AF598595

0 483 659 00 X0

THE RETINA

A Model for Cell Biology Studies

Part I

CELLULAR NEUROBIOLOGY: A SERIES

EDITED BY SERGEY FEDOROFF

THE NODE OF RANVIER
Edited by JOY C. ZAGOREN AND SERGEY FEDOROFF

THE RETINA: A MODEL FOR CELL BIOLOGY STUDIES, PART I
Edited by RUBEN ADLER AND DEBORA FARBER

THE RETINA: A MODEL FOR CELL BIOLOGY STUDIES, PART II
Edited by RUBEN ADLER AND DEBORA FARBER

THE RETINA

A Model for Cell Biology Studies

Part I

EDITED BY

RUBEN ADLER

The Michael M. Wynn Center for the Study of Retinal Degenerations
The Wilmer Ophthalmological Institute
The Johns Hopkins University
School of Medicine
Baltimore, Maryland

DEBORA FARBER

Jules Stein Eye Institute
UCLA School of Medicine
Los Angeles, California

1986

ACADEMIC PRESS, INC.

Harcourt Brace Jovanovich, Publishers
Orlando San Diego New York Austin
London Montreal Sydney Tokyo Toronto

ACADEMIC PRESS, INC.
Orlando, Florida 32887

United Kingdom Edition published by
ACADEMIC PRESS INC. (LONDON) LTD.
24–28 Oval Road, London NW1 7DX

Library of Congress Cataloging in Publication Data
Main entry under title:

The Retina : a model for cell biology studies.

Includes bibliographies and index.
1. Retina–Cytology. I. Adler, Ruben. II. Farber, Debora. [DNLM: 1. Retina–cytology. 2. Retina–physiology. WW 270 R43804]
QP479.R468 1986 599'.01823 85-23005
ISBN 0–12–044275–2 (Pt. I : alk. paper)

PRINTED IN THE UNITED STATES OF AMERICA

86 87 88 89 9 8 7 6 5 4 3 2 1

This book is dedicated to
Alfred J. (''Chris'') Coulombre,
a pioneer of the field of retinal cell biology
and an outstanding human being

CONTENTS

TROPHIC INTERACTIONS IN RETINAL DEVELOPMENT AND IN RETINAL DEGENERATIONS. *IN VIVO* AND *IN VITRO* STUDIES

Ruben Adler

CELL MOTILITY IN THE RETINA

Beth Burnside and Allen Dearry

MOLECULAR DYNAMICS OF THE ROD CELL

Paul A. Hargrave

CYCLIC NUCLEOTIDES IN RETINAL FUNCTION AND DEGENERATION

Debora B. Farber and Terrence A. Shuster

PHOTOSENSITIVE MEMBRANE TURNOVER: DIFFERENTIATED MEMBRANE DOMAINS AND CELL–CELL INTERACTION

Joseph C. Besharse

CONTRIBUTORS

Numbers in parentheses indicate the pages on which the authors' contributions begin.

Ruben Adler (1, 111), The Michael M. Wynn Center for the Study of Retinal Degenerations, The Wilmer Ophthalmological Institute, The Johns Hopkins University, School of Medicine, Baltimore, Maryland 21205

Joseph C. Besharse (297), Department of Anatomy and Cell Biology, Emory University School of Medicine, Atlanta, Georgia 30322

Beth Burnside (151), Department of Physiology–Anatomy, University of California, Berkeley, Berkeley, California 94720

Allen Dearry (151), Department of Physiology–Anatomy, University of California, Berkeley, Berkeley, California 94720

Debora B. Farber (1, 239), Jules Stein Eye Institute, UCLA School of Medicine, Los Angeles, California 90024

Christopher C. Getch (67), Department of Biology, Princeton University, Princeton, New Jersey 08544

Paul A. Hargrave (207), Department of Ophthalmology and Department of Biochemistry and Molecular Biology, College of Medicine, University of Florida, Gainesville, Florida 32610

Robert E. Marc (17), Sensory Sciences Center, University of Texas Graduate School of Biomedical Sciences, Houston, Texas 77030

Terrence A. Shuster (239), Jules Stein Eye Institute, UCLA School of Medicine, Los Angeles, California 90024

Malcolm S. Steinberg (67), Department of Biology, Princeton University, Princeton, New Jersey 08544

PREFACE

Retinal Cell Biology: Past, Present, and Future

It is difficult at times to define the field of cell biology. One need only attend the annual meeting of the American Society for Cell Biology to experience the enormous breadth of this field. Nonetheless, many would define cell biology in a very general way as a combination of microscopic anatomy, biochemistry, physiology, and pathology.

The confluence of these disciplines into a unifying field did not really occur until the 1950s, but we can trace the ancestry of the discipline to Matthias Schleiden, Theodor Schwann, and Rudolf Virchow, the botanist, histologist, and pathologist, respectively, who are generally credited with the development of the cell theory in 1839. Somewhat later in that century the study of retinal cells was initiated. Ultimately, Max Schultze (1867) set forth the duplicity theory for photoreceptor function through his observation that cones are the receptors for bright light and color and that rods function in dim light. Subsequently Franz Boll (1877) observed that the photochemical event underlying light detection was the bleaching of "visual purple" now known as rhodopsin. He also suggested that a carotenoid might be a precursor for this photopigment. His contemporary, Willy Kühne (1878) was the first to solubilize photopigments in bile salts thereby demonstrating their hydrophobic nature. Santiago Ramon y Cajal (1894) soon thereafter modified methods shared by fellow Nobel laureate Camillo Golgi and displayed the neuronal network of the retina in such breathtaking detail that his work serves as the standard even to this day. Therefore, by the turn of the century, the study of retinal cells by representatives of the separate disciplines of microscopic anatomy, biochemistry, and physiology was well established.

Retinal research was graced by investigators of equivalent intuition and creativity during the first half of this century. The disciplines of electrophysiology and biochemistry, exemplified by Ragnar Granit, Haldane Hartline, and George Wald, produced insights into retinal electrical activity, cell interactions, and photochemistry worthy of the Nobel Prize so appropriately bestowed upon them. The contributions of these gifted scientists and their contemporaries drew into the field a new generation of capable scientists armed with new and powerful techniques. Fritiof Sjostrand, Eichi Yamada, Adolph Cohen, Brian Boycott, John Dowling, and many others began the systematic examination of retinal cells by electron microscopy. The application of tissue autoradiographic technique to retinal studies by Richard Sidman, Richard Young, and Bernard Droz in the early to mid 1960s added a dynamic component to microscopy. It was possible for the first time to mark retinal birth dates and to study the metabolism of individual cell types.

Particularly suitable for autoradiographic studies were the photoreceptors with their highly compartmentalized structure and their prodigious biosynthesis and assembly of membranes. By virtue of this technique, an entirely new role was discovered for retinal pigment epithelial cells, the phagocytosis and photoreceptor outer segment membranes. Furthermore, a new research strategy began to emerge. Whereas scientists of the past had adhered rather strictly to their chosen disciplines each of which drew upon a rather limited constellation of methods, the new approach was to combine the methods of multiple disciplines in order to address problems of ever-increasing complexity. This is the essence of cell biology and, with the advent of this approach, retinal cell biology came into full bloom.

A subject that benefited greatly from this philosophy was the study of membrane biosynthesis in photoreceptors. Michael Hall and I combined radiobiochemical and autoradiographic methods in the study of rhodopsin biosynthesis and assembly into outer segment disk membranes, and David Papermaster and Barbara Schneider applied high-resolution immunocytochemical methods to plot membrane trafficking from sites of synthesis to sites of assembly within the cell.

Another area that was aided tremendously by this combined approach was that of retinal neurotransmitter studies. Relatively little progress was made on this subject until the introduction of autoradiographic and immunocytochemical techniques. Autoradiographic analysis of high-affinity-uptake sites for neurotransmitters as initiated by Berndt Ehinger and Dominic Lam combined with immunocytochemical localization of enzymes involved in neurotransmitter synthesis have been very illuminating as have studies introduced by Harvey Karten and Nicholas Brecha on the localization of the myriad of peptides that are now known to exist either as neurotransmitters or neuromodulators in the amacrine cells of the retina. Immunocytochemical techniques are now employed by a growing number of retinal cell biologists and, when applied at high resolution in particular, will add significantly to our understanding of the molecular division of labor among retinal organelles.

Research during the 1970s has featured another group of exciting topics. The subject of photoreceptor transduction has been hotly debated among proponents who favor the hypothesis that calcium serves as the internal messenger during signal amplification, as first proposed by William Hagins and Shuko Yoshikami, and the adherents of cyclic guanosine monophosphate, as initially suggested from the work of Mark Bitensky and collaborators. Whatever the outcome, it appears likely that the photoreceptors will be the first sensory cells in which the mystery of transduction is solved. Furthermore, the contributions of Debora Farber and Richard Lolley and their collaborators have shown how cyclic nucleotide metabolism, when perturbed, can lead to photoreceptor degeneration of the type observed in some animal models for retinitis pigmentosa.

Rhythmic phenomena in the retina, pioneered by Matthew LaVail through his observations on cyclic outer segment disk shedding, have also been a subject of intense investigation resulting in the discovery by several laboratories that these activities are controlled intraocularly rather than by remote tissues such as the pineal gland. The development of an *in vitro* preparation by Joseph Besharse for the study of the molecular basis of rhythmic activity has already moved us significantly toward an understanding of these processes. Superimposed upon these metabolic studies has been exciting progress in our understanding of the role of extracellular adhesion and matrix molecules in retinal development and trophic interactions. The recent work of Ruben Adler and associates now provides us with the capability to study the differentiation of photoreceptors in culture.

Another informational leap is now occurring in retinal cell biology, spawned by the technology that has impacted so favorably on other tissues as well, namely, recombinant DNA research. Some of the most exciting observations are so new that they have not yet appeared in print. The first application of recombinant DNA methods to retinal research has involved the amino acid sequencing of proteins involved in phototransduction. The sequencing of bovine rhodopsin by conventional methods was initiated by Paul Hargrave and his colleagues shortly after this protein was purified by Joram Heller in 1968.

About 15 years of intensive work were required before the primary structure of this hydrophobic membrane protein was solved. Utilizing this important sequence information, the fact that the vertebrate photopigments are well conserved, and rapid nucleotide sequencing developed independently by the laboratories of Frederick Sanger and Walter Gilbert, Jeremy Nathans and David Hogness have sequenced human rhodopsin and all of the human cone opsins in the space of just a few years. Additionally, the primary sequences of all of the subunits of the G protein and light-activated cyclic nucleotide phosphodiesterase thought to be involved in phototransduction will probably be solved in several laboratories by the time that these books (Parts I and II) appear in print. With this information in hand we will have, for the first time, the opportunity to determine the molecular basis of several forms of inherited photoreceptor degeneration that are thought to involve one or more of these proteins.

We are currently experiencing a fast-moving and exciting period for the retinal cell biologist and the opportunities appear almost limitless at this time. It is therefore most fitting that a two-part series entitled *The Retina: A Model for Cell Biology Studies* should be made available to those who already have an interest in this subject and to others who may be inspired to join us. We hope that you find the subject as inviting and rewarding as we have during the past two decades.

DEAN BOK

ISSUES AND QUESTIONS IN CELL BIOLOGY OF THE RETINA

DEBORA FARBER

Jules Stein Eye Institute
UCLA School of Medicine
Los Angeles, California

RUBEN ADLER

The Michael M. Wynn Center for the
Study of Retinal Degenerations
The Wilmer Ophthalmological Institute
The Johns Hopkins University
School of Medicine
Baltimore, Maryland

Most investigations of enzymes and other chemical constituents of the developing chick retina are of recent date. . . . thus far inadequate attention has been given to chemical differences between the pigmented epithelium and the neural retina. . . . it seemed reasonable to trace in more detail the histogenesis of this tissue and to make correlation of the findings with the data on retinal chemistry. [Coulombre, 1955]

> At any stage of embryonic development, each tissue of the eye may be viewed either as a source of influences which control the growth and differentiation of adjoining tissues, or as a target of influences from neighboring tissues. By systematically studying the ability of each tissue to influence or to be influenced at each stage in embryogenesis, it is possible to construct a flow sheet of the successive steps by which the eye is shaped. [Coulombre and Coulombre, 1965]

I. Introduction

That neurons are constantly regulated by molecular signals originating in other cells has become one of the central concepts of contemporary cellular neurobiology. The properties of a multicellular society such as the retina are the reflection of the properties of its constituent cells. Similarly, many attributes of each cell are controlled by its microenvironment. Retinal development, structure, biochemistry, function, and pathology can all be regarded as reflecting individual cell responses to the environmental conditions prevalent at each particular point in the life history of an organism. In addition, cells should be perceived as entities in which individual molecules become integrated into supramolecular structures capable of performing diverse functions. *Cell biology*, a relatively new multidisciplinary branch of the biological sciences, approaches the cell with a variety of powerful biochemical and morphological techniques. Its goal is to understand the structure and function of a cell through the investigation of its molecules, their supramolecular organization, and their function. Although many aspects of retinal biology had been explained in "cellular" terms before (i.e., Cajal, 1928), an experimental "cell biology approach" to the retina probably is only some 30 years old. The success of this discipline is reflected in its many recent contributions, which have increased our understanding of the retina as a whole by correlating development, organization, biochemistry, function, and pathology. By recognizing this impact of cell biology and emphasizing the usefulness of the retina as a model system for the investigation of cellular mechanisms of general relevance, these two volumes should serve as a modest tribute of admiration and recognition to the pioneers of this discipline. Alfred J. ("Chris") Coulombre, to whom these two volumes are dedicated, occupies a site of honor as one of the founding fathers of the cell biology of the retina.

Contributing authors have been asked to review the field of their expertise, with emphasis not only on established concepts but also on unanswered questions, emerging issues, and areas of controversy. They have also been asked to describe technical breakthroughs which have had, or might have, significant impact on research in their fields. Correlations between basic retinal cell biology and human retinal disease have been made whenever possible. Because of space limitations, some important topics (i.e., glial cells, cell proliferation) are briefly discussed in this Introduction but have not been alloted individual articles. To orient the reader who may not be familiar with the retina, we also present in the following pages some basic information about retinal structure and development. Moreover, we discuss the advantages of the retina as a model system for cell biology studies, and call attention to problems or questions which have not yet received proper attention.

II. The Adult Retina

A. Localization within the Eye

The retina is the innermost of the three coats that form the wall of the eyeball (Fig. 1). The outer coat is thick and tough and protects the delicate inner struc-

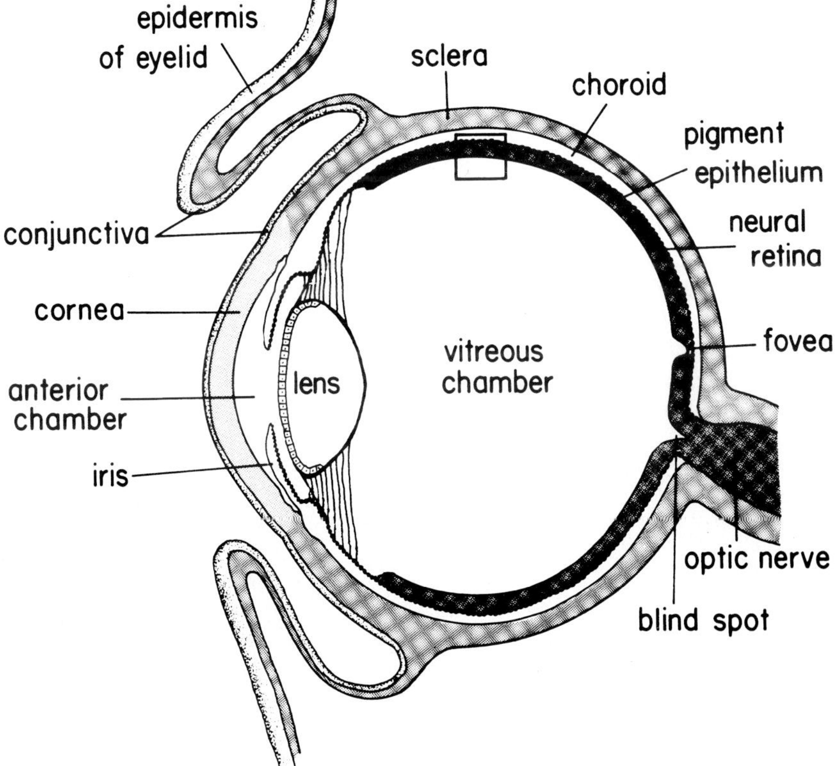

FIG. 1. Diagram of a horizontal section of a high vertebrate eye. The cellular organization of a portion of the neural retina (□) is represented in Fig. 2.

tures. It is opaque in the larger posterior portion, the sclera, and transparent in the smaller anterior section, the cornea. The middle vascular coat, the uvea or choroid, provides the nutrients to the ocular tissues, and its anterior segment, the ciliary body, is the muscular instrument for the process of accommodation carried out by the lens. The lens is suspended from the inner surface of the ciliary body. The iris is a continuation of the ciliary body and divides the anterior segment of the eye into two unequal parts, the anterior and posterior chambers, which are filled with transparent aqueous humor. The two chambers are connected by the opening of the iris, the pupil, which can be reduced or expanded through the contraction or relaxation of its constrictor and dilator muscles. The iris regulates in this way the amount of light entering the eye. The large posterior cavity of the eye is the vitreous chamber, which is filled with a viscous, transparent substance, the vitreous humor.

B. Cell Layers and Neuronal Networks

The neural retina is perhaps the central nervous system (CNS) tissue best suited for cell biology studies. Its limited number of cellular classes, which in the adult stage are organized in a perfectly layered structure, simplify the identification and localization of different biochemical and physiological functions (Fig. 2). The photoreceptor cells, lying outermost, are adjacent to the retinal pigment epithelium, a cellular lining which separates the retina from the blood capillaries of the choroid. The innermost retinal layer, which is closest to the vitreous humor, is formed by the ganglion cells and their axons, which bundle together to form the optic nerve fiber layer. An ''inner limiting membrane'' separates this layer from the vitreous humor. Between the photoreceptor and the ganglion cell layers is a cellular stratum, many times referred to as the ''inner nuclear layer,'' formed by horizontal, bipolar, interplexiform, and amacrine neurons and the glial-like Müller cells. Synaptic contacts also occur in defined layers of the retina. An ''outer plexiform layer,'' between photoreceptors and the inner nuclear layer, contains synapses between photoreceptors, bipolar, and horizontal cells. In the ''inner plexiform layer,'' which is located between the inner nuclear layer and the retinal ganglion cells, the latter make contacts with bipolar and amacrine cells, which also interact with each other. Interplexiform cells have processes extending into the inner plexiform and the inner nuclear layers. The latter processes make synaptic contacts with horizontal and bipolar cells. Müller cells span the retina from the inner limiting membrane through the outer nuclear layer which contains the nuclei of photoreceptors. At that level, junctions between Müller cells create the light microscopy image of an ''outer limiting membrane.'' Light is transmitted through cornea, aqueous humor, pupil, lens,

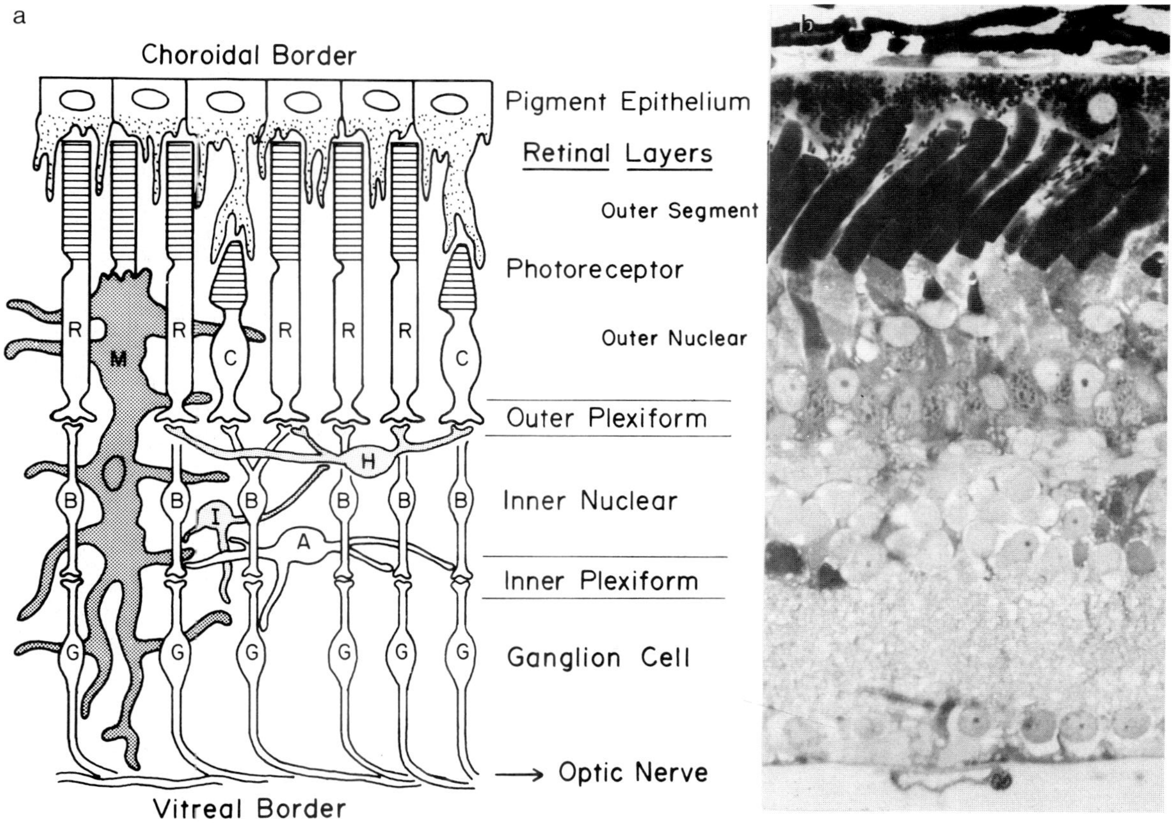

a

Choroidal Border

Pigment Epithelium

Retinal Layers

Outer Segment

Photoreceptor

Outer Nuclear

Outer Plexiform

Inner Nuclear

Inner Plexiform

Ganglion Cell

→ Optic Nerve

Vitreal Border

b

FIG. 2. The adult vertebrate retina. (a) Diagram of the retinal layers and their synaptic relationships. The main cell types are designated as R, rods; C, cones; H, horizontal cells; B, bipolar cells; I, interplexiform cells; A, amacrine cells; G, ganglion cells; M, Müller cells. (b) Light micrograph of the retina of the toad *Xenopus laevis,* showing the different layers. ×641. (Courtesy of Dr. John Flannery.)

vitreous humor, and inner retinal layers to the photoreceptors, which convert it into electrical impulses that are sent to the brain via the optic nerve. This means that before leaving the eye, the electrical signals generated by the photoreceptors have to pass through the neural network of the inner retina, in which they follow specific synaptic pathways.

The retina offers an optimal system to study neuronal circuitry in the CNS. Mechanisms controlling the integration and encoding of the visual signals can be studied by psychophysical, neurophysiological, morphological, and neurochemical techniques. Intracellular recordings combined with dye injections have made it possible to correlate the structure and function of different inner retinal neurons. In addition, many types of neurotransmitters and neuromodulators are now known to act at the various synapses, and the presence of their synthesizing enzymes and uptake and release mechanisms have been demonstrated in the inner retinal layers. The ultrastructural organization of the photoreceptor synaptic terminal suggests that these visual cells also release chemical transmitters; some studies suggest further that the transmitter is released continuously in the dark and that this process is diminished in the presence of light. An article by Robert E. Marc (Part I) presents a discussion of retinal networks, and in Part II P. Michael Iuvone discusses work that has been carried out on presynaptic and postsynaptic neurotransmitter systems that act as markers of neurochemical differentiation in the retina.

One of the recurring principles emerging from contemporary cell biology is that, in spite of the extreme diversity of cell types found in the vertebrate organism, many basic mechanisms are shared by the vast majority of the cells. Retinal cells are particularly well suited for the investigation of several of these activities. For example, a large number of cellular mechanisms have been investigated using as an experimental system just *one* of the cell types present in the retina, namely, the photoreceptor cell (Figs. 2 and 3). Photoreceptor outer segments are an excellent system for the investigation of the assembly, structure, and function of biological membranes. Disk membranes are arranged with striking regularity within the outer segment and undergo a continuous process of renewal. Every day some disks are assembled at the outer segment base while others are shed at its tip and phagocytized by the pigment epithelium. Thus, the phenomenon of disk renewal allows detailed investigation of basic cell activities, which are discussed in detail in several articles in these volumes, for example (1) synthesis, vectorial transport, and membrane insertion of specific proteins (Joseph C. Besharse, Part I), (2) phagocytosis (Virginia M. Clark, Part II), and (3) circadian rhythms (Joseph C. Besharse, Part I; P. Michael Iuvone, Part II). Photoreceptor outer segments are also obviously well suited for studies of the phototransduction process. Photopigments are an integral part of disk membranes, and a whole set of "light-sensitive" enzymes are associated with either

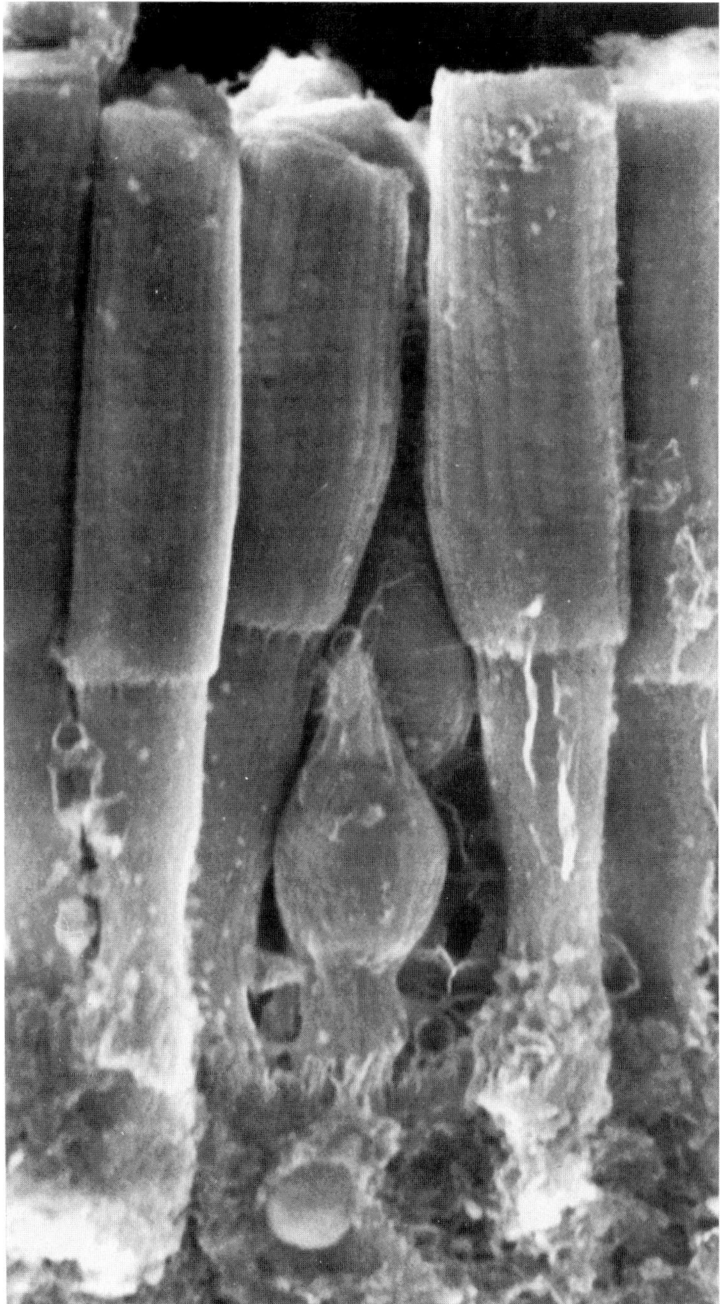

FIG. 3. Scanning electron micrograph of *Xenopus* photoreceptors from the rod outer segments to the outer nuclear layer. The retinal pigment epithelium has been detached from the neural retina to expose the photoreceptor outer segments. Note a single cone with its spherical inner segment and short, conical outer segment. ×4000. (Courtesy of Dr. John Flannery.)

the membranes or the cytosol compartments. Issues which have attracted many investigators during the last few decades include measurement of changes in the spectral properties of the visual pigments upon illumination, the capture and identification of short-lived intermediaries, the kinetics of these reactions, and biochemical processes activated by light. Contributions by Paul A. Hargrave (Part I) and by Debora B. Farber and Terrence A. Shuster (Part I) deal with the biochemical aspects of the visual process. Robert W. Massof (Part II), on the other hand, describes psychophysical and physiological approaches to the study of visual pigments and phototransduction. Finally, the extreme polarization and asymmetry of the photoreceptors offer many advantages for cell motility studies, as discussed in detail by Beth Burnside and Allen Dearry in an article (Part I) that deals largely with the role of the cytoskeleton in photomechanical movements (the elongation and contraction of photoreceptors in response to environmental light). Recently developed cell culture systems for embryonic photoreceptors should allow further investigation of cytoskeletal involvement in photoreceptor organization and function (see article by Ruben Adler, Part I).

The retinal pigment epithelium (RPE) serves multiple functions which are essential for proper retinal function (see article by Virginia M. Clark, Part II). The RPE is vital to the integrity of rod and cone photoreceptors and participates in the continuous renewal of their outer segments through the process of phagocytosis. The RPE acts as a barrier between the choriocapillaris and the extracellular space surrounding the retina, by restricting the entrance of unwanted molecules to the retina and controlling the composition of fluids present in the extracellular space. Specific ions and molecules such as amino acids and vitamin A reach the retina through (and in some cases after processing by) the RPE. After the absorption of light occurs in the outer segment disks, causing rhodopsin bleaching, the derivatives of rhodopsin's chromophore are sent for storage to the retinal pigment epithelium. In this tissue they may be converted into 11-*cis*-retinal, which then would return to the photoreceptors, hypothesis not proven yet. Melanin granules, also synthesized by the pigment epithelium, absorb light energy and reduce light scatter and, as a result, improve the resolution of visual images. Furthermore, the pigment epithelium is involved in the synthesis of glycosaminoglycans, which are found in the subretinal space and may contribute to adhesion of the retina to the pigment epithelium (see article by Tyl Hewitt, Part II).

C. *Glia*

Glial cells have the dubious privilege of being among the most enigmatic cells in the entire organism. It is unwarranted to regard them as ancillary components of the nervous system, whose existence would be almost exclusively justified by the services they provide to neurons. However, the investigation of glial cell function in the intact organism has proved very difficult. Perhaps the only glial function clearly established *in vivo* is the production of a myelin sheath by Schwann cells in the peripheral nervous system and by oligodendrocytes in the CNS. Possible functions of astrocytes, on the other hand, can be extrapolated from an extensive series of *in vitro* studies. For example, cultured astroglial cells display a variety of high-affinity uptake mechanisms for neurotransmitters and can also regulate potassium concentrations in their microenvironment (reviews by Kuffler *et al.*, 1984; Ripps and Witkovsky, 1985). Similar *in vivo* functions have been suggested on the basis of these *in vitro* studies, but there is little direct evidence supporting this extrapolation.

The situation is even more complicated in the case of retinal Müller cells, which have some properties resembling astrocytes, radial glial cells, and even oligodendrocytes (Ripps and Witkovsky, 1985). Müller cells share with astroglial cells the capacity to take up different neurotransmitters by means of high-affinity mechanisms. Important enzymatic systems are restricted in the retina to Müller cells, including glutamine synthetase (required for glutamate metabolism) and carbonic anhydrase (which is involved in the control of sodium and chloride movements and in fluid balance). Their developmental regulation has been extensively studied (reviewed by Moscona and Linser, 1983). The investigation of Müller cell properties using *in vitro* techniques has been more difficult than similar studies with other retinal cell types (see article by Ruben Adler, Part I). To complicate the issue even further, Müller cells are apparently not the only glial element in the retina. Hume *et al.* (1983) recently described the invasion of the mouse retina in early postnatal life by circulating macrophages which participate in the phagocytosis of neurons that undergo developmental death. After completing their debris-removing activities these cells appear to remain in the retina and assume microglial identity. In addition, more typical astrocytic cells, clearly different from Müller cells, can be found in the layer of the optic nerve fibers in the mammalian retina. The presence of myelin-forming glial elements has also been described in the retina of several species of birds (Smith, 1982), but their identity has not been established. In summary, glial cells remain the least well understood and the least studied cell type in the retina. However, their apparent importance is such that they certainly deserve more attention than they have received so far.

III. The Developing Retina

As described above, the mature neural retina is a multilayered structure characterized by a striking degree of order and organization. In the mature retina

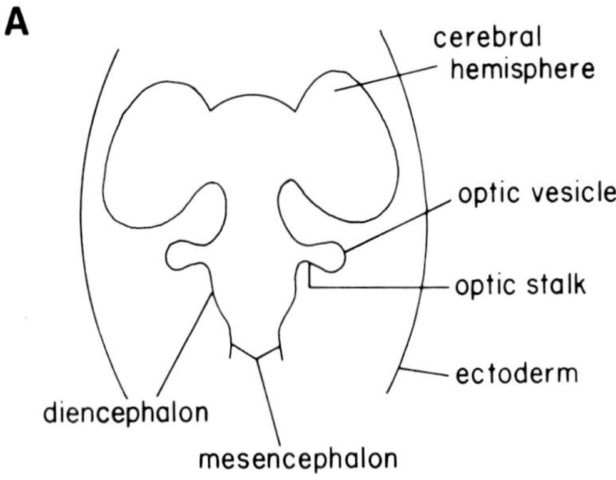

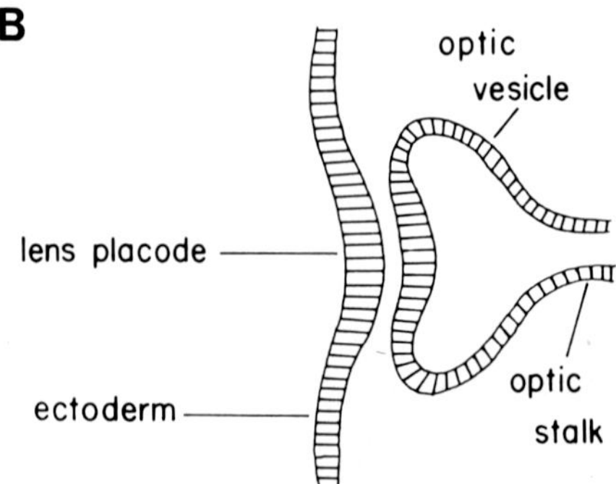

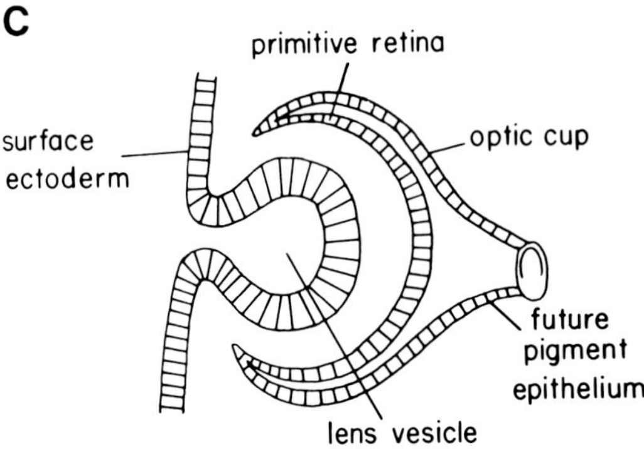

FIG. 4. Diagrammatic representation of the development of the retina and related eye structures in vertebrates. Note that a hollow optic vesicle appears by evagination of the wall of the diencephalon (A and B). In turn, the invagination of the optic vesicle leads to the development of an optic cup consisting of two layers (C). The inner layer gives rise to the neural retina, and the outer layer to the pigment epithelium. The lens has a separate origin. See text for further details.

most cells are postmitotic and can be identified by their position, structure, biochemical properties, and function. Neurons are selectively interconnected with other neurons and, in the case of ganglion cells, also with postsynaptic elements located in the brain. These interconnections depend on the existence of appropriate numbers of neurons of the different varieties. Also, special mechanisms are necessary to determine that synaptic connections only form and/or persist between "correct" neuronal partners. The magnitude of the developmental phenomena necessary for the genesis of this orderly complexity can be easily appreciated by comparing the retina in the mature eye (Fig. 2) with the optic cup from which this organ is derived (Fig. 4C). The optic cup is an evagination from the wall of the diencephalon (Fig. 4A,B). It consists of an external layer which gives rise to the retinal pigment epithelium and an internal layer which develops into the neural retina itself. Both layers are continuous at the level of the future ora serrata (where the iris will later develop) and in the region of the optic stalk (future optic nerve). The neuroepithelium, which forms the wall of the optic cup in the young embryo, appears as a homogeneous population of epithelial cells capable of mitotic division, which extend from the vitreal to the scleral surfaces of the future retina (Fig. 5).

It is bewildering to think of the precision of the mechanisms which are necessary to transform a simple neuroepithelium into a structure as complex and yet as accurately organized as the retina. Many of these mechanisms still remain obscure, and the description that follows is therefore presented largely to highlight some questions which await to be investigated. The transition from the simplicity and homogeneity characteristic of the early optic cup to the complexity featured by the mature retina appears to be accomplished through the harmonious occurrence of a series of developmental phenomena. Cell proliferation leads to increases in cell numbers which, together with cell movements and with the development of nerve fibers, are apparently responsible for retinal growth and for the development of a layered pattern. At very precise times in development (see below) cells located in specific regions of the retina become postmitotic and migrate toward the future vitreal surface of the retina (Fig. 5). After reaching a defined location in one of the developing layers, cells begin expressing characteristic differentiated properties (formation of axons and dendrites, acquisition of enzymes for neurotransmitter synthesis, development of exocytotic mechanisms for neurotransmitter release, electrical excitability, etc.). Axons grow along highly selective pathways and synapse upon specific targets. Retinal ganglion cells send their axons toward the point of origin of the optic nerve and, through this nerve, to specific postsynaptic elements in the brain. As in many other regions of the nervous system, the adjustment of cell numbers between the retina and postsynaptic centers in the brain (and perhaps also between different populations of retinal neurons) is accomplished through the developmental neuronal death phenomenon.

Even the most superficial analysis of retinal development results in the for-

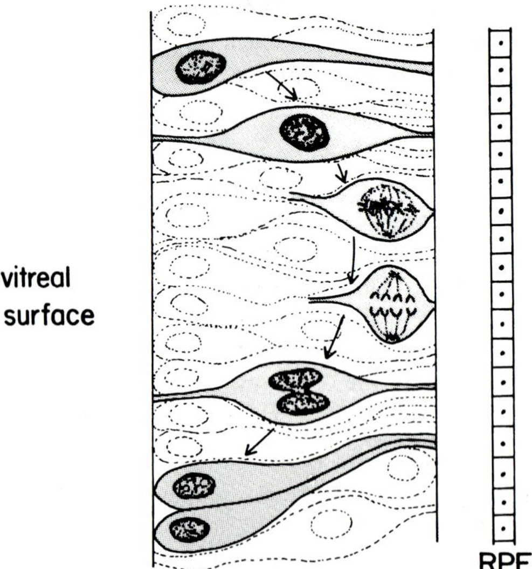

Fig. 5. Diagrammatic representation of the inner wall of the optic cup (future neural retina) during early embryonic development. The cells span the entire width of the pseudostratified neuroepithelium. Mitotic figures are found at the surface closer to the retinal pigment epithelium (RPE). This peculiar distribution is explained by nuclear migration during the cell cycle (see arrows), which is repeated until the cells become postmitotic. Postmitotic cells migrate toward the vitreal surface of the retina and undergo differentiation after reaching their definitive location in one of the retinal layers.

mulation of many questions regarding cellular mechanisms of retinal differentiation. It is known, for example, that in different regions of the retina different cell types are "born" (i.e., undergo their last mitotic division) at specific times in development. Is each cell predetermined to develop in a particular direction *before* becoming postmitotic, or is there a set of signals which dictates its pattern of differentiation *after* it becomes postmitotic? The importance of this question is highlighted by the observation that retinal neurons become postmitotic according to highly stereotyped spatial and temporal gradients. For example, retinal ganglion cells tend to be the first and photoreceptors the last neuronal elements to be generated in each retinal region. At the same time, neuronal "birth" occurs earlier in the more posterior than in the peripheral regions of the retina. These developmental rules seem to be basically preserved throughout the vertebrate species which have been studied, although some interspecific differences have been found (for review see Raymond, 1985). There is in all cases a clear-cut correlation between the time when a cell becomes postmitotic, its final location within the retina, and the pattern of differentiation that it will follow. How are these different phenomena interrelated? Are there molecular signals which control the stereotyped sequence of these events? Is there a cause–effect correlation between the different phenomena which make up this sequence?

As described above, the inner nuclear layer of the retina contains the glial Müller cells and four types of neurons, namely the horizontal, bipolar, amacrine, and interplexiform cells. Moreover, recent studies using immunocytochemical and autoradiographic techniques have shown that these cell types are not homogeneous. In light of substantial evidence that suggests that selective gene expression during cell differentiation is largely controlled by microenvironmental cues, it is not very easy to understand intuitively how cells which originate from common ancestors, and which are located in such intimate proximity to each other, can achieve such a diverse pattern of differentiation. Which molecular signals direct some cells to differentiate as neurons and which instruct other cells to become glial? Even more puzzling is the fact that the four types as well as many subtypes of neurons present in the inner nuclear layer share a common basic pattern of organization while showing many differences superimposed upon that common pattern. For example, amacrine, interplexiform, bipolar, and horizontal cells all develop nerve fibers, but their neuritic processes are so different in morphology that they serve as criteria for their identification. They all use neurotransmitters, but there are differences in the neurotransmitter systems used by each neuronal type (see articles by Robert E. Marc, Part I, and P. Michael Iuvone, Part II). The morphology of their synaptic terminals is also characteristically different. The investigation of the cellular and molecular mechanisms controlling complex patterns of cell differentiation in the CNS is one of the most challenging tasks faced by contemporary neurobiology. Because of its accessibility and its peculiar organization, the retina appears as the most suitable CNS organ in which these questions may be studied.

Axonal growth and guidance are cell motility phenomena important for the development of retinal networks and the retinotectal projections. The latter require the polarized outgrowth of ganglion cell axons toward the vitreal surface of the retina, their oriented elongation toward the choroid fissure (where the optic nerve emerges from the eye), the orderly arrangement of ganglionic cell axons within the optic nerve, and the development of topographically specific connections with tectal cells. Several articles in these volumes reflect the diversity of approaches that can be used to study these phenomena. Christopher C. Getch and Malcolm S. Steinberg (Part I) discuss the role of contact-mediated cell–cell interactions and cell surface macromolecules involved in cell adhesion. Karl Herrup and Jerry Silver (Part II) provide us with some insight on the utilization of genetic chimeras for the investigation of these issues and emphasize the role that preformed pathways might play in mechanically guiding growing axons. Complementary to these two articles is that by Tyl Hewitt (Part II) in which extracellular matrix molecules are shown to be important constituents of the neural retina, capable of affecting neuronal survival as well as the production and growth of axonal fibers.

IV. The Pathological Retina

The retina also offers an attractive system for the investigation of cellular mechanisms of importance in pathological conditions. In many instances, fundamental information regarding normal processes has been obtained from the comparison of healthy and abnormal tissues.

The rather peripheral location of the retina makes it one of the CNS components more directly exposed to trauma. In addition, the retina can be affected by genetic disorders and by lesions induced by drugs or environmental factors. When this occurs, many specific abnormal responses are observed in some of the cell types, making them ideal targets for cell biological investigations.

Several genetic disorders of the retina have been identified in human beings, for example retinitis pigmentosa, and others are being studied in different species of animals. In fact, a wealth of information has been accumulated during the last decade about the histopathology, physiology, and biochemistry of retinas affected with inherited degenerations. Generally, these diseases are transmitted as autosomal recessive traits and are expressed at different ages in development, with characteristic times of onset and rates of progression. When the photoreceptor cells are the target of the genetic lesion, these disorders result in blindness. A common end state is then observed, a photoreceptorless retina, with the pigment epithelium juxtaposed to the inner neuronal layers which remain intact and functional.

It has been suggested that there are periods in the life of a photoreceptor cell in which it becomes particularly vulnerable to genetic or environmental insults (Farber and Lolley, 1977). In some of the animals affected by inherited retinal degenerations (i.e., *rd* mice and Irish setter and collie dogs), the critical event appears to occur at the time of photoreceptor differentiation. In other cases, such as the RCS rat, photoreceptors differentiate normally but the renewal of their outer segments appears disrupted. These degenerative disorders are being investigated with different techniques. Genetic mosaics are used to study gene action in both normal and abnormal retinal development (see article by Karl Herrup and Jerry Silver, Part II). Biochemical studies have revealed abnormalities in cyclic nucleotide metabolism which precede the onset of morphological signs of retinal degeneration in some animal mutants as well as in some experimentally induced degenerations (see article by Debora B. Farber and Terrence A. Shuster, Part I). As discussed by Ruben Adler in Part I, an entirely different approach to retinal degeneration stems from studies of the developmental neuronal death phenomenon. This phenomenon, which is normal during embryonic development, led to the concept that neuronal survival is regulated by "trophic factors" originating in the microenvironment surrounding the neurons (postsynaptic cells, glial cells, etc.). The possible role of altered trophic mechanisms in retinal degenerations deserves further investigation. Yet another approach to retinal degeneration is presented by Robert W. Massof (Part II), who discusses the use of psychophysical techniques to evaluate the progression of visual loss in human beings affected by these diseases. This article also proposes a challenging correlation between psychophysical observations and cellular aspects of the visual process.

Endothelial cells provide a clear example of a cell type which is not particularly conspicuous in the normal retina but which can assume a protagonistic role in some pathological conditions. The phenomenon of "neovascularization" is characterized by the development of new blood vessels and has the potential to impair retinal function to such an extent that it can lead to total blindness. It is therefore of paramount importance to investigate the possible existence of factors which inhibit the development of new blood vessels, as well as those which, under abnormal circumstances, can stimulate endothelial cells to proliferate and migrate to form new blood vessels. The last few years have witnessed an explosion of new knowledge in this area, as reviewed and discussed in the article by Bert M. Glaser (Part II).

The importance of basic developmental mechanisms for the well-being of the adult organism goes beyond the requirements which must be met to achieve full maturation of the different organs. As in other parts of the CNS, the injured neural retina could be repaired if it were possible for its cells to recapitulate in the adult the developmental mechanisms which generate the retina during embryogenesis. The capacity for retinal regeneration (and, in particular, for the regeneration of a retinotectal projection) does exist in some lower vertebrate species but has been lost in higher vertebrates. The article by Bernice Grafstein (Part II) reviews this work and proposes research avenues which will hopefully teach us how to intervene in the retina, to stimulate its regeneration in animal species where the phenomenon does not occur spontaneously.

V. Concluding Remarks

Our understanding of the cell biology of the retina has advanced dramatically in the 30 years that have elapsed since researchers found ways to establish correlations between retinal structure and biochemistry. As our research tools increase in power, we can ask ever more refined questions about intimate molecular aspects of retinal (and other) cells. "Progress," however, cannot be taken at face value without some critical evaluation of the criteria used in its determination. We certainly need an overall sense of direction for our investigations, and general biological principles and questions should not be forgotten even if most of our energy is devoted to the study of the specifics of one or another molecular interaction. It is relevant in this context that the formulation of those general principles and questions took place well before we acquired the tools for cell biology investigations. Therefore, it appears to us warranted, as well as necessary, to put contemporary studies in perspective, with respect both to the past (from which much inspiration can still be drawn) and to the future (when the success or failure of our research strategies will become evident). The Preface, by Dean Bok, presents such a link between past, present, and future of the cell biology of the retina.

References

Cajal, S. R. (1928). "Degeneration and Regeneration of the Nervous System" (R. M. May, trans.). Hafner, New York, 1959.

Coulombre, A. J. (1955). Correlations of structural and biochemical changes in the developing retina of the chick. *Am. J. Anat.* **96,** 153–189.

Coulombre, J. L., and Coulombre, A. J. (1965). Regeneration of neural retina from the pigmented epithelium of the chick embryo. *Dev. Biol.* **12,** 79–92.

Farber, D. B., and Lolley, R. N. (1977). Influence of visual cell maturation or degeneration on cyclic AMP content of retinal neurons. *J. Neurochem.* **29,** 167–170.

Hume, D. A., Perry, V. H., and Gordon, S. (1983). Immunohistochemical localization of a macrophage-specific antigen in developing mouse retina: Phagocytosis of dying neurons and differentiation of microglial cells to form a regular array in the plexiform layers. *J. Cell Biol.* **97,** 253–257.

Kuffler, S. W., Nicholls, J. G., and Martin, A. R. (1984). "From Neuron to Brain." Sinauer, Sunderland, Massachusetts.

Moscona, A. A., and Linser, P. (1983). Developmental and experimental changes in retinal glial cells: Cell interactions and control of phenotype expression and stability. *Curr. Top. Dev. Biol.* **18,** 155–188.

Raymond, P. A. (1985). The unique origin of rod photoreceptors in the teleost retina. *TINS* **8,** 12–17.

Ripps, H., and Witkovsky, P. (1985). Neuron–glia interactions in the brain and retina. *Prog. Ret. Res.* **4,** 181–219.

Smith, R. L. (1982). Retinal myelination in birds. *In* "The Structure of the Eye" (J. G. Hollyfield, ed.), pp. 191–204. Elsevier, Amsterdam.

THE DEVELOPMENT OF RETINAL NETWORKS

ROBERT E. MARC

Sensory Sciences Center
University of Texas Graduate School of Biomedical Sciences
Houston, Texas

I. Introduction

This article was designed as a review, for cell biologists and vision scientists alike, of some of the major events in the maturation of the retina pertinent to the assembly of neuronal circuits. How do unique cell types arise and how do they "wire" themselves? We are still unsure, but over the past decade there has been a significant accumulation of observations that provide insight to these processes and that clearly chart a course for future experiments. I will attempt to characterize the task of mapping developmental processes in terms relevant to the anatomy and physiology of vision. One should appreciate, however, that a vast literature centers on descriptive approaches to retinal development, and many of those works have little immediate significance for understanding the maturation of retinal networks. Consequently, I have been somewhat selective of references (and perhaps uneven in coverage) in an attempt to consolidate central issues about functional development. In particular I have not discussed tissue culture methods and avian development in great detail. Though these are important approaches to development, it is difficult to integrate them into a general assessment of circuitry. The findings crucial to our understanding of functional retinal assembly are really quite recent and though they are often no more than chronologies of events (rather than demonstrations of mechanisms), they have allowed the partitioning of development into apparently independent phases.

A. *Types of Retinal Neurons*

Describing the development of retinal networks is a neuroanatomical effort. What is known of morphogenetic changes of retinal neurons has been derived primarily from light microscope examinations of Golgi-impregnated neurons and serial-section electron microscopic studies. There have been some notable electrophysiological contributions, but as we shall discover, many of the events associated with the specification of cellular participants in certain networks significantly predate synapse formation and subsequent synaptic activity. More recently, various biochemical markers have become available, and in combination with anatomical methods, constitute a powerful approach to neuronal development.

There are two initial goals in studying the development of networks: (1) to characterize the individual classes of neurons and (2) to specify the synpatic relationships among classes. These are enormous tasks which have not been adequately achieved for the differentiated state. With the help of neurochemical probes, classical neuroanatomy, and combined intracellular recording/dye injection, it has been possible to define many subcomponents of small chains of neurons in the vertebrate retina. These small networks are often referred to as "microcircuits," to recognize the fact that a physiological network involving a set of photoreceptors, intermediate neurons, and an individual ganglion cell is actually composed of many microcircuits. Moreover, different classes of retinal networks may share some microcircuits, so we view microcircuits as building blocks.

We are now aware of six broad categories of retinal neurons: (1) photoreceptors, (2) bipolar cells (BCs), (3) horizontal cells (HCs), (4) amacrine cells (ACs), (5) interplexiform cells (IPCs), and (6) ganglion cells (GCs). Each of these categories is now known to come in several subvarieties, and while the members of each category share many properties, the differences among members are dramatic and often absolutely diagnostic. Upon consideration of this diversity in development, we may reach the conclusion that one cannot use the broad categories for much more than general discussion, and that a subset of a category may have different developmental rules than its relatives.

The diversity of cell types is substantial. Among the photoreceptors there are rods and cones. Depending on species, however, cones may be subdivided on structural grounds and/or visual pigment content into two, three, or as many as seven subtypes (see reviews by Liebman, 1972; Munz and MacFarland, 1974; Marc, 1982a). Likewise, two or three HC types have been characterized in mammals (Boycott and Kolb, 1973; Boycott *et al.*, 1978; Wassle *et al.*, 1978a,b; Kolb *et al.*, 1980), and three or four in various fishes, reptiles, and avians (Stell, 1967; Stell and Lightfoot, 1975; Leeper, 1978; Gallego, 1983). The diversity of BC types is incompletely documented, but nine have been described in cat retina (Kolb *et al.*, 1981), seven in grey squirrel retina (West, 1978), 11 in rabbit retina (Famiglietti, 1981), and five types of mixed rod–cone BCs with an unknown number of pure cone BCs in fishes (Scholes, 1975; Ishida *et al.*, 1980). AC and GC types have been extensively studied, and although documentation is incomplete, it is currently thought that the tally in some vertebrates may be over 20 subtypes of each variety (e.g., Kolb *et al.*, 1981; Brecha, 1983). IPCs are poorly understood, some species may have no IPCs, others only one, and most teleost fishes clearly possess two types (Dowling and Ehinger, 1978; Marc and Lam, 1981; Marc, 1982b; Marc and Liu, 1984). To make matters more complicated, Mariani (1982) has described a new neuron that combines the features of GCs and IPCs: the biplexiform cell. Thus in a particular vertebrate retina, one may encounter 30–60 structurally and/or neurochemically unique neurons. An additional caveat is that we do not understand the evolutionary relationships among the retinas of the seven vertebrate classes, or even within a class, as some dramatic differences exist among orders. This fact should strongly temper extrapolations across species regarding developmental events.

B. Recognition of Cell Types

A primary requisite of charting the development of networks or microcircuits is the ability to recognize cell types or processes associated with certain cell types as maturation progresses. Until recently this effort has been exclusively anatomical and predominantly ultrastructural. In the gross sense, the maturation of elements of most of the broad cell categories has been followed in terms of the laminar positions the cells occupy in the mature retina: photoreceptors in the outer or distal retina, GCs in the inner or proximal retina, and the somas of HCs, BCs, ACs, and IPCs distributed in a large nuclear band between the two. As will be discussed, the process of lamination is one stage of network analysis, albeit coarse in terms of the circuitry we wish to comprehend.

On a finer scale, some ultrastructural studies have attempted to define how the broad cell categories mature from apparently elongated neuroblast-like elements to appropriately shaped and sited neurons. These studies rely heavily on recognition of ultrastructural features associated with each of the cell types beyond simple location and general shape. For example, photoreceptors are recognized not only by position but by their progressive elaboration, at the interface of the neural retina and the pigment epithelium, of stacks of membrane comprising the outer segment. At their proximal ends, the synaptic terminals of photoreceptors contain characteristic presynaptic specializations, synaptic ribbons. The outer plexiform layer is filled with the dendritic processes of BCs and HCs and these form postsynaptic contact with photoreceptors in characteristic positions: the HCs lateral to the ribbon synapse, and the BCs in a variety of positions depending on functional variety (invaginating ribbon synapses at the central position, and both invaginating and noninvaginating nonribbon synapses). The inner plexiform layer is the nexus of AC, GC, and BC processes and is itself highly structured. BCs form ribbon synapses onto various ACs and/or GCs, while ACs provide conventional synaptic input to both BCs and GCs. It is thus possible to use electron microscopy to observe the development of specific kinds of contacts associated with the broad cell categories.

Without question, the most powerful methods in the study of retinal network maturation involve the use of biochemical, specifically neurochemical, markers of cellular identity. The major methods now in use are (1) light and electron microscope autoradiography of neurotransmitter-specific high-affinity uptake, (2) immunocytochemistry of peptide-like immunoreactivity or synthetic enzyme immunoreactivity, and (3) histochemistry for monamines. The application and significance of most of these methods will become apparent as we review individual cases. It is thus possible to probe for the presence of neurochemical markers characteristic of mature cell types during histogenesis.

C. Basic Retinal Networks

Prior to embarking on a survey of developmental studies, it may be useful for those unfamiliar with retinal structure, chemistry, and function to survey the general organization of one specific retina; aficionados of circuitry may be satisfied to bypass this section.

I will employ the retina of the goldfish (*Carassius auratus*), a teleostean cyprinid fish, as an exemplar for this review and note relevant variations in other species elsewhere. The basic set of neurons with which we have to contend is shown in Fig. 1, although it is admittedly incomplete. The goldfish retina is known to possess eight types of photoreceptors: rods and seven kinds of cones composed of six morphological and three spectral types (Marc and Sperling, 1976a,b; Stell and Hárosi, 1976).

This array of photoreceptors possesses stereotyped patterns of connections with the second-order elements in the network: HCs and BCs. Goldfish HCs are of four types: three cone horizontal cells (H1, H2 and H3 HCs) and the rod HCs (Stell and Lightfoot, 1975). BCs of the goldfish have either mixed rod and cone inputs or pure cone inputs (Stell, 1967; Scholes, 1975). The five cell types shown in Fig. 1 are the mixed variety and they play a predominant role in our investigations of connectivity and function in the fish retina.

The explanation of the types requires a slight diversion. Anatomical and physiological studies of the cat retina produced evidence of an organizing principle for the seemingly elaborate and incomprehensible maze of the inner plexiform layer. Famiglietti and Kolb (1976) proposed that the IPL was subdivided into a distal (sublamina a) and a proximal (sublamina b) portion. Specifically, they proposed that the mammalian cone BCs that hyperpolarized in response to light and passed this signal directly to GCs (i.e., OFF-center GCs) made their synaptic connections in sublamina a; conversely, cone BCs that depolarized in response to light and conveyed that signal directly to GCs (i.e., ON-center GCs) were confined to sublamina b. As expected, this model has some exceptions and has undergone some requalification, but its general truth is well established in most classes of vertebrates, including teleost fishes (Famiglietti *et al.*, 1977). Thus, goldfish mixed BCs can be subdivided into two categories based on the sublamination of their synaptic terminals in the IPL and their physiological properties (Famiglietti *et al.*, 1977; Stell *et al.*, 1977; Ishida *et al.*, 1980): (1) center-depolarizing or ON-center BCs have their synaptic terminals in sublamina b, are of three varieties with respect to proportion of rod, red cone, and green cone inputs, and are thus designated Bb1, Bb2, and Bb3 BCs; (2) center-hyperpolarizing or OFF-center BCs have their synaptic terminals in sublamina a, are of two known varieties regarding rod and cone proportion, and are designated Ba1 and Ba2 BCs. Their forms, inputs, and outputs are partially summarized in Fig. 1. Cone BCs are more complicated yet and are not included in this review.

ACs are a diverse group of interneurons involved in the lateral disposition of information among the vertical cone/rod → BC → GC chains. While our knowledge of photoreceptor and BC morphology and function is largely (not wholly) independent of neurochemical markers, our knowledge of AC connectivity, form, and function is almost exclusively due to the localization of neurochemical markers in unique subsets of the AC population. A notable exception is the

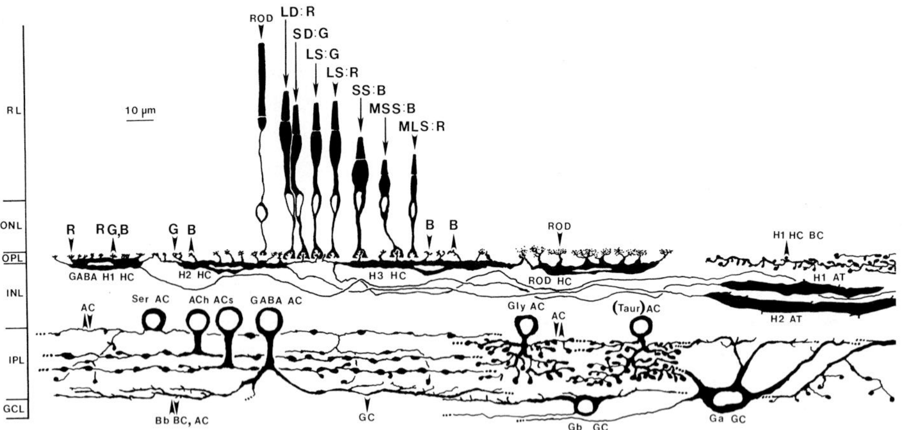

FIG. 1. Scale silhouette drawing of some of the unique classes of neurons known for the goldfish retina. This is part of the diversity for which we must account during the maturation of the retina. Arrowheads indicate direction of synaptic transfer; parentheses indicate provisional neurotransmitter identity; question marks indicate that the inputs–outputs are unknown. Most of the drawings are based on the following references: Ishida *et al.* (1980), Marc and Sperling (1976a), Stell and Lightfoot (1975). Large type abbreviations: B, blue-sensitive cone; G, green-sensitive cone; LD, long member of a double cone; LS, long single cone; MLS, miniature long single cone; MSS, miniature short single cone; R, red-sensitive cone; SD, short member of a double cone; SS, short single cone. Small type abbreviations: a1, type 1 OFF-center bipolar cell; a2, type 2 OFF-center bipolar cell; AC, amacrine cell; ACh, acetylcholine; b1, type 1 ON-center bipolar cell; b2, type 2

elegant Golgi impregnation and electron microscopy done by Kolb and colleagues (e.g., Kolb, 1979; Kolb *et al.*, 1981) in the cat retina. The best-known neurochemical varieties of ACs are indicated in Fig. 1. ACs utilizing γ-aminobutyric acid (GABA) as their transmitter have been identified through autoradiographic studies of high-affinity [^{3}H]GABA uptake (Marc *et al.*, 1978) and immunocytochemical localization of binding sites for antibodies against fish glutamic acid decarboxylase (GAD) (Lam *et al.*, 1979; R. E. Marc, D. M. K. Lam, and W.-L. S. Liu, unpublished data). They synapse directly onto Bb BCs and receive direct input from them (Marc *et al.*, 1978); GABAergic ACs make contacts with other ACs and GCs as well (J. F. Muller and R. E. Marc, unpublished data). There are two systems of unistratified acetylcholine-utilizing ACs identified by high-affinity [^{3}H]choline uptake (R. E. Marc and W.-L. S. Liu, unpublished data); little is known of their distributions or contacts. Putative serotoninergic ACs are primarily unistratified cells confined to the most distal part of the IPL with some dendritic trailers extending into other layers. They have been identified by high-affinity [^{3}H]serotonin uptake and serotonin-like immunoreactivity (Marc, 1980, 1982b; Tornquist *et al.*, 1983; Marc and Liu,

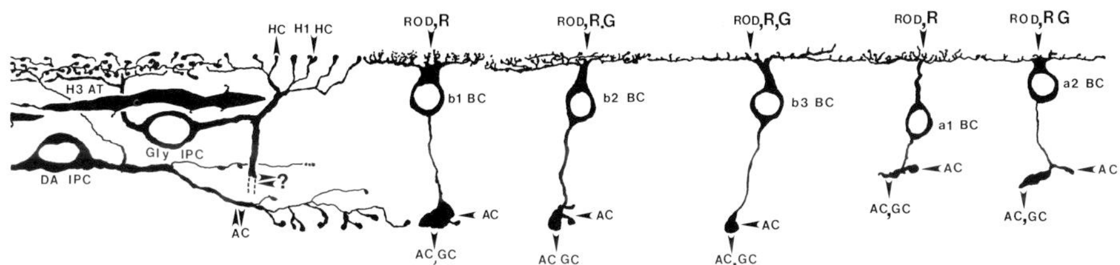

ON-center bipolar cell; b3, type 3 ON-center bipolar cell; Bb, ON-center bipolar cell; BC, bipolar cell; DA, dopamine; GABA, γ-aminobutyric acid; Ga, OFF-center ganglion cell; Gb, ON-center ganglion cell; GC, ganglion cell; GCL, ganglion cell layer; Gly, glycine; HC, horizontal cell; H1, red-dominated spectrally monophasic horizontal cell; H2, red-depolarizing/green-hyperpolarizing, spectrally biphasic horizontal cell; H3, green-depolarizing/blue + red-hyperpolarizing, spectrally triphasic horizontal cell; H1 AT, axon terminal of H1 horizontal cell; H2 AT, axon terminal of H2 horizontal cell; H3 AT, axon terminal of H3 horizontal cell; INL, inner nuclear layer; IPC, interplexiform cell; IPL, inner plexiform layer; ONL, outer nuclear layer; OPL, outer plexiform layer; RL, receptor layer; ROD HC, rod horizontal cell; Ser, serotonin; Taur, tuarine.

1985b), and monoamine fluorescence techniques (Ehinger and Florén, 1978). Their primary connections are with other ACs, some cone BCs, and some GCs. High-affinity [^{3}H]glycine labeling is the sole technique available for the structural characterization of glycinergic neurons. Glycinergic ACs of the goldfish retina are numerous, with small, diffuse arbors and contacts largely restricted to ACs and GCs (Marc and Lam, 1981; Marc, 1982b). Other ACs abound in the fish IPL but we will not discuss them here.

There are two varieties of IPCs in the goldfish retina. The first, dopaminergic IPCs, have been extensively characterized by Dowling and Ehinger (1975, 1978). Their somas are located in the AC layer. In the IPL, dopaminergic IPCs receive input from ACs and provide output to ACs; in the OPL they synapse upon H1 HCs and some BCs. Glycinergic IPCs were discovered by Dominic Lam and myself (Marc, 1980, 1982b; Marc and Lam, 1981). Their somas are found within the INL and they have processes that extend upward to the OPL where they receive extensive conventional synaptic input from H1 HCs and are apparently presynaptic to some HC dendrites. Furthermore, they are postsynaptic to conventional synapses made by the axon terminals of cone horizontal cells

(Marc and Liu, 1984). From this dendritic field, a stout process descends into the IPL. The connections of glycinergic IPCs in the IPL are not known but it is clear that they constitute a signal pathway to the IPL that bypasses bipolar cells.

Finally, there are several types of GCs, but only two are shown in Fig. 1 to indicate the generic types with which we will be concerned: ON-center GCs that ramify in sublamina b and are referred to as Gb GCs, and OFF-center GCs that ramify in sublamina a and are referred to as Ga GCs.

There are significant interspecies variations on the types, numbers, and forms of these cells and they will be noted where necessary. Given this array of elements and some of their connective patterns (there are more than have been indicated), how are we to unravel the differentiation of specific cell types and their ultimate connectivity? I believe we must simply focus on a few well-described neurons and their connections. Indeed, this has been the strength of neurochemical markers: they allow us to ignore the vast, complex background of differentiations, thus providing a "figure versus ground" context to observe the successive alterations in a single (or at least restricted) population of neurons as development progresses. Physiological measurements have much of this character but are, as previously noted, largely constrained to studying postconnection development.

FIG. 2. Some of the major classes of microcircuits contributing to the center-surround organization of retinal GCs. Abbreviations as in text and Fig. 1. Each fundamental microcircuit is delineated by a dashed border and numbered. Large arrowheads are presynaptic terminals associated with the vertical or "straight-through" transmission line to GCs. Medium-sized arrowheads are presynaptic elements linked to or arising from IPCs. Small arrowheads are presynaptic elements involved in local feedback loops. The solid dots are postsynaptic sites, unlabeled dots are sites of presumed sign-conserving or conventional "excitatory" synaptic transfer, and dots indicated by "i" are sites of sign-inverting or conventional "inhibitory" synaptic transfer. Synapses marked by * are of unknown sign and the dopamine IPC to H1 HC transfer is marked by a star to indicate that its synaptic effect is primarily mediated by a second-messenger system (adenyl cyclase). (a) Organization of three microcircuits contributing to the ON-center/OFF-surround system. Circuit 1 represents the passage of photic signals from cones directly to GCs. The important feature of this network is the sign-inverting synapse between the cone and the Bb BC. Circuit 2 is the opponent surround network for the OPL, composed of a layer of HCs coupled by gap junctions (three horizontal lines connecting the cells). The HCs engage in local feedback circuits with cones as described in the text. Circuit 3 is another surround path composed of GABA ACs in the IPL. They form local feedback networks with Bb BCs in the same fashion as HCs feed back onto cones. (b) Partial schematic of the organization of the OFF-center/ON-surround pathway. The vertical microcircuit for this pathway is identical to circuit 1 except that the synaptic transfer from cones to Ba BCs is sign conserving. Circuit 2 contributes to this path in the same manner as the ON pathway. Circuit 4 is a network similar to circuit 3, except that it involves a different morphological class of ACs (possibly taurinergic). (c) The two known IPC microcircuits. Circuit 5 shows the communication of signals arising in the IPL from ACs to dopaminergic IPCs that are presynaptic to H1 HCs. The primary effect of this pathway seems to be control of coupling among H1 HCs. Circuit 6 shows the organization of glycinergic IPCs, receiving input from H1 HCs and transferring it to the IPL.

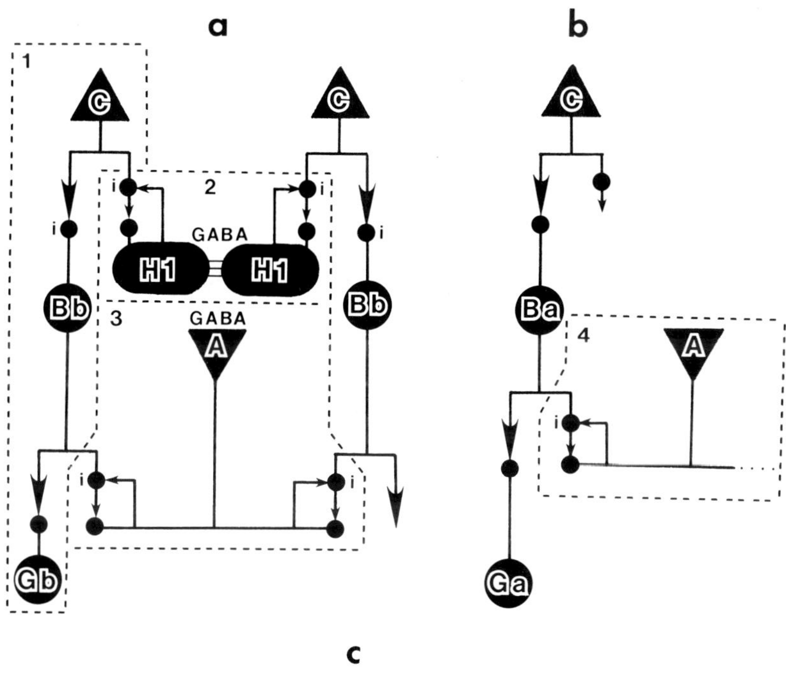
a
b
1
C
C
C
2
i
GABA
H1
H1
Bb
Bb
Ba
3
GABA
A
4
A
Gb
Ga

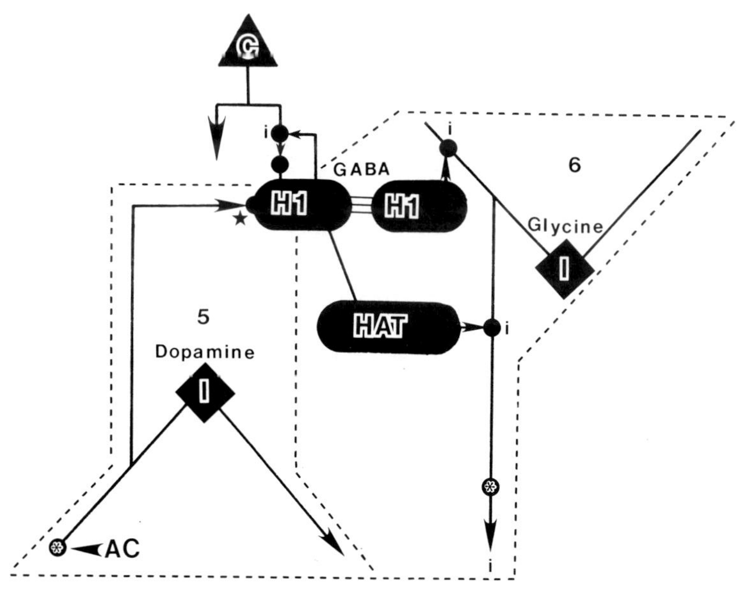
c
C
i
GABA
H1
H1
6
Glycine
I
HAT
5
Dopamine
I
AC

In any event, the goldfish retina possesses basic microcircuits that serve as target networks for the differentiating fish retina, and as general models (with modifications) for other vertebrate systems (Fig. 2). One example involves the vertical pathway of cones → Bb BCs → Gb GCs (Fig. 2a, circuit 1) and two lateral pathways of cones → H1 HCs → cones (Fig. 2a, circuit 2) and Bb BCs → GABAergic ACs → Bb BCs (Fig. 2a, circuit 3). The physiological features of the network are as follows. Red-sensitive cones are presynaptic to both Bb BCs and H1 HCs. The synaptic transfer from cones to Bb BCs is sign inverting and that to H1 HCs is sign conserving. This means that a light stimulus falling on a cone will hyperpolarize it and consequently *depolarize* the Bb BC and *hyperpolarize* the H1 HC. Since horizontal cells are electrically coupled to each other by low-resistance gap junctions, a light falling on a distant cone may influence the central cone (i.e., the cone in circuit 1, Fig. 2a) in the following manner: a strong stimulus to the distant cone will hyperpolarize it and the sign-conserving property of cone → H1 HC transmission will lead to a hyperpolarization in the HC. The signal will spread decrementally through the electrically coupled HC layer and the H1 HC presynaptic to the central cone will hyperpolarize. The sign-inverting H1 HC → cone path will lead to a small depolarization in the cone which, by a second sign inversion, leads to a hyperpolarization in the Bb BC. This generates, in part, the well-known ON-center/OFF-surround property of BCs. The second lateral path is quite similar. A Bb BC distant from the central Bb BC is depolarized by illumination of the cones providing its input. In fishes, the outputs of all mixed rod–cone BCs appears to be sign conserving (Naka, 1977). Thus the deplarized Bb BC will effect the depolarization of the GABAergic ACs to which it is presynaptic. The GABAergic AC will in turn hyperpolarize the central Bb BC, once again leading to the same kind of ON-center/OFF-surround property. That the Bb BC transfers its voltage responses directly to the Gb GC through a sign-conserving synapse means that the Gb GC is endowed with ON-center/OFF-surround properties as well. It is also likely that the Gb GCs receive direct GABAergic AC input (V. F. Muller and R. E. Marc, unpublished data). This is a basic circuit, repeated over the extent of the the retina as a matrix of circuits. Note that it is decomposable into three overlapping microcircuits as indicated in Fig. 2a.

Figure 2b depicts the fundamental elements of the OFF-center/ON-surround channel that differ from those in Fig. 2a. The organization of the vertical channel is similar to circuit 1 and the lateral channel (circuit 2) is also used. The GABAergic AC may not be here and another, yet unknown, transmitter system used, as depicted in Fig. 2b, circuit 4. Finally, Fig. 2c indicates how the horizontal cell serves as a nexus for signal transfer to and from the IPL via dopaminergic (circuit 5) and glycinergic (circuit 6) IPCs, respectively. With these concepts in hand, we will review our current knowledge regarding how some of these associations might develop.

II. The Development of Retinal Layers

The retina develops by the invagination of the distal half of the optic vesicle. The apposition of the inner faces of the distal and proximal halves of the collapsing vesicle forms the presumptive retinal–pigment epithelium interface, slightly separated by the remainder of the ventricular space. Detailed descriptions of lens–retina–pigment epithelium interactions and growth are available in many reviews and need not be repeated here. One of the first gross histological indicators of specialization in retinal differentiation is the process of separating cell bodies into layers destined to become the outer nuclear layer (ONL), inner nuclear layer (INL), amacrine cell layer (ACL), and ganglion cell layer (GCL). The immature retina is composed of proximodistally elongated neuroepithelial cells (NCs), many of which have processes contacting the vitreal remnant, and proximally directed processes of varying lengths. The NCs simultaneously transform and migrate; during this process, and long before any synaptic contacts are made, NCs apparently make neurochemical commitments as well as select their general cell category.

A. *Cell Sorting*

We do not know the signals that promote the differentiation of layers, but there is good reason to suspect that many of them involve interactions among neighboring cells. The process of "sorting" among cell types has been examined extensively in the embryonic chick retina, stemming originally from the efforts of Moscona (1965), who described the reaggregation of dissociated embryonic chick retinal cells into histotypic structures. Sheffield (1982) has demonstrated that dissociated and density-gradient fractionated cells from the 14-day embryonic chick retina (which has well-defined retinal layers) may be rhodamine or fluorescein labeled to mark their positions in a variety of sorting assays. Some fractions appear to reassociate *in vitro* with characteristic patterns depending on the mix of cells; the cell fraction presumedly derived from GCs seems to prefer assuming a cortical position around a core of cells from other fractions. This is reminiscent of the histotypic structures observed in retinoblastomas and suggests that there are interfacial binding preferences as well as homotypic binding preferences. Ignoring, for the moment, issues of extracellular matrix anisotropy and mechanisms of cell mobility, it is quite plausible that space and time variations in the ratios of heterocellular binding events (binding among unlike members) to homocellular binding events (binding among like members) can confer a spectrum of configurations from fully mixed through sorted with interfacial zones to fully segregated (Steinberg, 1962; Steinberg and Poole, 1981). Cell-surface components that may participate in cell adhesion are thus of great interest.

Thiery *et al.* (1977) and Rutishauser *et al.* (1976, 1978a,b) described the isolation of a protein from chick neural tissue that apparently subserves some adhesive function in development. Subsequently, a few other putative cell-surface adhesion molecules have been described. Of interest in retinal development are (1) neural-cell adhesion molecule (N-CAM; Thiery *et al.*, 1977; Buskirk *et al.*, 1980; Edelman, 1983); (2) cognin (Hausman and Moscona, 1975, 1976, 1979); (3) calcium-dependent adhesion mechanisms (Brackenbury *et al.*, 1981). These three, at least, seem to possess different time frames of action, with the calcium-dependent system functioning in early chick embryos (6–7 days). This is the age at which lamination is beginning, but it significantly predates synaptogenesis.

Antibodies against N-CAM are able to disrupt the later process of lamination in cultured chick retinas (Buskirk *et al.*, 1980) and to inhibit axon fasiculation in chick dorsal root ganglion cells (Rutishauser *et al.*, 1978). As Buskirk *et al.* (1980) have noted, the disruption of layer formation by antibodies against N-CAM is only partial, leading to blurred borders and misplaced cells, but not preventing the process of layer formation itself. Thus some additional mechanism must be invoked to mediate the cell-category-specific process of layering. The detailed consideration of glycoconjugate biochemistry associated with retinal lamination is clearly outside the scope of this review, but this is an area of investigation which may be one of the most important for the elucidation of cell connectivity. Nevertheless, lamination is not connectivity per se, and the processes of cellular transformation, neurite extension, and contact are more proximal issues in understanding the emergence of networks.

B. Cell Specification

The major cell categories appear to arise in a stereotyped pattern in most vertebrates, as assayed by their withdrawal from S-phase activity. [^{3}H]Thymidine-labeling experiments in the frog (*Rana pipiens;* Hollyfield, 1968), African clawed frog (*Xenopus laevis;* Jacobson, 1968; Hollyfield, 1971), killifish (*Fundulus heteroclitus;* Hollyfield, 1972), chick (*Gallus domesticus;* Fujita and Horii, 1963; Kahn, 1974), goldfish (*Carassius auratus;* Ungar and Sharma, 1980), and mouse (Sidman, 1961; Blanks and Bok, 1977; Carter-Dawson and LaVail, 1979) indicate that future ganglion cells are the first to withdraw from the cell cycle and the remaining cell types follow a species-specific sequence. The exact profile for all cell types is not quite as clear. Part of this confusion may arise from the relatively crude way in which cells are classified in these studies (according to broad categories).

In the mouse retina, a population of HCs withdraws from S phase soon after the GCs (Sidman, 1961; Blanks, 1982), while BCs and photoreceptors continue to proliferate. Conversely, HCs in the fish retina seem to be among the last to withdraw (Hollyfield, 1972; Ungar and Sharma, 1980). This timing difference may have something to do with the fact that mammalian and fish horizontal cells are not biochemically equivalent cell types. As we shall see, the rabbit retina appears to harbor the early development of a set of HCs that may not survive to maturity (D. A. Redburn, personal communication). This may imply that there are retinal cell types whose appearance and disappearance may be cryptic, revealed only by the use of chemical probes. This is different from the phenomenon of cell death, which involves cells presumably equivalent to some surviving varieties. This is another issue we have little space to discuss herein but which is undoubtedly a critical event in retinal maturation.

How do cells transform themselves in the process of forming layers? This is a particularly difficult problem to address, and has been approached with both ultrastructural and Golgi-impregnation techniques to reveal alterations in form and position. The primary limitation on these approaches is that they are static. Interpreting sequential changes requires a great deal of inference where identifying a particular cell line is concerned and is constrained to evaluating four or five broad categories of cells rather than identifying one of the many dozen of unique types known to populate the adult retina. Although a developmental sequence is obvious from [^{3}H]thymidine studies, it is equally apparent that sufficient overlap occurs in withdrawal from S phase by the major cell categories (cf. Hollyfield, 1972), that a subset of any one category could easily be considered to have its own schedule, and that the gross scheme of a proximal → distal gradient in differentiation could be largely epiphenomenal with respect to subsequent connectivity.

Hinds and Hinds (1974, 1978, 1979) have examined the prenatal maturation of several cell types in the mouse retina. Most of their work has focused on embryonic days 13–17 (E13–E17). By utilizing reconstruction of long series of thin sections, the shapes, positions, and patterns of transformation of four types of NCs were charted. Most if not all NCs possess a ventricular process forming the distal border of the presumptive retina. As various classes of future second-order neurons transform and migrate, many release their ventricular contacts, and depending upon their destinies, extend or shorten their proximal or vitreal process.

Future photoreceptors retain their ventricular contact and possess a centriole pair that moves to a distal location in the cell beneath a knob or swelling that contains the beginnings of a cilium. Future HCs eventually release their ventricular contacts, migrate proximally, and withdraw their initially short vitreal processes (Hinds and Hinds, 1979). A more complicated sequence was proposed for ACs and GCs (Hinds and Hinds, 1974, 1979). The NCs destined for the inner retina appear to release their ventricular contacts and migrate proximally. The ventricular process is dramatically shortened and the vitreal process remains elongated. The future GC is finally positioned in the presumptive GC layer, wherein it begins to extend distally oriented dendrites and the vitreal process extends tangentially (i.e., in the plane of the retina) to begin forming an axon. This is not unlike the impression gained by the examination of Golgi-impregnated material (Morest, 1970; Nishimura, 1980), in which future GCs were seen to detach from the ventricular surface, migrate proximally, and extend a basal axon, followed by the slow growth of their dendritic arbor.

The differentiation of ACs is more complex and less well understood. Hinds and Hinds (1978) propose that future ACs cotransform with future GCs, migrate into the presumptive GC layer, and perhaps even form axons and initial dendrites. A secondary migration and transformation then occurs, in which future ACs move distally and withdraw their nascent axons. The notion that future ACs and GCs are intimately related cell types is supported by their general morphologies, the incidence of displaced GCs and ACs, and some comparative anatomy. The retinas of certain lampreys and hagfishes lack separate AC and GC layers. Instead, both ACs and GCs form the proximal limit of the INL and both cell groups send processes into a proximal plexiform layer (Walls, 1935); this could represent the primordial format. As the GC layer migrates across the IPL in other species, however, the intimate association of ACs may not have been lost, and only through a secondary process would most ACs regain their appropriate positions at the border of the INL and IPL. This is only one scenario, however, and we have no rigorous means to test it. A particularly difficult aspect of subsequent development is that the many subvarieties of GCs and ACs have dendritic terminations in particular layers of the IPL. Do these arbors arise simultaneously among matching populations of cells, or are they diffuse, poorly stratified, and unsynchronized, depending on secondary processes to direct their growth and connectivity? Such questions require more specific labels to reveal their mechanisms. In the next section we will consider the expanded view of layering and cell specification that neurochemical markers provide.

III. Neurochemical Differentiation in the Retina

Some of the major breakthroughs in charting retinal development have involved the use of biochemical and structural probes of neurochemical identity. The ontogenetic appearance of synthetic enzymes, release mechanisms, clearance mechanisms, neurotransmitter content, and receptor binding have been studied in several vertebrates, most intensively in nonmammalians. Recently, it has been possible to discern the future neurochemical statuses of many cell lines long before they synthesize or release a given neurotransmitter, because stereospecific presynaptic clearance mechanisms for some neurotransmitters are "installed" first. Thus, an autoradiographic means exists to chart the appearance of specific cell types, rather than simply follow a broad cell group. In this section we will examine the development of markers for several neurotransmitters, primarily nonpeptide systems. Much of the data will bear on both the cellular transformations that occur prior to synaptogenesis and the subsequent maturation of the networks that subserve adult visual function. Each transmitter substance will be considered separately because unique cell types are thus specified. However, the maturation of one neurochemical system may eventually be shown to be linked to another. Not all transmitter systems have been evaluated in depth and some important ones, such as acidic amino acid systems, will not be discussed.

A. γ-Aminobutyric Acid

γ-Aminobutyric acid (GABA) is one of the most prominent inhibitory transmitters in the retina; it is present in all vertebrate retinas examined so far and serves as a pathway for the transmission of surround information. Its distribution among cell types seems to vary among species (Marc and Lam, 1986), but ultrastructural evidence is needed to interpret apparent differences. GABAergic HCs are present in the adult retinas of almost all bony nonmammalians, but not in chondrichthyan or mammalian retinas, as assayed by autoradiographs (ARGs) of high-affinity [^{3}H]GABA uptake. All vertebrates possess some form of GABAergic AC system; some seem to have only one, whereas other may have several.

The retinas of several species have also been examined for the presence of glutamic acid decarboxylase (GAD), the enzyme responsible for GABA synthesis. The original and definitive experiments were performed by Lam (1975), who demonstrated that isolated goldfish HCs were capable of synthesizing [^{14}C]GABA from [^{14}C]glutamic acid *in vitro*. Subsequently, [^{3}H]GABA-accumulating neurons of the goldfish retina were also shown to possess GAD-like immunoreactivity (Lam *et al.*, 1979). Double-label experiments in my laboratory have shown, conclusively, that all [^{3}H]GABA-accumulating neurons in the goldfish possess GAD-like immunoreactivity (R. E. Marc, D. M. K. Lam, and W.-L. S. Liu, unpublished data). All of these efforts have culminated in the view that neuronal uptake of [^{3}H]GABA during *in vitro* experiments is a powerful and valid approach to identifying GABAergic neurons in the retina.

The first indication of GABAergic function during retinal development is the appearance of high-affinity GABA uptake (Hollyfield *et al.*, 1979; Lam *et al.*, 1982; Kong *et al.*, 1983; Marc and Summerall, 1980; Marc and Liu, 1986). The embryos of ectotherms may be isolated soon after fertilization and the optic vesicle exposed to saline solutions containing micromolar amounts of [^{3}H]GABA. When this is done, high-affinity uptake of [^{3}H]GABA may be observed as early as stage 32 in *Xenopus* (Hollyfield *et al.*, 1979) and stage 24 in the goldfish (Marc and Liu, 1986). In both these cases, the appearance of [^{3}H]GABA uptake is noticeable in the presumptive OPL within cells that appear to be horizontally elongated. It is reasonable to assume that the labeled cells are NCs recently transformed into HCs.

In the goldfish retina, [^{3}H]GABA uptake is associated with several cell groups at stage 24/25; new HCs, somas in the AC layer, diffuse NC-like processes and somas poorly localized in the INL, and a few cells in the GC layer (Figs. 3–6). The latter observation is important because the goldfish has sparse displaced GABAergic ACs. Thus, some of the future GABAergic ACs could have migrated through the GC layer, as proposed by Hinds and Hinds (1978) for mouse retina. By stage 25, usually the day of hatching, diffuse label in the INL has disappeared and only HCs and ACs remain labeled (Figs. 7–10). In both *Xenopus* and the goldfish, at the first appearance of HCs, there are no synaptic contacts among cell types (Figs. 4–6). Although the OPL has begun to form and there are some processes coursing through it, no synaptic ribbons are present in the photoreceptor cells and the photoreceptors are just beginning to construct outer segments, as indicated by the protrusion of a cilium into a small space between the pigment epithelium and the neural retina.

Even though cell division is slowing down in the fundus, and [^{3}H]thymidine studies indicate that most cells are finished with S phase, dividing cells are common in the ONL and INL while the HCs are maturing (e.g., Fig. 4). It is not possible to say whether some of these mitotic cells have or have not made commitments before their final division. Indeed, some of the mitotic figures in the OPL appear to be labeled, but only slightly. Since their volumes are considerably greater than those of new HCs, one cannot assert that the total label content of large dividing cells is less. A more rigorous test of this would be pulse-label studies using ^{3}H-labeled neurotransmitters, since it has been demonstrated that cell division can occur after an initial neurochemical commitment in sympathetic neurons (Patterson, 1981; Black, 1983). HCs are among the last to complete S phase in ectotherms (Jacobson, 1968; Hollyfield, 1972; Sharma and Ungar, 1980), and if their neurochemical specification is postmitotic, the commitment of some of them to GABAergic function would have to be made rapidly after the final division.

After photoreceptors have clearly matured and possess well-developed outer segments and synaptic ribbons (Figs. 8–10), GABAergic HCs can easily be characterized by electron microscope autoradiography: their horizontally elongated somas are quite evident (Figs. 9 and 10), though they lack the size of mature HCs, and label can be shown to be associated with processes occupying lateral positions in the terminals of cone photoreceptors (Fig. 10) just as in the adult. Importantly, these young H1 HCs possess no axons or axon terminals for

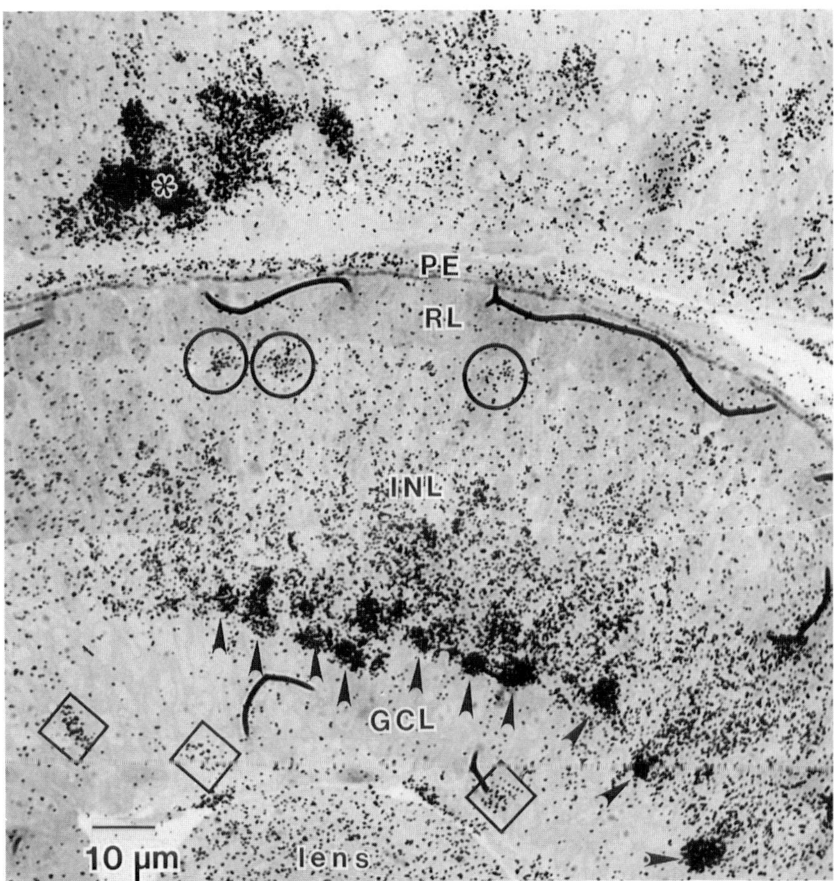

FIG. 3. High-affinity [^{3}H]GABA uptake by retinal cells in the stage 24/25 goldfish embryo. Retinal cells have segregated into three major layers: RL, INL, and GCL. There is no distinctive OPL or IPL, yet certain [^{3}H]GABA-accumulating neurons position themselves near the presumptive borders of these yet-unformed layers. Moderately labeled, horizontally elongated cells (circles) are positioned at the bases of the photoreceptors (PE, pigment epithelium). Many moderately to heavily labeled, vertically elongated cells seem to be streaming through the INL to form a band of densely labeled presumptive ACs (indicated by arrowheads). A few labeled cells are found in the GCL (diamonds); these are of interest since some displaced GABAergic ACs appear in the adult. Neurons in the brain are also maturing at roughly the same rate as retinal neurons. This is indicated by the accumulation of [^{3}H]GABA in a cluster of cells visible just beyond the pigment epithelium and choroid layers (*). The cluster later dissipates as the cells spread apart, probably due to the elaboration of neuropil. Light microscope autoradiograph, 0.5-μm section, 14-day exposure.

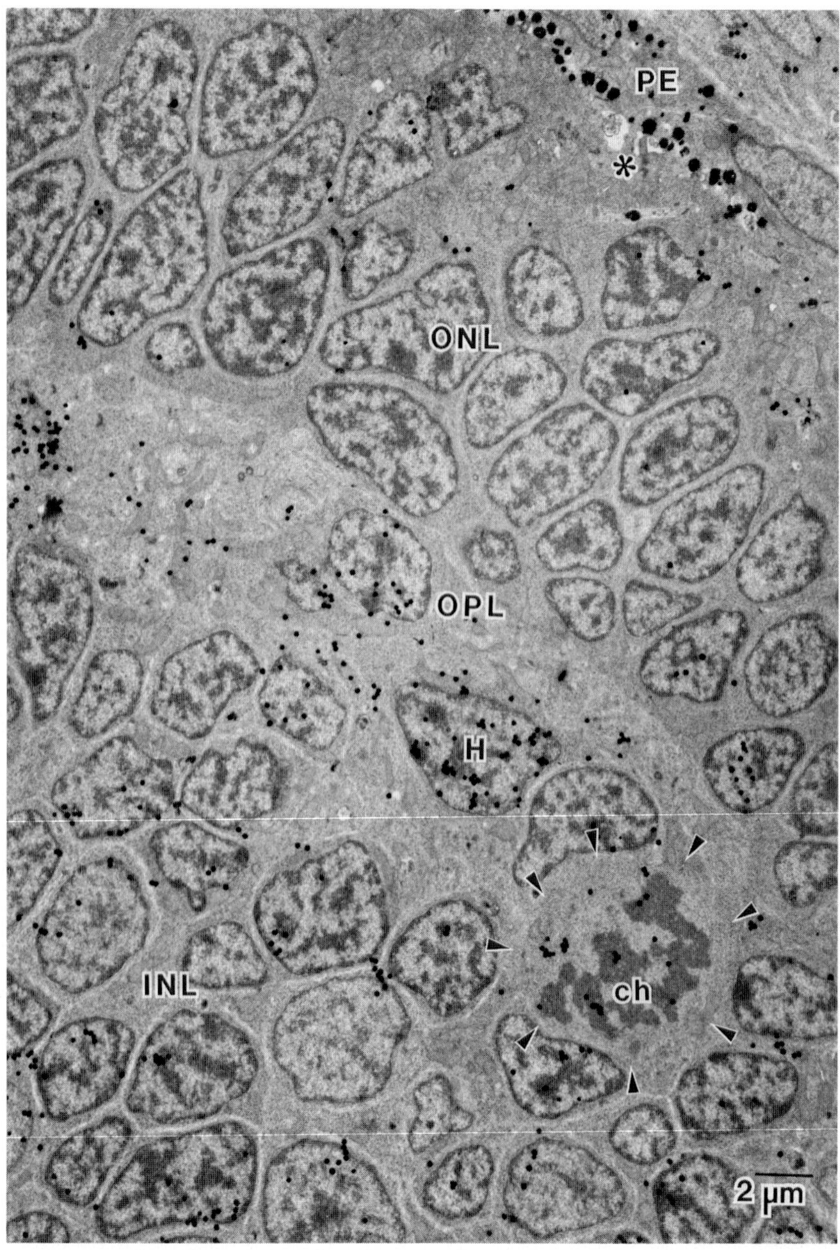

FIG. 4. Electron microscope autoradiograph of [^{3}H]GABA uptake in a stage 24/25 goldfish retina. At the ultrastructural level, the nascent OPL is visible as an interrupted band of processes about 2 μm thick. Positioned near this border is the soma of a presumed GABAergic H1 HC (H). A dividing cell (encircled by arrowheads) is located nearby and is lightly labeled. It may be that GABAergic commitment is made prior to final division. Ventricular remnants (*) between the PE and ONL are the regions into which the photoreceptors are beginning to extend ciliary elements in preparation for outer segment assembly. This micrograph is continuous with those in Figs. 5 and 6. Three-month exposure; ch, chromosome.

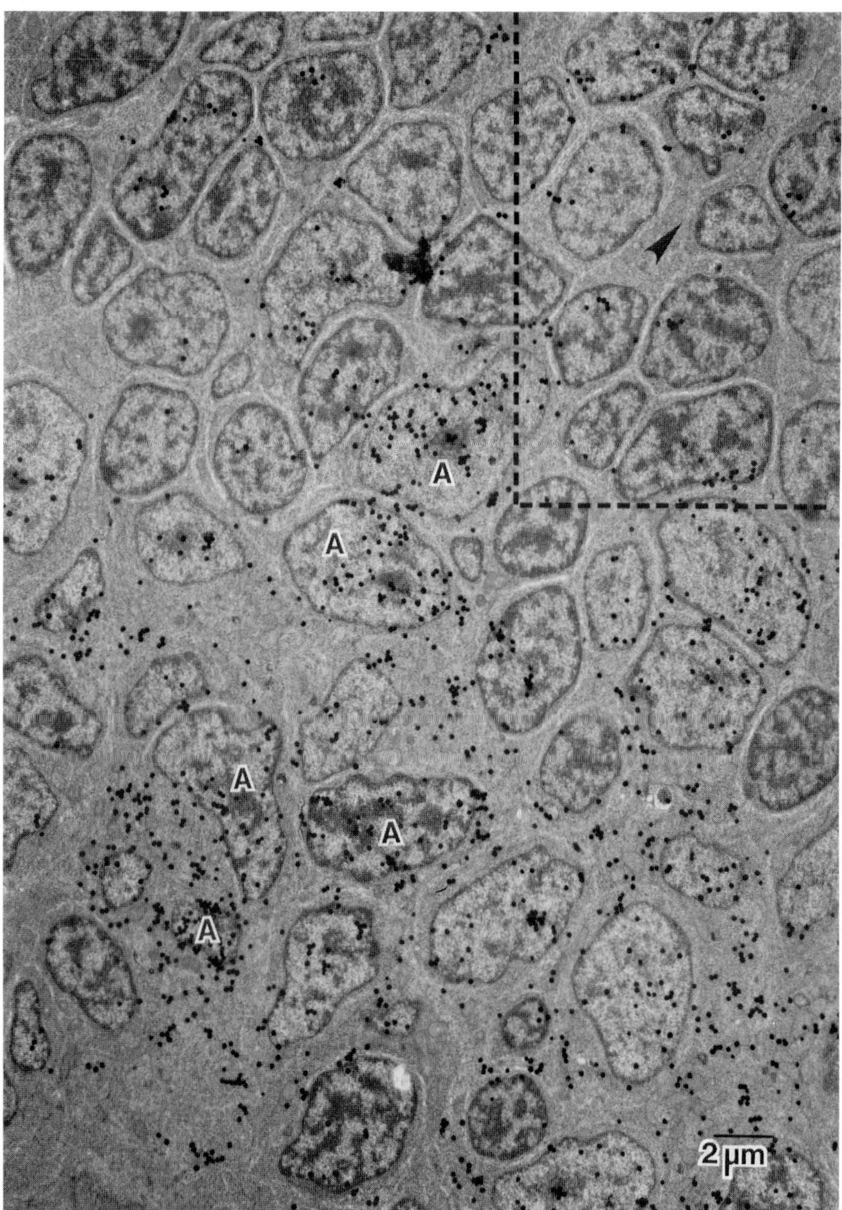

FIG. 5. A continuation of the field shown in Fig. 4. The dashed border in the upper right corner indicates the region of overlap with the lower left corner of Fig. 4. The heavily labeled somas of a number of ACs (A) are distributed through the INL. There are also many unlabeled cells which are probably the precursors of BCs. The arrow indicates the direction of the ONL. Three-month exposure.

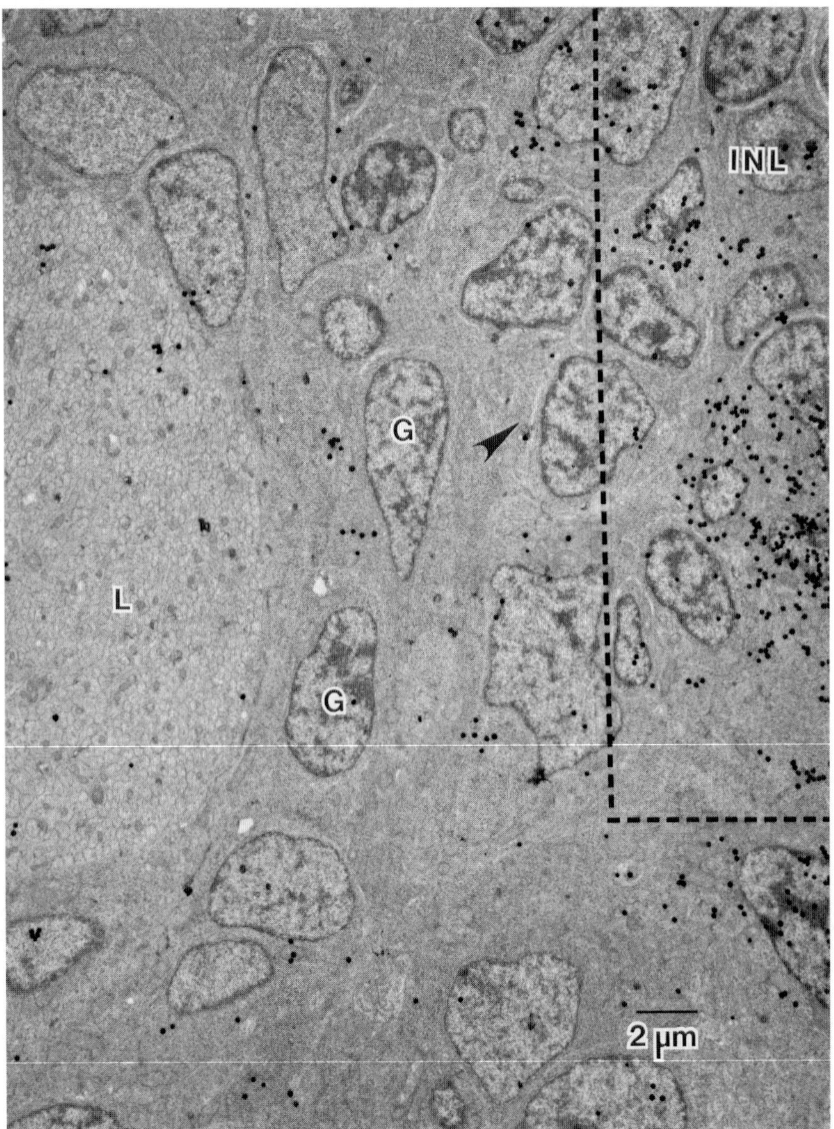

FIG. 6. A continuation of the field shown in Fig. 5. The dashed border in the upper right corner indicates the region of overlap with the lower left corner of Fig. 5. Note that there is no evidence of the IPL and the abundance of unlabeled GCs (G). Even in the absence of a barrier visible in electron micrographs, the ACs appear to stop migrating at a well-defined point. This does not preclude them having rapidly moved through the GC layer, as suggested by Hinds and Hinds (1979) for mouse ACs, but nevertheless the evidence seems more consistent with a model in which the ACs are specified early, migrate through the INL to a given level, and stop. The arrowhead indicates the direction of the ONL. L, Lens. Three-month exposure.

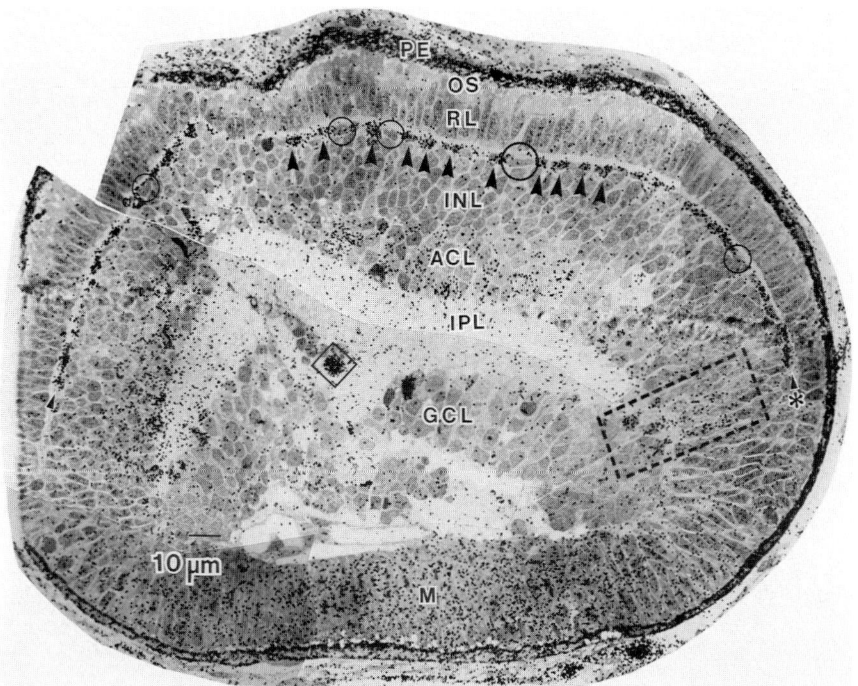

FIG. 7. Photomontage of GABAergic HCs and ACs in the stage 25/26 goldfish retina. The pigment epithelium (PE) is considerably thickened and contains dense pigment. Photoreceptors (RL) have now produced outer segments (OS), and the OPL is a distinct border that runs along the bases of the photoreceptors for most of the width of the eye (small arrowheads denote the two ends of the border). Well-labeled HCs line the OPL (medium arrowheads) interspersed with a few unlabeled HCs (circles), presumed to be non-GABAergic. The INL in the fundus is generally free of label, while many somas in the ACL are labeled. The section is somewhat oblique, since these eyes must be punctured to allow entrance of the label, and therefore the GCL has been cut obliquely. The GCL is mostly unlabeled but for a few isolated cells, one of which is indicated by the diamond. These older eyes afford a useful glimpse of the transition from immature retina associated with the margin (M) to mature retina at the fundus. At the right side (*) the OPL begins at the position where the photoreceptor cells are just beginning to form outer segments. Beneath that region is a section of immature retina (dashed border) exhibiting the same pattern of labeling as a stage 24 retina; elongated NC-like elements apparently increase in label density as they migrate closer to the border between the ACL and GCL. The GCL remains unlabeled, similar to the stage 24 GCL. Light microscope autoradiograph, 0.5 μm, 3-week exposure.

some time and do not exhibit any gap junctions with each other. The importance of the latter observation will be discussed in the section on dopamine (Section III,C).

In *Xenopus* the clearance mechanism for GABA is in place prior to the maturation of GAD activity and GABA release (Hollyfield *et al.*, 1979). The first indication of GABA synthesis is observed at stage 37/38 and it increases until it

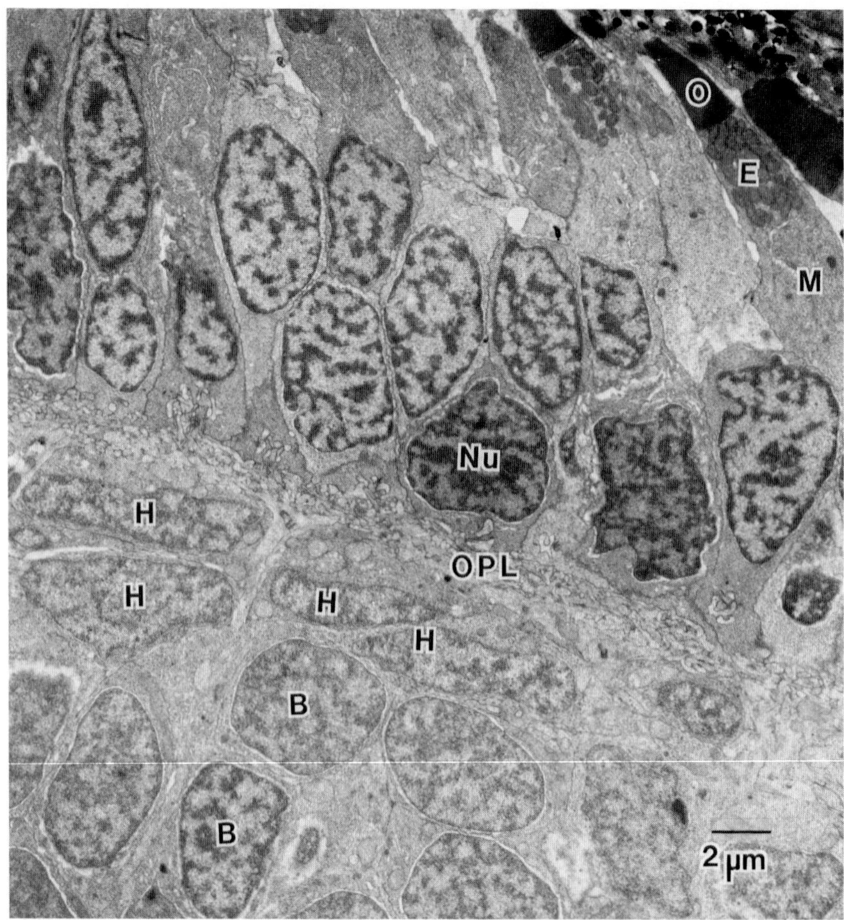

FIG. 8. Electron micrograph of a stage 26/27 goldfish retina showing the nearly mature form of many of the cells in the outer retina. Cones now have distinct outer segments (O), an ellipsoid (E), and a myoid (M), which separates the distal portions of the cone from its nucleus (Nu). The OPL is distinct and contains many laterally oriented processes. Several layers of small HC somas (H) are evident, suggesting that the distal–proximal segregation of GABAergic and non-GABAergic HCs is nearly complete. Compare with Fig. 7, in which the unlabeled and labeled HCs form a single row. B, Bipolar cells.

plateaus around stage 40 at nearly 75% of adult activity. In a similar manner, potassium-evoked release of preloaded [^{3}H]GABA cannot be observed until stage 37/38. This is somewhat puzzling, since recent evidence indicates that GABAergic ectotherm HCs can release some GABA via a nonvesicular mechanism that is relatively insensitive to extracellular calcium ion depletion or most

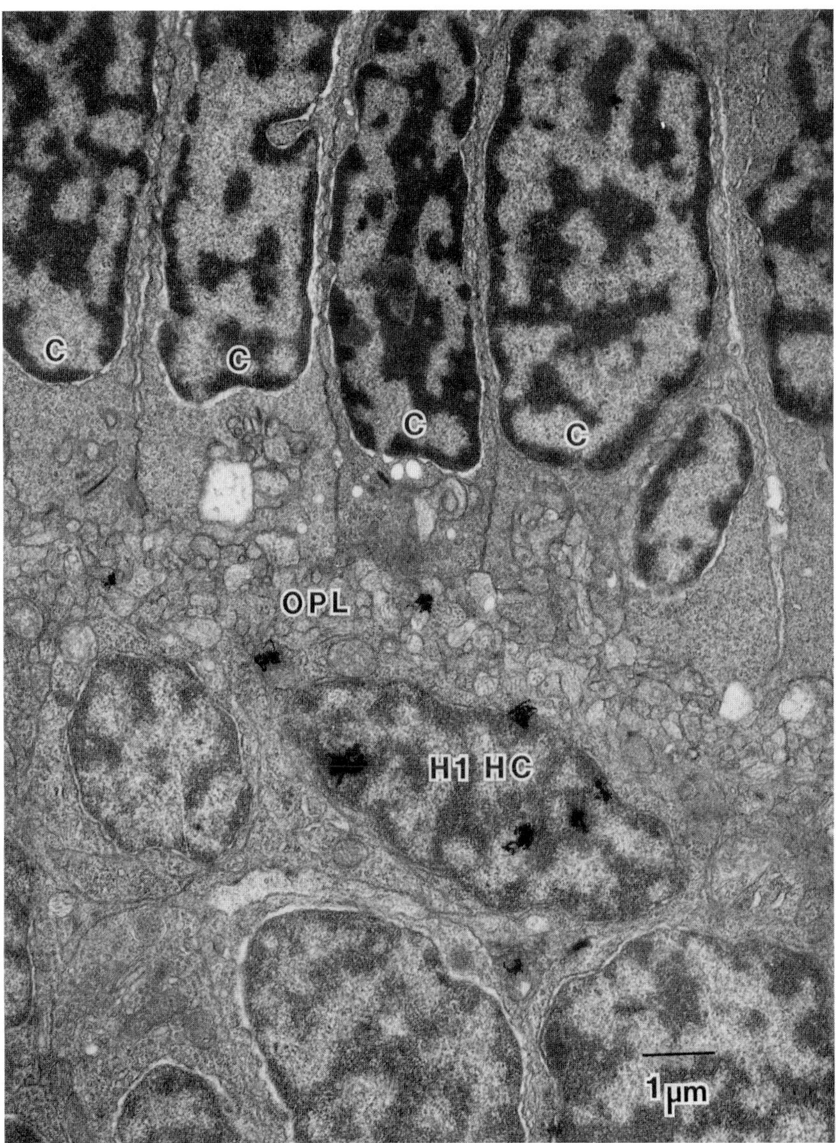

FIG. 9. A row of cones (C) exhibiting synaptic ribbons and many invaginating processes in a stage 26/27 goldfish. A single labeled cell (H1 HC) is present at the OPL border and it has the characteristic gross morphology of a GABAergic HC. Several aspects of fine structure are still absent, however, there are no conventional synapses from the HC to BCs or IPCs, nor are any gap junctions visible. Furthermore, the HCs have yet to elaborate axons and axon terminal expansions. One-month exposure.

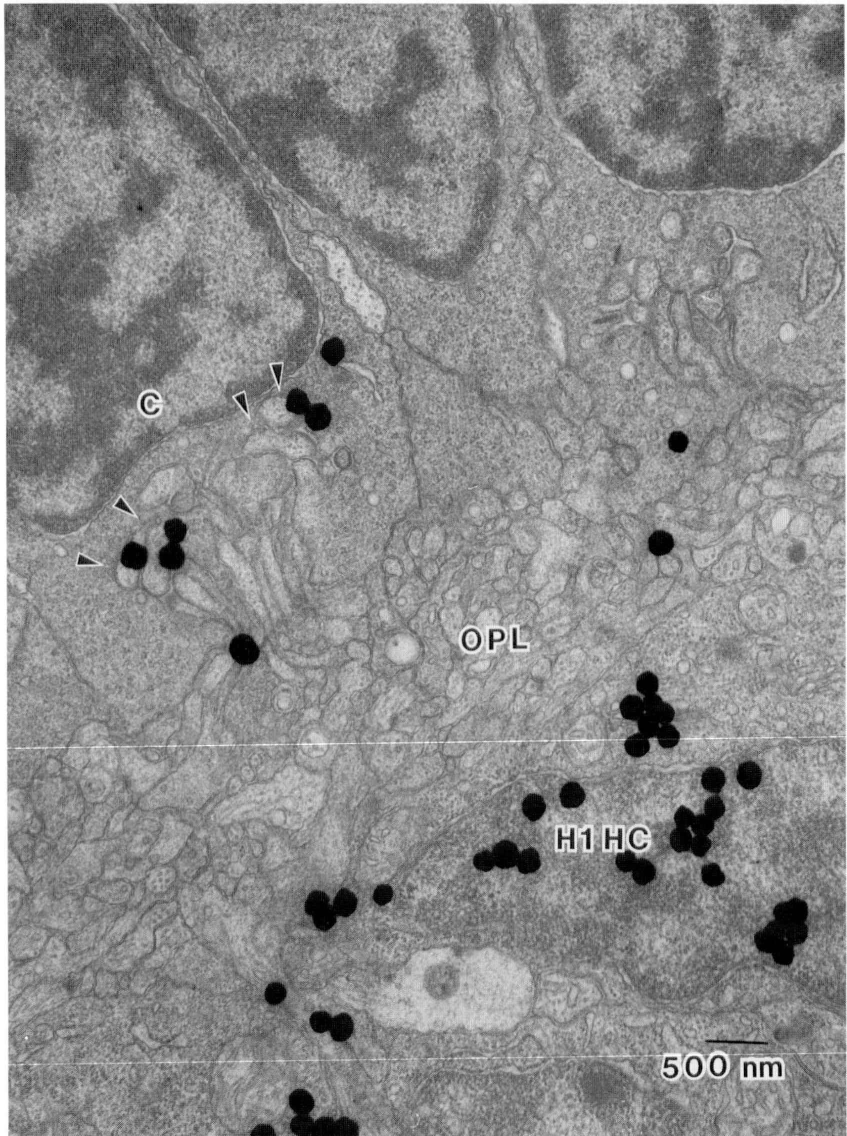

FIG. 10. Three-month exposure electron microscope autoradiograph of a stage 26/27 goldfish, demonstrating label preferentially associated with the lateral processes (arrowheads) of HCs in the cone invagination. Cones now possess abundant synaptic vesicles and ribbons, suggesting that the local feedback network of circuit 2 (Fig. 2a) is intact at this stage.

blockers of calcium function (Schwartz, 1982; Yazulla and Kleinschmidt, 1983; Ayoub and Lam, 1984). Extremely high levels of cobalt ions (i.e., concentrations in excess of 3 m*M*) can block this apparently calcium-insensitive mechanism, but magnesium ions cannot (Ayoub and Lam, 1984). Since the potassium-evoked release in developing *Xenopus* retinas was blocked by 10 m*M* cobalt, it is not possible to say whether a conventional vesicular mechanism was interrupted or the transport process contravened. Hollyfield *et al.* (1979) showed, however, that potassium-evoked release did arise from the HCs, and not ACs, by observing label loss autoradiographically. Further evidence supporting the notion that nonvesicular release plays a role at this stage comes from the inability of Witkovsky and Powell (1981) to find synapses in *Xenopus* horizontal cells until after stage 46.

Why should there be a discrepancy between the onset of uptake and release? We cannot be sure at present, but a possible cause is rapid metabolism of accumulated [^{3}H]GABA in the young HCs. Since there should be no intracellular pool of GABA prior to the synthesis of GAD, any accumulated GABA could be rapidly degraded if GABA transaminase (GABA-T) were present. Thus, little labeled GABA would be expected to be available for efflux until GAD levels were elevated. No one has demonstrated the presence of GABA-T in embryonic nonmammalian retinas at these early stages, although it is known that rabbit retinas show considerable basal GABA-T activity prior to the onset of significant GABA synthesis (Lam *et al.*, 1980).

Mammalian GABAergic neurons are ACs; no GABAergic HCs exist in the adult mammalian retina. In the rabbit retina, *in vivo* injections of [^{3}H]GABA will label a subset of the presumptive AC population even at birth, when photoreceptors have yet to form outer segments and the OPL is just barely defined (Lam *et al.*, 1980). Indeed, these presumptive neurons are evident in the embryonic rabbit retina at about embryonic day 22 (E22), about 9 days before birth (Fung *et al.*, 1982). As the retina matures, more well-defined labeled cells are evident. In qualitative agreement with results in the ectotherm retina, however, release is restricted to a low basal rate until after endogenous GABA levels rise. Rabbit GABAergic neurons, as characterized by GAD immunoreactivity (Brandon *et al.*, 1979), form three or four distinct strata, only weakly evident with *in vivo* [^{3}H]GABA-labeling metbods, so it is difficult to equate the cell types. The approximate numbers and distributions of the labeled cells appear to change little during development, other than some gradual thinning, thus giving some confidence that the same kinds of cells are labeled through maturation.

Are all such cells GABAergic? We candidly cannot say. The *in vivo* injection technique is used instead of *in vitro* approaches because of some peculiarities in mammalian retinal neurotransmitter uptake. The retinal glia, Müller's cells, avidly take up GABA, and thus constitute a formidable spatial buffer of GABA. The first cells to be labeled *in vitro* or *in vivo* are always the Müller's cells.

However, after some time delay (usually hours), the glia become unlabeled and the neurons labeled (Ehinger, 1977). The loss of label from glia is probably due to the action of glial GABA-T and the subsequent production of labeled substrates which constitute a pool for relabeling neurons. How much of the remaining activity is found as GABA? According to Lam *et al.* (1980), about three-quarters of the potassium-releasable tritium comigrates with GABA in high-voltage electrophoresis. Not knowing the actual partitioning of label and the relative terminal densities of various neurons hampers interpretation somewhat, but it is probably safe to say that at least half to three-quarters of the labeled cells in the developing rabbit retina are probably GABAergic.

At the time GABA release first appears, synaptic maturation of the rabbit IPL is only fractional (McArdle *et al.*, 1980). Conventional synapses are present at low density from birth, but ribbon synapses and a concomitant increase in the number of conventional synapses become apparent between days 7 and 9. The increase in release of preloaded [^{3}H]GABA closely parallels the maturation of conventional synapses and both GAD and GABA content (McArdle, 1980; Fung *et al.*, 1982), indicating that the bulk of GABAergic neurons are associated with the postnatal rise in synapse numbers.

As a final comment on putative GABAergic cells, D. A. Redburn (personal communication) has observed that rabbit retinas at postnatal day 1 (P1) possess a unique population of [^{3}H]GABA-accumulating cells that seem identical to HCs; there is even evidence of release from these cells. Furthermore, GAD immunoreactivity (IR) has been found in E17–P3 mouse retinas in HC-like elements, with a gradual decline to no detectable immunoreactivity in 4-week-old mice (Rusoff and Schnitzer, 1984). This transient and otherwise cryptic process may imply that there is indeed some "recapitulation" of phylogenetic structures during embryogenesis.

B. *Glycine*

The evidence for glycine as a retinal neurotransmitter is voluminous and has been exhaustively reviewed (Voaden, 1976; Marc, 1985). The sole method for the structural characterization of glycinergic neurons is autoradiography of cells labeled through high-affinity [^{3}H]glycine uptake. In all vertebrates examined so far (see Marc, 1985, for a review), glycinergic neurons are represented by a large population of ACs, usually with diffuse or weakly stratified arborizations. Most vertebrates possess a "second" glycine-labeled system that is thought to be glycinergic. In teleost fishes and anuran amphibians, glycinergic IPCs engage in synaptic interactions with horizontal cells and the inner plexiform layer (Marc and Lam, 1981; Rayborn *et al.*, 1980; Marc and Liu, 1984). In many other vertebrates either IPCs and/or BCs are labeled by [^{3}H]glycine (Marc, 1985).

Mammals clearly possess an assortment of [^{3}H]glycine-labeling BCs (McGuire, 1980; Sterling, 1982; Marc and Liu, 1984; Marc, 1985).

Rayborn *et al.* (1980) examined the maturation of glycinergic systems in *Xenopus* retina and discovered a pattern of development remarkably similar to that of GABAergic systems. At stage 31, uptake was poor and not well localized, but by stage 32 some concentration into a subset of the NC-like population was evident. AC-like morphology began to appear at stage 37/38. Note that this is many stages later than the time for clear visualization of GABAergic HCs by [^{3}H]GABA uptake autoradiography. Using [^{3}H]glycine preloading, Rayborn *et al.* (1980) also demonstrated the onset of glycine release capabilities at stage 42, much later than that seen for GABA release. Again, it appears that neuronal clearance mechanisms are activated prior to maturation of release processes. Conventional synapses in the *Xenopus* IPL appear around stage 40 and rapidly increase in number thereafter (Fisher, 1976), and the timing of glycine release is in good correspondence with that observation.

The goldfish retina exhibits a similar sequence of maturation of glycinergic ACs, with the first appearance of distinctively labeled elements somewhere around stage 25 (Fig. 11). As in the *Xenopus* retina, this system seems to develop a bit more slowly than GABAergic systems. However, we find no evidence of glycinergic IPCs at the time when glycinergic ACs are evident, and well past the time when GABAergic HCs have layered and made their contacts with photoreceptors (stages 25–27). We know that glycinergic IPCs constitute a significant alternate pathway for HC signals to reach the IPL; glycinergic IPCs receive direct synaptic input from HC somas and axon terminals (Marc and Liu, 1984). Furthermore, we have found no evidence of HC synapses, and there are no HC axon terminals by stage 27.

It is difficult to characterize the transformation of a given cell type because an intermediate stage may be short-lived. Thus a temporal sample may be inadequate. It has been demonstrated, however, that many developing ectotherm retinas display a gradient of maturation from margin to fundus (e.g., Hollyfield 1968, 1969). Figure 11 displays such a gradient in the maturation of glycinergic ACs. NCs near the margin exhibit nondescript, uniform labeling, but a border exists about where (presumably when, as well) the OPL forms: at this border, some future ACs exhibit enhanced [^{3}H]glycine uptake and beyond the border, some ACs are heavily labeled while the remainder of the cells in the INL decrease their uptake. One important aspect of this gradient is that the labeled cells have already assumed their positions in the AC layer, unlike early GABAergic ACs.

Mammalian glycinergic development has been examined in the rabbit retina by Kong *et al.* (1980) and Fung *et al.* (1982). The first evidence of glycinergic commitment occurs at about E25, once again later than putative GABAergic neurons (Fung *et al.*, 1982). The labeled cells are not well formed and it is difficult to assert that they will all preserve their identities to become the full

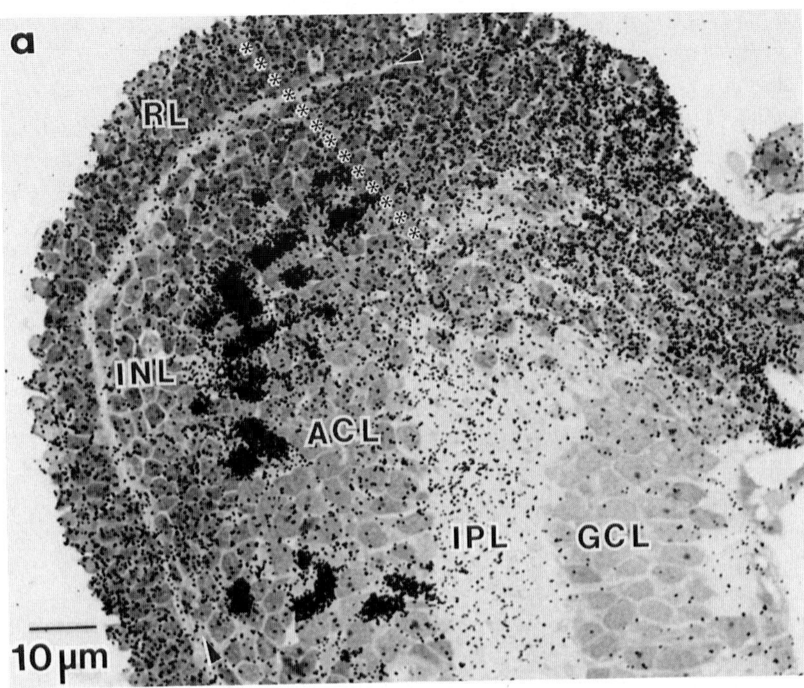
a
RL
INL
ACL
IPL
GCL
10 µm

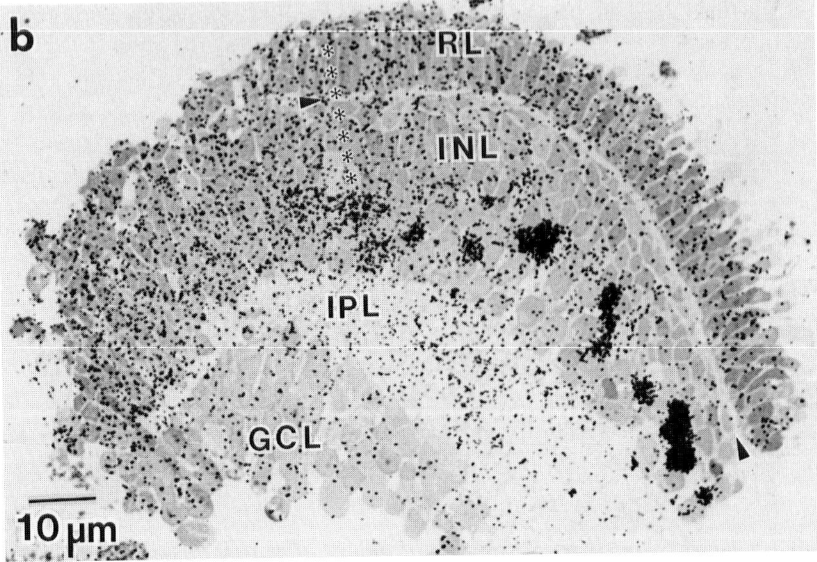
b
RL
INL
IPL
GCL
10 µm

cohort of adult glycinergic neurons. In truth, the distinctive diffuse but bistratified aborizations characteristic of the glycinergic neurons is particularly difficult to observe, and a number of cells in the GC layer also label profusely, which is uncharacteristic of the adult retina. This result points out the risks of intravitreal injections of ligands, especially for a common metabolite such as glycine. Calcium-dependent release from [^{3}H]glycine prelabeled cells does not occur until after birth, beginning at about day P7 and reaching 90% of adult levels by P12 (Kong *et al.*, 1980). This corresponds with the maturation of conventional synapses in the rabbit IPL (McArdle *et al.*, 1977) and is practically identical to the onset of release from [^{3}H]GABA prelabeled cells.

C. *Dopamine*

Catecholamine-containing neurons have been identified in virtually every vertebrate retina thus far examined, and in most cases it is clear that the dominant catechole, if not the only one, is dopamine. Methods used to characterize dopaminergic neurons include the use of aldehyde-based derivatization methods to document the presence of a catechole-derived fluorophor (e.g., Haggendal and Malmfors, 1965; Ehinger and Falck, 1969; Dowling and Ehinger, 1975, 1978), high-affinity [^{3}H]dopamine uptake autoradiography (Kramer *et al.*, 1970; Sarthy and Lam, 1979; Marc, 1980, 1982; Frederick *et al.*, 1982; Floren, 1983), monoamine-analog cytotoxicity (also based upon the selectivity of high-affinity uptake) to induce ultrastructurally recognizable synaptic degeneration (Dowling and Ehinger, 1975, 1978), and recently, localization of tyrosine hydroxylase-like immunoreactivity (TH-IR) (Brecha *et al.*, 1984; Nguyen-Legros *et al.*, 1981, 1983). Each of these techniques has been applied to the characterization of monoaminergic cell development in the vertebrate retina.

In *Xenopus* retina, Sarthy *et al.* (1981) have reported a pattern of maturation for dopaminergic ACs that is very different from that noted for GABA and

FIG. 11. High-affinity [^{3}H]glycine uptake in stage 25/26 goldfish retinas. These retinal fragments were extruded through slits in the eye and consisted of the marginal zones and a transition region leading to a mature area. (a) All layers are distinct in the mature area to the left of the border of asterisks (*****). Many ACs demonstrate heavy labeling, but the GCL is free of label. The label in the IPL also resembles the adult form, where the processes of glycinergic ACs are largely excluded from the most proximal region of the IPL (see Fig. 1). The extent of the OPL is indicated by arrowheads. The OPL begins in the immature region to the right of the asterisk border, and there is no evidence of *specific* [^{3}H]glycine uptake by any subset of the cells. There is diffuse labeling of all cells however. (b) A second fragment in which the immature zone is to the left of the asterisk border. In this preparation it is possible to see a gradation in selective [^{3}H]glycine uptake from no selectivity at the left margin, through partial selectivity in several cells immediately beneath the asterisk border, to strong selectivity in clusters of ACs in the mature region. There is no evidence for glycinergic IPCs at this stage. Light microscope autoradiograph, 0.5-μm section. 3-week exposure.

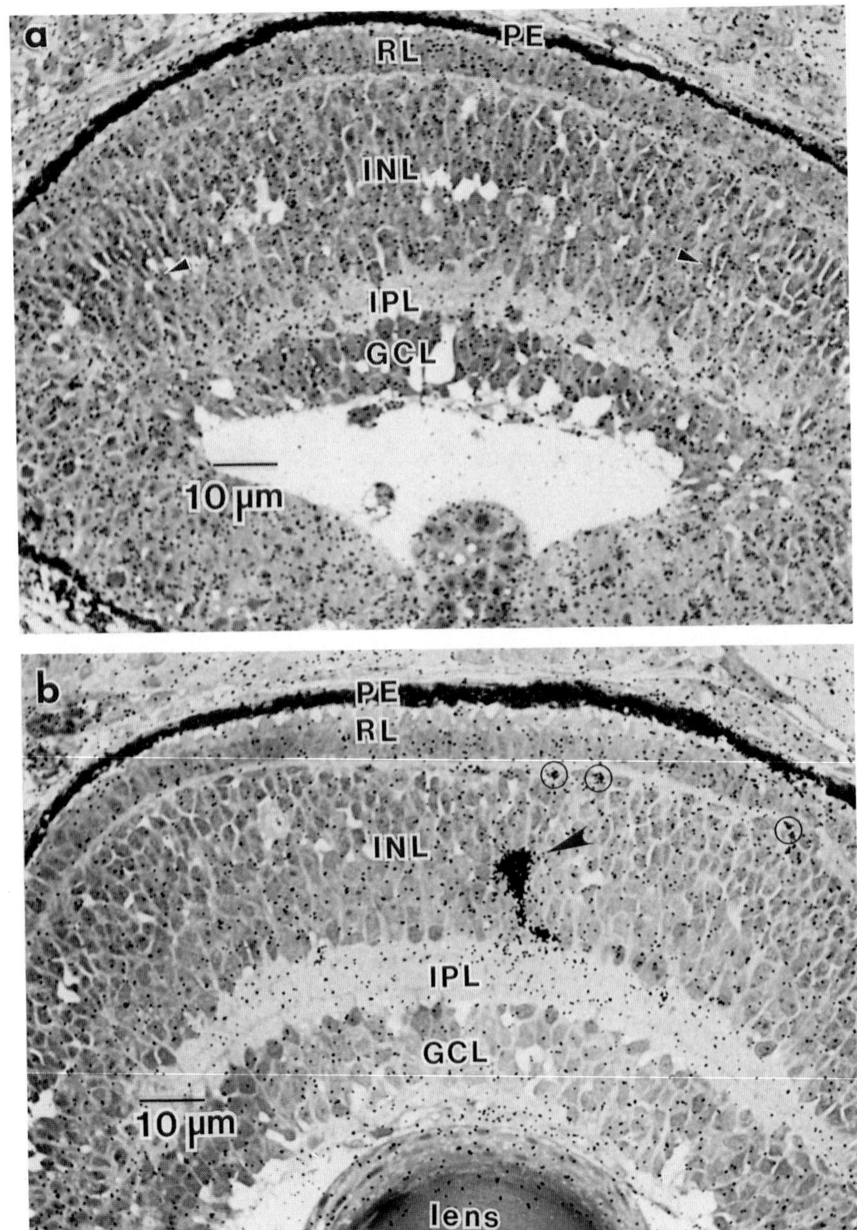
a
PE
RL
INL
IPL
GCL
10 µm
b
PE
RL
INL
IPL
GCL
10 µm
lens

glycine. The earliest indication of a putative dopaminergic system arises at stage 35/36, when the ability to synthesize [^{3}H]dopamine from [^{3}H]tyrosine appears. Much later, at stage 43, [^{3}H]dopamine uptake is detectable by autoradiography. Finally, release is measurable at stage 46. While the temporal separation between dopamine synthesis and release is partially explicable in terms of lack of maturation of the synaptic specialization necessary for exocytosis, the late differentiation of uptake mechanisms was unexpected. In fact, the late development of uptake is not supported by observations in the goldfish and Siamese fighting fish retinas (R. E. Marc, W.-L. S. Liu, and R. D. Summerall, unpublished data; Fig. 12), where IPC-like morphologies are detectable with [^{3}H]dopamine uptake at goldfish stages 25–27, just after the appearance of high-affinity [^{3}H]glycine uptake.

Exactly the same phenomenon was observed by Negishi *et al.* (1983) in goldfish retina using monoamine fluorescence methods. By taking advantage of the selective ability of monoamine neurons to transport and accumulate catechole and analogs, they were able to visualize catechole-like and indole-like fluorescence at stage 25/26. However, retinas at this stage did not possess any detectable endogenous fluorescence, indicating an ability to transport but not synthesize catecholes and indoles. Evidence of endogenous content was evident by stage 27. In sum, it seems that more in-depth evluation of monoamine maturation is needed to reconcile the ectotherm sequence, but the results for fishes are consistent.

In spite of this, one feature of dopaminergic function seems likely to have late development: dopaminergic control of HC coupling through gap junctions. Teranishi *et al.* (1983, 1984) have demonstrated that the intercellular low-resistance pathway among H1 HCs of a cyprinid fish (carp), mediated by extensive gap junctions, is negatively modulated by dopamine, dopamine agonists, cyclic AMP analogs, and agents which activate cyclic AMP synthesis (such as forskolin). These compounds increase membrane resistance and restrict the diffu-

FIG. 12. Maturation of high-affinity [^{3}H]dopamine uptake in the maturing Siamese fighting fish retina (*Betta splendens*), which has the same progression of maturation as the goldfish retina. (a) At a stage comparable to goldfish stage 25, the *Betta* retina has photoreceptors with small outer segments, a distinct lamination of the retina, including the IPL. The arrowheads indicate the width of the mature zone in this specimen. At this stage, GABAergic and glycinergic cells are abundant yet there is no evidence of high-affinity [^{3}H]dopamine accumulation anywhere in the eye. (b) At stage 26, the IPL is thicker and the mature zone comprises more of the retinal width. A [^{3}H]dopamine-accumulating cell is now present (arrowhead), although it appears to have just begun forming processes. No distinct terminals are present in the IPL and only a few lightly labeled clusters are in the OPL (circles). Mature dopaminergic IPCs possess profuse terminal arbors in the OPL, so that the the dopaminergic IPC shown here must be considered quite immature. Indeed, there is no evidence of dopamine synthesis in goldfish dopaminergic cells until stage 27 (Negishi *et al.*, 1983). Circuit 5 (Fig. 2c) may not be functional until quite late in development. Light microscope autoradiograph, 0.5-μm section, 3-week exposure.

sion of small dye molecules among cells. Conversely, dopamine antagonists or neurotoxins that selectively kill catechole-accumulating neurons lead to decreased resistance and massive diffusion of small molecules among HCs (Cohen and Dowling, 1983; Teranishi *et al.*, 1984).

As noted earlier, gap junctions are not evident among goldfish HCs at stage 25–27 and must arise later. Whether they arise when the dopaminergic IPCs exhibit synaptic release is not known. One of the more interesting questions is whether the synthesis of gap junction protein is developmentally modulated by dopamine. It is now clear that the primary effect of dopamine on fish HCs is to activate adenyl cyclase through type D1 dopamine receptors. In avians the adenyl cyclase system is evident by day E7 (DeMello, 1978) and dopamine-activated, D1-mediated increases in cyclic AMP appear by days E10–11 (Ventura *et al.*, 1984), well before synaptogenesis. The avian dopaminergic system is an AC network, not an IPC network, but the analogy may still hold; that receptors and second messenger systems are installed long before effectors (e.g., gap junctions) are present.

Dopaminergic neuron maturation in mammalian retinas offers yet another picture of the relations among uptake, release, and content. In the rabbit retina, both uptake and K^+-evoked release of preloaded radiolabeled neurotransmitter appear prenatally, around E27, but significant dopamine synthesis does not appear until P8 and does not reach adult levels until after P20. This suggests that functioning of dopaminergic neurons in the rabbit is delayed significantly and occurs only after a number of adult-like circuits are detectable. The phenomenon of early release is quite surprising, since major IPL synaptogenesis in the rabbit begins around day P7, much later than the release. However, some synapses are present in the IPL of the newborn rabbit (McArdle *et al.*, 1977); thus they could be future dopaminergic synapses. Either sparse, cryptic dopaminergic synapses appear early or nonsynaptic release is occurring. Aldehyde-induced catecholamine fluorescence in the rat retina is not present until P10, with visible processes not detectable until P16 (Kato *et al.*, 1980).

The fluorescence technique is rather insensitive, however, and immunocytochemistry for TH-IR (Nguyen-Legros *et al.*, 1983) allows visualization of putative dopaminergic somas in the AC layer as early as P3 with almost mature forms by P7, long before catecholamine fluorescence is detectable. The immunocytochemistry agrees nicely with the biochemical data of Morgan and Kamp (1982) that indicate that synthesis of dopamine in the rat retina commences between days P2 and P4. After inhibition of aromatic amino acid decarboxylase, these authors were able to monitor the accumulation of dihydroxyphenylalanine as a measure of light-induced dopamine turnover. Though dopamine content in the P7 rat retina is only 36% of adult levels, light-induced dopamine turnover is significant at that time, indicating that some of the synaptic inputs to dopaminergic neurons are in place. Thus, the results in rat retina further support the phenomenon of late maturation of mammalian dopaminergic systems.

What do mammalian dopaminergic ACs do? No one is really sure, although the results of Morgan and Kamp (1980, 1982) and Proll *et al.* (1982) would lead one to believe that light leads to depolarization and release, thus making dopaminergic ACs ON-type neurons of sustained or transient varieties. It is quite intriguing that one of the synaptic targets of mammalian dopaminergic ACs is a variety of glycinergic AC (Pourcho, 1982) known as the type AII AC (Famiglietti and Kolb, 1975) or the gly2 AC (Pourcho, 1980; Marc and Liu, 1985). The AII ACs are known to be coupled to each other and certain cone BCs by gap junctions (Famiglietti and Kolb, 1975; McGuire *et al.*, 1980; Marc and Liu, 1985) and to receive direct inputs from rod BCs (Famiglietti and Kolb, 1975). Thus, one of the targets of dopaminergic ACs is a neuron possessing extensive coupling with other cells. Does the mammalian dopaminergic AC perform functions similar to the fish dopaminergic IPC? When do the gap junctions of AII ACs mature? The glycinergic ACs of rabbit are present at birth and should bear, at some time, dopamine receptors which will activate adenyl cyclase. In accord with this scenario, J. A. Pachter and D. M. K. Lam (personal communication) have found dopamine-activated adenyl cyclase activity at birth in the rabbit retina.

D. *Acetylcholine*

The study of cholinergic neurons has been constrained by technical difficulties. Only recently have antibodies against the synthetic enzyme choline acetyltransferase (ChAT) proved to be of good specificity and sensitivity. Localization of neurons possessing high-affinity [^{3}H]choline uptake requires freeze-dry tissue preparation, anhydrous embedding, and dry contact autoradiography, all quite difficult procedures. Furthermore, none of these methods has been applied systematically to retinal maturation. ChAT-IR has been demonstrated in goldfish ACs that ramify in two bands in the IPL (Tumosa *et al.*, 1984), and an identical pattern is observed with dry-contact autoradiography of high-affinity [^{3}H]choline uptake (R. E. Marc, unpublished data). Autoradiography of [^{3}H] choline uptake in the chicken retina (Baughman and Bader, 1977) and the rabbit retina (Masland and Mills, 1979) yields an identical bistratified system, currently interpreted as two sets of unistratified ACs. Thus, a variety of species seem to have a significant conservation of form among cholinergic neurons. We know little of their connectivity, although the physiological studies of Masland and Ames (1976) indicate that many rabbit retinal ganglion cells receive direct input from cholinergic ACs.

No anatomical data exist for the development of cholinergic neurons, and most of the biochemical data have been obtained from avian retinas. ChAT activity in the embryonic chick retina begins to increase on E6 and reaches a plateau at about 40% of adult level by E11 which holds until hatching. Thereafter ChAT activity progressively approaches adult levels. Acetylcholine content does not mimic this pattern, however, since it exhibits a major increase near E19 (Bader *et al.*, 1978), which correlates with a major increase in V_{max} values for both high-affinity and low-affinity [^{3}H]choline uptake to near-adult levels on E19. In the rabbit retina, [^{3}H]choline uptake is present on the day of birth (prenatal function was not determined), and K^+-stimulated release of acetylcholine appears by day P4 (Lam and Pu, 1982). ChAT activity is low at birth but increases significantly near day P3. This is similar to the sequence found for the other conventional neurotransmitters.

A feature of cholinergic function that is accessible to structural investigation is the ultrastructural localization of cholinoreceptive sites. Adult and embryonic avian distributions of nicotinic binding sites have been studied with α-bungarotoxin conjugated to horseradish peroxidase (Vogel *et al.*, 1977; Daniels and Vogel, 1980). As early as E12, when some synapses are already forming in the immature IPL, α-bungarotoxin binding sites were found, and by E16 were localized to junctions involving ACs or BCs. α-Bungarotoxin-binding sites involved roughly 8% of encountered synapses, a fraction similar to that found in the adult chicken. Thompson (1982) demonstrated K^+-evoked acetylcholine release in the chick retina on day E12, when cholinergic postsynaptic markers are already present. Thus both presynaptic and postsynaptic markers for cholinergic systems seem to develop early, in parallel with the bulk of synaptogenesis.

E. Peptides

The burgeoning numbers of amacrine cells exhibiting peptide-like immunoreactivity portend an equal number of studies on their development. At present, few studies of import to our goals are available, and most of those studies are in avians. One exception is the demonstration of substance P-like immunoreactivity (SP-IR) in the retinas of mature and developing goldfishes (Brecha *et al.*, 1981). The adult goldfish retina seemingly possesses a single population of ACs that exhibits SP-IR, with somewhat pyriform morphology and a unistratified dendritic arbor localized to mid-IPL. The first instance of SP-IR was detected at stage 27 and the cells observed were similar to the adult form. The absence of immunoreactivity in a transitional form indicates that the mRNA required for synthesis of SP precursors is delayed until after the AC reaches nearly adult shape. This is contrary to the observation that high-affinity GABA uptake in goldfish ACs is operative while the presumptive ACs are still quite elongated in shape and apparently migrating through the INL. However, GABA synthesis in *Xenopus* is delayed until a stage at which significant AC maturation has been attained. Thus, synthetic mechanisms may be among the last of presynaptic physiology to be installed, along with the formation of presynaptic release mechanisms.

There has also been investigation of vasoactive intestinal peptide-like immunoreactivity (VIP-IR) in ACs of neonatal rats (Terubayashi *et al.*, 1982). VIP-IR cells do not appear until long after birth, about P12, and even then have weak immunoreactivity. By P21 the cells have attained adult form and immunoreactivity. Thus, their chronologies seem somewhat delayed, like mammalian dopaminergic neurons.

F. Summary

We now have abundant evidence that the neurochemical maturation of retinal cells occurs very early in development for most of the "conventional" neurotransmitters and with various delays for monoamines and peptide systems. In most cases, however, the neurochemical specification seems to occur quite early, soon after the development of layers is evident and long before the onset of significant synaptogenesis and functional activity. This implies that there is a great degree of predetermination in the retina, and that the identities of cells have little to do with electrical activity. The same may not be true of synaptogenesis or subsequent physiological maturation.

IV. Maturation of Retinal Networks

The aspect of retinal development least accessible to analysis is the maturation of physiological responses to photic stimulation. In this section I wish to address some of the anatomical and physiological data that pertain specifically to the issues of when (and perhaps how) specific modes of connectivity arise. I will not attempt an encyclopedic listing of electrophysiological results in developing retinas but will rather seek to integrate results from model systems that are most commonly used in developmental work and are thus likely to serve as good frameworks for further study of other vertebrate retinal models. One aspect of maturation is clear, however: the neurochemical specification of conventional transmitter systems in the vertebrate retina is complete far in advance of the appearance of light-driven responses.

A. Retinal Networks in Xenopus

In some respects it is unfortunate that most of the important physiology and connective anatomy relevant to maturation has been done in *Xenopus,* since the connectivity, neurochemistry, and physiology of the mature retina in *Xenopus* is less well characterized than in other model systems. The major principles of organization are likely to be similar, however, and the needed data are forthcoming steadily (e.g., Witkovsky and Stone, 1983; Hassin and Witkovsky, 1983; Stone and Witkovsky, 1984). Some of the initial functional data were provided by extracellular single-unit recordings from incoming retinal ganglion cell axons in the tectum of larval *Xenopus* (Pomeranz and Chung, 1970; Pomeranz, 1972; Chung *et al.*, 1973). Chung *et al.* (1972) noted progressive alterations in the photoresponsive properties of GCs in larval development and compared them to the morphologies of Golgi-impregnated GCs from comparable stages. Neurons dominated by ON responses were recorded earliest, about stage 43. OFF responses became evident at stage 45 and gradually increased in strength until they were roughly equal in prominence to ON responses by stage 49.

It is not clear, however, that these responses are fully comparable to those generated by what have become conventionally known as ON-center or OFF-center channels. Early larval *Xenopus* GCs lack the precise receptive field properties of mature neurons and they fatigue rapidly. Furthermore, most of the responses are of the transient variety, sustained responses only appearing later near stage 61. It is currently thought that the ON responses of both transient and sustained GCs are driven through ON-center BCs (Miller and Dacheux, 1976). A strong test of the origin of early larval GC responses would be an attempted blockade with 2-amino-4-phosphonobutyric acid, a glutamate analog known to selectively inhibit the photoresponses of ON-center BCs (Slaughter and Miller, 1981). Morphological analysis of developing synapses in *Xenopus* OPL does not provide any strong evidence for changes in photoreceptor–BC synapse morphology that would imply the sequential maturation of two different functional classes of BCs, such as ON-center before OFF-center (Witkovsky and Powell, 1981). Considering that the differential development of ON versus OFF only spans a stage or two, that spontaneous activity is low, and that immature GCs fatigue readily, there may be no significant difference in the maturation of these two parallel channels.

Chung *et al.* (1973) attempted to compare the morphologies of Golgi-impregnated neurons appearing at various stages with the patterns of ganglion cell responses observed at the same time. They realized there were severe limits to such interpretations and, indeed, the possible correlations were only partially consonant with what we now expect for the laminar distributions of retinal ganglion cell dendrites. At the stages manifesting transient responses dominated by ON responses, bistratified GCs and GCs with arbors largely constrained to sublamina a were observed. There is evidence in fishes that transient GCs are bistratified (Murakami and Shimoda, 1975; Famiglietti *et al.*, 1977), partially supporting the congruence of a receptive field type with dendritic morphology. However, the bisublaminar model proposed by Famiglietti and Kolb (1976) has been shown to be valid for achromatic cone channels (i.e., the channel dominated by long-wave cones) in many vertebrates. Thus, the finding that certain early GCs arborize in the future sublamina a, the OFF sublamina, while ON responses dominate tectal recordings demands explanation. The pharmacological approach suggested above is probably the most direct. It may be that GC arbors undergo considerable pruning and regrowth in achieving their final forms.

Mature ganglion cell receptive fields in all vertebrates are characterized by various "trigger" properties; those photic conditions that most strongly evoke responses. It is generally conceded that these properties arise from different forms and strengths of lateral interactions in both plexiform layers. Certainly HCs provide significant contributions as demonstrated by Naka and Witkovsky (1972) and Naka (1977). The major pathway for these actions appears to be HC → cone → BC or direct HC → BC transmission. The BC signals are then conveyed directly to the GCs. Witkovsky and Powell (1981) note that conventional synapses between HCs and their targets were not evident in *Xenopus* at stage 46 or earlier, yet HC release of [^{3}H]GABA is evident by stage 37/38, albeit at roughly one-third of juvenile levels. Surround effects are really not strongly evident in *Xenopus* GCs until postmetamorphic stages (Chung *et al.*, 1975). Thus, the roles subserved by HCs and ACs as lateral interneurons are certainly cryptic in larvae, despite the clear maturation of functional neurotransmitter markers and some GC responses. The events transpiring at metamorphosis are incompletely understood and certainly beyond the scope of this review. They do appear, however, to be maturational changes of the pure "network" variety and not those defining the essential biochemical character of any given cell type. Cell death, intermittent synaptic contacts, futile contacts, dendritic growth and pruning, and even photic exposure may play some role in the proper wiring of a retina.

B. *Retinal Networks in the Rabbit*

The efforts of a small group of dedicated workers have recently afforded an intriguing view of postnatal retinal maturation in the rabbit. At birth (P1), all cell layers of the rabbit retina are evident, though immature. The IPL is thick and does contain conventional synapses at roughly one-quarter the density of the mature retina (McArdle *et al.*, 1977), though there are no outer segments. In this sense the mammalian retina differs from that of ectotherms, in which photoreceptor maturation begins at roughly the same time as synaptogenesis in the IPL. No photic responses are detectable at birth, as expected, but GCs seem to be electrically competent and responsive to exogenously applied agents, such as acetylcholine (Masland, 1977). This is an important observation because cholinergic neurons are not mature at birth (Pu and Lam, 1983). Thus, GCs synthesize and insert acetylcholine receptors into their membranes without prior exposure to acetylcholine, it would seem. The first evidence of light-driven behavior is a cornea-negative potential resembling the isolated mass photoreceptor potential or PIII (Masland, 1977; Dacheux and Miller, 1981a) occurring on P6, which corresponds to the initial outpouching and membrane elaboration of photoreceptor cilia (McArdle *et al.*, 1977). By P8, about the time that BC ribbon synapses appear (McArdle *et al.*, 1977), initial small HC, BC, AC (Dacheux and Miller, 1981a,b), and GC (Masland, 1977) responses can be recorded. The small size of the responses is probably partially attributable to incomplete synaptogenesis and small outer segment size. Furthermore, most responses fatigue quickly. Somewhere around P10–P12, some major network properties appear to mature, as evidenced by the presence of some adult-like GCs, the development of conventional synapses in the OPL, and appearances of surround properties in BCs. Likewise, this is the period in which GABA and glycine release reach near-adult levels, implying completion of the release mechanisms in those groups of ACs. Thereafter, all properties seem to gradually mature, and GC responses become more and more like adult responses, as if some fine tuning of connectivity were occurring.

Several developmental studies have involved the postnatal maturation of the cat retina, and in many respects these have more intensively characterized GC properties. In truth, the fundamental processes determining connectivity and the ordinal progression of maturation seems to be similar for rabbit retina. Cat retinal GCs seem to have the same dimensions and arborization patterns at 3 weeks of age (P21) as adult GCs (Russoff and Dubin, 1977, 1978; Russoff, 1979), remarkably similar to the plateau in maturation reported for many aspects of retinal structure, chemistry, and physiology in the rabbit. The major changes after P21 and continuing until about P63 involve optical rearrangements leading to smaller receptive fields, although some synaptic alterations may occur. In particular, Hamaski and Sutija (1979) have noted progressive changes in the encounter frequencies of X-type and Y-type GCs from 3 to 12 weeks, implying that significant maturation of specific connections occur. There are structural evidences of postnatal plasticity in the retina. Recently, Perry and Linden (1982) showed that the GCs of rat retinas exhibited dendritic competition, as indicated by the growth of GC dendrites into zones vacated by mechanical lesions on postnatal day 1. Certain classes of retinal GCs are also purported to undergo alterations in physiological properties due to deprivation (Ikeda and Tremain, 1979). No matter the outcome of these issues, it remains that the bulk of retinal maturation involves self-assembly independent of environmental experience.

V. Summary

Over the past decade we have observed that the maturation of functional properties of the vertebrate retina is separable into four major phases: (1) the proliferative phase, in which the laminar destination and ultimate cell class of NCs is determined; (2) the neurochemical phase, in which the biochemical components of most of the neurons are synthesized and installed; (3) the synaptic phase, a brief period in which synaptic release mechanisms mature, roughly in parallel or slightly preceding the onset of detectable photoreceptor function; and (4) the physiological phase, in which the appropriate properties of receptive fields are progressively attained.

The relationships among these phases are not understood, except in the sense that the functional destinies of most cells seem determined in the proliferative phase. It is thus unlikely that we will learn much about the assembly of networks without detailed knowledge of the molecular biology of NCs. Different ACs clearly have different developmental programs. Conventional neurotransmitter types (GABA, glycine, acetylcholine) appear early, often associated with cells that have little resemblance to the adult form, whereas other systems (monoamines, peptides) show evidence of commitment only when the cells have already taken on the "amacrine" shape. The next generation of probes required to characterize neurochemical specification at earlier stages in maturation will be cDNA–mRNA *in situ* hybridization. For example, Gee *et al.* (1983) have demonstrated the localization of mRNA coding for pro-opiomelanocortin (the precursor molecule for adrenocorticotropic hormone, β-endorphin, and melanotropins) to neurons in mammalian hypothalamus with cDNA probes. In the short term, these methods are likely to be most useful for peptide-producing cells, rather than those using conventional neurotransmitters, because the gene sequences for the peptides are more easily obtained.

Nevertheless, the approach is important because some evidence indicates that neurochemical specification can be made prior to the termination of mitotic activity. Goldfish GABAergic HCs show selective [^{3}H]GABA uptake as soon as the OPL is first identifiable, and some dividing cells seem significantly labeled (e.g., Fig. 4). The commitment to synthesis of the GABA transporter must thus have been made very early in retinal histogenesis. So far, the retina is unlike neural tissue such as sympathetic ganglia, where cells have been shown to be able to switch commitments between cholinergic and adrenergic function (Patterson, 1979). There have been no attempts to test this apparent immutability.

In addition to investigating the onset of neurochemical commitment with new technologies, we have only begun to tap the utility of electron microscope autoradiography. Our own work is incomplete, especially regarding the inner plexiform layer. One issue that remains puzzling is that the appearance of mature receptive fields seems significantly later than much of synaptogenesis. The implication is that the establishment of synaptic contacts or at least the maturation of release mechanisms is not the final step in making functional connections among cells. In culture systems, embryonic chick retinal and spinal cholinergic cells will synapse with rat striated muscle cells, but the retinal cells form only transient synapses (Ruffolo *et al.*, 1978). This supports the notion that contact

TABLE I

EVENTS IN GOLDFISH RETINAL DEVELOPMENT

Stage	Event
18	Optic vesicle and ventricle form. All cells in vesicle exhibit DNA synthesis
19	Optic cup forms. Pigment epithelium and lens rudiment are evident. Some future GCs withdraw from the cell cycle
20	Cell proliferation continues. More future GCs and some future ACs withdraw from cycle
21	Most future GCs, more future ACs, and some future INL cells withdraw from cycle
23	Cells elongated, NC-like. First pigment granules evident in pigment epithelium. All future GCs in central retina withdraw from cycle. Most ACs, BCs, and some photoreceptors withdraw. Future HCs still cycling
23–24	Initial segregation into layers. No OPL or IPL evident, but photoreceptor layer, INL, and GC layer distinguishable. Some future HCs still cycling. First evidence of selective [^{3}H]GABA uptake into HCs, ACs, and NC-like cells
24	Distinct layering. Last future-HC mitoses in fundus. Selective [^{3}H]GABA uptake by HCs, ACs, and NCs. No IPL labeling, although a thin IPL has formed between GCs and ACs. First evidence of photoreceptor ciliary protrusion into ventricular remnant. No photoreceptor synaptic ribbons formed yet
25	Fishes hatch. GC layer thins, and IPL thickens. ACs form distinct layer of pale cells. All cells in fundus withdrawn from cycle. Confluent [^{3}H]GABA labeling of all distal HCs. No HC axons or axon terminals present. Photoreceptor terminals possess flat bases, rudimentary ribbons. HCs make superficial contact with photoreceptors. First distinct outer segments formed. [^{3}H]GABA labeling in IPL. BC and AC synapses evident. First selective [^{3}H]glycine uptake into ACs. First evidence of [^{3}H]dopamine uptake and uptake-induced catecholamine and indoleamine fluorescence
27	Nearly mature form of retina. Abundant OPL and IPL synapses. Endogenous catecholamine and indoleamine fluorescence. First evidence of substance P-like immunoreactivity in ACs

formation in the retina may be a dynamic process in which presynaptic elements are influenced by postsynaptic status with regard to permanence of contacts. Thus, the actual generation of receptive field properties may not be wholly independent of photic history (for example, see Tucker and Hollyfield, 1977). If so, ultrastructural localization of terminals accumulating ^{3}H-labeled neurotransmitters will be seriously hampered by the inability to determine whether the contact is futile. This also stresses the importance of understanding how cell death in neurons distal to GCs is regulated. Developing ACs possessing a large proportion of futile contacts may be destined to die. Culture systems will play a great role in determining the trophic factors and contact events that regulate neuronal survival; much has already been achieved (see article by Adler, this volume), but the correlation of culture systems with functional networks is still incomplete, especially with regard to associations among different neurochemical classes of ACs.

TABLE II

EVENTS IN *Xenopus* RETINAL DEVELOPMENT

Stage	Event
32	Onset of specific high-affinity [^{3}H]GABA uptake by HC-like elements
33/34	Specific [^{3}H]GABA uptake by HC-like and AC-like elements. Onset of specific high-affinity [^{3}H]glycine uptake
35/36	Onset of [^{3}H]dopamine synthesis from [^{3}H]tyrosine
37/38	Onset of glutamic acid decarboxylase (GAD) activity. First evidence of GABA release. Contacts of photoreceptors with HCs present
40	GAD activity reaches plateau levels
41	Onset of major increase in IPL synapses
42	Onset of glycine release
43	Onset of high-affinity [^{3}H]dopamine uptake. Earliest GC responses, dominated by ON-responses
45	GC OFF-responses clearly present; ON-responses still dominate
46	First evidence of dopamine release
48	First evidence of conventional HC synapses
49	GC ON- and OFF-responses roughly equal
58	Gap junctions between HCs present (onset not determined)
61	Sustained GC responses appear
Postmetamorphosis	Appearance of clear surround responses

Some of the major events in the maturation of goldfish, *Xenopus,* and rabbit retinas have been summarized in Tables I, II, and III, respectively. What have we learned with regard to the adult cell forms and circuitries discussed in the introduction? First, it seems that each type of retinal neuron has its own developmental program determined early in histogenesis, before or near the onset of layer formation. Second, ectotherm GABAergic and non-GABAergic HCs are separate entities throughout postmitotic development. They form contacts with apparent cone photoreceptors and do not seem to make errors in their targets, such as contacting rod terminals. Thus, the lateral circuit 2 in Fig. 2a seems partially complete very early in development. The gap junctions between horizontal cells are not evident, even at stage 27, when the retina seems quite mature. Indeed, the onset of gap junction synthesis involving any retinal neuron is not known in any species, but appears to be quite late in *Xenopus* (Witkovsky and Powell, 1981) and chick (Cooper and McLaughlin, 1981).

The recent evidence implicating dopaminergic neurons in the modulation of gap junction coupling among horizontal cells may be important in this context, and an interesting question is whether there is an obligate relationship between monoamine development and gap junction assembly. Although we have no electron microscope autoradiography of dopaminergic neurons in development, the dopaminergic IPCs are present and capable of dopamine synthesis by stage

TABLE III

EVENTS IN RABBIT RETINAL DEVELOPMENT

Stage	Event
E22	Onset of [^{3}H]GABA uptake into future ACs
E25	Onset of [^{3}H]glycine uptake into future ACs
E27	Onset of [^{3}H]dopamine uptake into future ACs. Onset of dopamine release
P1	GAD and GABA levels at about 20% adult levels. Conventional synapses present at about 25% of adult levels. Spontaneous GC action potentials; responsiveness to acetylcholine. Protrusion of photoreceptor cilia into ventricular space. Presynaptic ribbon specializations undergoing assembly in photoreceptors. High-affinity [^{3}H]choline uptake present (probably develops prenatally)
P3	Photoreceptor ribbon synapses abundant. Choline acetyltransferase activity increases
P4	Onset of acetylcholine release
P6	Onset of GABA and glycine release. Outer segment develops initial membrane elaborations. First evidence of light-driven activity: PIII-like potential
P8	Onset of major increase in all IPL synapses. GABA and GAD levels and GABA release reach 80% of adult levels. Initial increase in dopamine content (minor). Initial, weak light-driven responses in a minor portion of the GC population. Small photic responses from HCs and BCs. Transient AC responses dominated by ON-components
P9	BC ribbon synapses present (probably also at P8)
P10	Most GCs light responsive, some with mature properties
P11	Conventional synapses (presumably from HCs) present in OPL
P12	Glycine release reaches 80% of adult levels. Surround responses present in some BCs
P16	Clear ON- and OFF-components in transient ACs
P20	GC properties largely mature. Outer segments and synaptic frequency nearly mature
P22	Secondary increase in dopamine content (major)

27 in the goldfish. Thus, circuit 5 (Fig. 2c) does seem in place at this time, although appearing later than GABAergic or glycinergic ACs. Circuit 6 (Fig. 2c) is apparently incomplete until times later than we have assayed, indicated by the absence of glycinergic IPCs and conventional synapses in GABAergic HCs. Supporting this concept is the late onset of conventional synapses in *Xenopus* HCs (stage 48) relative to the first appearance of glycine uptake (stage 33).

We have not discriminated the vertical channels in development with neurochemical markers (circuit 1, Fig. 2a). However, by the time we observe GABAergic HCs with well-invaginated profiles in cone terminals, there are abundant nonlabeled profiles present, indicating the elaboration of BC dendrites. Witkovsky and Powell (*Xenopus;* 1981) and McArdle *et al.* (rabbit; 1977) indicated that indentation of the early photoreceptor terminal by HC processes slightly predates BC dendrite contact. There is no evidence to indicate that ON- or OFF-center channels mature in a particular order or that, if they do, the order has functional consequences.

The networks we understand the least from a functional aspect, the AC circuits (e.g., circuit 3, Fig. 2a; circuit 4, Fig. 2b), are quite easy to reveal by neurochemical probes. The conventional transmitter systems (GABA, glycine, acetylcholine) mature quickly and develop demonstrable release properties well in advance of GC photoresponses. The surround properties presumably mediated by these neurons are slow to develop however, which is somewhat surprising. If (1) the neurochemical mechanisms are mature, (2) abundant synapses are found, (3) postsynaptic responsivity is present (e.g., Masland, 1977), and (4) bipolar channels to the IPL are apparently intact, why are antagonistic surrounds and/or trigger feature responses difficult to demonstrate in both ectotherms and mammals until much later? Such late development would make sense if the neurochemical properties of ACs were slow to appear, but this is not the case at all. This prolonged maturation requires detailed ultrastructural and physiological investigation. Presumably, like immature cells in culture, retinal ACs are making and breaking contacts, searching for the appropriate postsynaptic entity. Witkovsky and colleagues have shown the *Xenopus* retina to be well suited for pharmacological investigation, so that the maturation of connections mediated by neurochemically identified cells may be charted. Thus, it may be possible to describe the maturation of GC properties in terms of the acquisition of a proper neurochemical spectrum of inputs.

Our understanding of circuit maturation is largely an assemblage of loosely associated observations, with few causal relationships established. Nevertheless, our course should be quite clear. We can propose a number of hypotheses linking (1) embryonic microenvironment, genetic status, and neurochemical destiny, (2) neurochemical identity and interactions with development of neighboring cells, and (3) photic history and generation of stable synaptic contacts. The technologies required for these investigations are well established.

References

Ayoub, G., and Lam, D. M. K. (1984). The release of γ-aminobutyric acid from the horizontal cells of the goldfish (*Carassius auratus*) retina. *J. Physiol. (London)* **355,** 191–214.

Bader, C. R., Baughman, R. W., and Moore, J. L. (1978). Different time course of development for high-affinity choline uptake and choline acetyltransferase in the chick retina. *Proc. Natl. Acad. Sci. U.S.A.* **75,** 2525–2529.

Baughman, R. W., and Bader, C. R. (1977). Biochemical characterization and cellular localization of the cholinergic system in the chicken retina. *Brain Res.* **138,** 469–485.

Black, I. B. (1983). Stages of neurotransmitter development in autonomic neurons. *Science* **215,** 1198–1204.

Blanks, J. C. (1982). Cellular differentiation in the mammalian retina. *In* "The Structure of the Eye" (J. G. Hollyfield, ed.), pp. 237–246. Elsevier, Amsterdam.

Blanks, J. C., and Bok, D. (1977). An autoradiographic analysis of postnatal cell proliferation in the normal and degenerative mouse retina. *J. Comp. Neurol.* **174,** 317–328.

Boycott, B. B., and Kolb, H. (1973). The horizontal cells of the rhesus monkey retina. *J. Comp. Neurol.* **148,** 115–139.

Boycott, B. B., Piechl, L., and Wässle, H. (1978). Morphological types of horizontal cell in the retina of the domestic cat. *Proc. R. Soc. London Ser. B* **203,** 229–245.

Brackenbury, R., Rutishauser, U., and Edelman, G. M. (1981). Distinct calcium-independent and calcium-dependent adhesion systems of chick embryo cells. *Proc. Natl. Acad. Sci. U.S.A.* **78,** 387–391.

Brandon, C., Lam, D. M. K., and Wu, J. Y. (1979). The γ-aminobutyric acid system in the rabbit retina: Localization by immunocytochemistry and autoradiography. *Proc. Natl. Acad. Sci. U.S.A.* **76,** 3557–3561.

Brecha, N. C. (1983). Retinal neurotransmitters: Histochemical and biochemical studies. *In* "Chemical Neuroanatomy" (P. C. Emson, ed.), pp. 85–129. Raven, New York.

Brecha, N. C., Sharma, S. C., and Karten, H. J. (1981). Localization of Substance P-like immunoreactivity in the adult and developing goldfish retina. *Neuroscience* **6,** 2737–2746.

Brecha, N. C., Oyster, C. W., and Takahashi, E. S. (1984). Identification and characterization of tyrosine hydroxylase immunoreactive amacrine cells. *Invest. Ophthalmol. Visual Sci.* **25,** 66–70.

Buskirk, D. R., Thiery, J.-P., Rutishauser, U., and Edelman, G. M. (1980). Antibodies to a neural cell adhesion molecule disrupt histogenesis in cultured chick retinae. *Nature (London)* **285,** 488–489.

Carter-Dawson, L. D., and LaVail, M. M. (1979). Rods and cones in the mouse retina. II. Autoradiographic analysis of cell generation using tritiated thymidine. *J. Comp. Neurol.* **188,** 263–272.

Chung, S. H., Stirling, R. V., and Gaze, R. M. (1975). The structural and functional development of the retina of the larval *Xenopus*. *J. Embryol. Exp. Morphol.* **33,** 915–940.

Cohen, J. L., and Dowling, J. E. (1983). The role of the retinal interplexiform cell: Effects of 6-hydroxydopamine on the spatial properties of carp horizontal cells. *Brain Res.* **264,** 307–310.

Cooper, N. G. F., and McLaughlin, B. J. (1981). Gap junctions in the outer plexiform layer of the chick retina: Thin section and freeze fracture studies. *J. Neurocytol.* **10,** 515–529.

Dacheux, R. F., and Miller, R. F. (1981a). An intracellular electrophysiological study of the ontogeny of functional synapses in the rabbit retina. I. Receptors, horizontal and bipolar cells. *J. Comp. Neurol.* **198,** 307–326.

Dacheux, R. F., and Miller, R. F. (1981b). An intracellular electrophysiological study of the ontogeny of functional synapses in the rabbit retina. II. Amacrine cells. *J. Comp. Neurol.* **198,** 327–334.

Daniels, M. P., and Vogel, Z. (1980). Localization of α-bungarotoxin binding sites in synapses of the developing chick retina. *Brain Res.* **201,** 45–56.

DeMello, F. G. (1978). The ontogeny of dopamine-dependent increase of adenosine 3′–5′ cyclic monophosphate in the chick retina. *J. Neurochem.* **31,** 1049–1053.

Dowling, J. E., and Ehinger, B. (1975). Synaptic organization of the amine-containing interplexiform cells of the goldfish and cebus monkey retinas. *Science* **188,** 270–273.

Dowling, J. E., and Ehinger, B. (1978). The interplexiform cell system. I. Synapses of the dopaminergic neurons of the goldfish retina. *Proc. R. Soc. London Biol.* **201,** 7–26.

Edelman, G. M. (1983). Cell adhesion molecules. *Science* **219,** 450–457.

Ehinger, B. (1977). Glial and neuronal uptake of GABA, glutamic acid, glutamine and glutathione in the rabbit retina. *Exp. Eye Res.* **25,** 221–234.

Ehinger, B., and Falck, B. (1969). Adrenergic neurons of some new world monkeys. *Z. Zellforsch.* **100,** 364–375.

Ehinger, B., and Florén, I. (1978). Indoleamine accumulating neurons in the retina of rabbit, cat and goldfish. *Cell Tissue Res.* **175,** 37–48.

Famiglietti, E. V., Jr. (1981). Functional architecture of cone bipolar cells in mammalian retina. *Vision Res.* **21,** 1559–1563.

Famiglietti, E. V., Jr., and Kolb, H. (1975). A bistratified amacrine cell and synaptic circuitry in the inner plexiform layer of the retina. *Brain Res.* **84,** 293–300.

Famiglietti, E. V., Jr., and Kolb, H. (1976). Structural basis for ON- and OFF-center responses in retinal ganglion cells. *Science* **194,** 193–195.

Famiglietti, E. V., Jr., Kaneko, A., and Tachibana, M. (1977). Neuronal architecture of on and off pathways to ganglion cells in carp. *Science* **198,** 1267–1269.

Fisher, L. J. (1976). Synaptic arrays in the inner plexiform layer in the developing retina of *Xenopus*. *Dev. Biol.* **50,** 406–412.

Frederick, J., Rayborn, M. E., Laties, A. M., Lam, D. M. K., and Hollyfield, J. G. (1983). Dopaminergic neurons in the human retina. *J. Comp. Neurol.* **210,** 65–79.

Fujita, S., and Horii, M. (1963). Analysis of cytogenesis in chick retina and tritiated thymidine autoradiography. *Arch. Histol. Jpn.* **23,** 359–366.

Fung, S. C., Kong, Y. C., and Lam, D. M. K. (1982). Prenatal development of GABAergic, glycinergic and dopaminergic neurons in the rabbit retina. *J. Neurosci.* **2,** 1623–1632.

Gallego, A. (1983). Horizontal cells of the tetrapoda retina. *In* "The S-Potential" (B. D. Drujan and M. Laufer, eds.), Progress in Clinical and Biological Research, Vol. 113, pp. 9–29. Liss, New York.

Gee, C. E., Chen, C.-L. C., Roberts, J. L., Thompson, R., and Watson, S. J. (1984). Identification of proopiomelanocortin neurones in the rat hypothalamus by *in situ* cDNA–mRNA hybridization. *Nature (London)* **306,** 374–376.

Glickman, R. D., and Adolph, A. R. (1982). Acetylcholine and substance P: Action via distinct receptors on carp retinal ganglion cells. *Invest. Ophthalmol. Visual Sci.* **22,** 804–808.

Haggendal, J., and Malmfors, T. (1965). Identification and cellular localization of the catecholamines in the retina and choroid of the rabbit. *Acta Physiol. Scand.* **64,** 58–66.

Hamasaki, D. I., and Sutija, V. G. (1979). Development of X- and Y-cells in kittens. *Exp. Brain Res.* **35,** 9-23.

Hausman, R. E., and Moscona, A. A. (1976). Isolation of retina specific cell-aggregating factor from membranes of embryonic neural retinal tissue. *Proc. Natl. Acad. Sci. U.S.A.* **3,** 410–411.

Hausman, R. E., and Moscona, A. A. (1979). Immunological detection of retina cognin on the surface of embryonic cells. *Exp. Cell Res.* **119,** 191–207.

Hinds, J. W., and Hinds, P. L. (1974). Early ganglion cell differentiation in the mouse retina: An electron microscopic analysis utilizing serial sections. *Dev. Biol.* **37,** 381–416.

Hinds, J. W., and Hinds, P. L. (1978). Early development of amacrine cells in the mouse retina: An electron microscopic, serial section analysis. *J. Comp. Neurol.* **179,** 277–300.

Hinds, J. W., and Hinds, P. L. (1979). Differentiation of photoreceptors and horizontal cells in the embryonic mouse retina: An electron microscopic, serial section analysis. *J. Comp. Neurol.* **187,** 495–512.

Hollyfield, J. G. (1968). Differential addition of cells to the retina in *Rana pipiens* tadpoles. *Dev. Biol.* **18,** 164–179.

Hollyfield, J. G. (1971). Differential growth of the neural retina in *Xenopus laevis* larvae. *Dev. Biol.* **24,** 264–286.

Hollyfield, J. G. (1972). Histogenesis in the retina of the killifish, *Fundulus heteroclitus*. *J. Comp. Neurol.* **144,** 373–380.

Hollyfield, J. G., Rayborn, M. E., Sarthy, P. V., and Lam, D. M. K. (1979). The emergence, localization and maturation of neurotransmitter systems during development of the retina in *Xenopus laevis*. I. γ-Aminobutyric acid. *J. Comp. Neurol.* **188,** 587–598.

Ikeda, H., and Tremain, K. E. (1979). Amblyopia occurs in retinal ganglion cells in cats reared with convergent squint without alternating fixation. *Exp. Brain Res.* **35,** 559–582.

Ishida, A. T., Stell, W. K., and Lightfoot, D. O. (1980). Rod and cone inputs to bipolar cells in goldfish retina. *J. Comp. Neurol.* **191,** 315–335.

Jacobson, M. (1968). Cessation of DNA synthesis in retinal ganglion cells correlated with the time of specification of their central connections. *Dev. Biol.* **17,** 219–232.

Kahn, A. J. (1974). An autoradiographic analysis of the time of appearance of neurons in the developing chick neural retina. *Dev. Biol.* **38,** 30–40.

Kato, S., Nakamura, T., and Negishi, K. (1980). Postnatal development of dopaminergic cells in the rat retina. *J. Comp. Neurol.* **191,** 227–236.

Kolb, H. (1979). The inner plexiform layer in the retina of the cat: Electron microscopic observations. *J. Neurocytol.* **8,** 295–329.

Kolb, H., Mariani, A., and Gallego, A. (1980). A second type of horizontal cell in the monkey retina. *J. Comp. Neurol.* **189,** 31–44.

Kolb, H., Nelson, R., and Mariani, A. (1981). Amacrine cells, bipolar cells and ganglion cells of the cat retina: A golgi study. *Vision Res.* **21,** 1081–1114.

Kong, Y.-C., Fung, S.-C., and Lam, D. M. K. (1980). Postnatal development of glycinergic neurons in the rabbit retina. *J. Comp. Neurol.* **193,** 1127–1135.

Kramer, S. G., Potts, A. M., and Magnall, Y. (1970). Dopamine: A retinal neurotransmitter. II. Autoradiographic localization of H3-dopamine in the retina. *Invest. Ophthalmol.* **10,** 617–624.

Lam, D. M. K. (1975). Biosynthesis of γ-aminobutyric acid by isolated axons of cone horizontal cells in the goldfish retina. *Nature (London)* **254,** 345–347.

Lam, D. M. K., and PU, G. A.-W. (1982). Postnatal development of cholinergic function in the rabbit retina. *Invest. Ophthalmol. Visual Sci. Suppl.* **22,** 81.

Lam, D. M. K., Su, Y. Y. T., Swain, L., Marc, R. E., Brandon, C., and Wu, J.-Y. (1979). Immunocytochemical localization of L-glutamic acid decarboxylase in the goldfish retina. *Nature (London)* **278,** 565–567.

Lam, D. M. K., Fung, S.-C., and Kong, Y.-C. (1980). Postnatal development of GABA-ergic neurons in the rabbit retina. *J. Comp. Neurol.* **193,** 89–102.

Leeper, H. F. (1978a). Horizontal cells of the turtle retina. I. Light microscopy of Golgi preparations. *J. Comp. Neurol.* **182,** 777–794.

Leeper, H. F. (1978b). Horizontal cells of the turtle retina. II. Analysis of interconnections between photoreceptor cells and horizontal cells by light microscopy. *J. Comp. Neurol.* **182,** 795–810.

Liebman, P. (1972). Microspectrophotometry of photoreceptors. *In* "Photochemistry of Vision. Handbook of Sensory Physiology VII/1" (H. J. A. Dartnall, ed.), pp. 481–528. Springer-Verlag, Berlin and New York.

McArdle, C. B., Dowling, J. E., and Masland, R. H. (1977). Development of outer segments and synapses in the rabbit retina. *J. Comp. Neurol.* **175,** 253–274.

McGuire, B., Stevens, J. K., and Sterling, P. (1980). Beta ganglion cells receive convergent input from 2 types of cone bipolars. *Soc. Neurosci. Abstr.* **6,** 347.

Marc, R. E. (1980). Retinal colour channels and their neurotransmitters. *In* "Colour Vision Deficiencies V" (G. Verriest, ed.), pp. 15–29. Hilger, Bristol, England.

Marc, R. E. (1982a). Chromatic organization of the retina. *In* "Cell Biology of the Eye" (D. McDevitt, ed.), pp. 435–473. Academic Press, New York.

Marc, R. E. (1982b). The spatial organization of neurochemically classified interneurons in the goldfish retina. I. Local patterns. *Vision Res.* **22,** 589–608.

Marc, R. E. (1985). The role of glycine in retinal circuitry. *In* "Retinal Transmitters and Modulators: Models for the Brain" (Wm. Morgan, ed.), Vol. 1., pp. 119–158. CRC Press, Boca Raton, Florida.

Marc, R. E., and Lam, D. M. K. (1981). Glycinergic pathways in the goldfish retina. *J. Neurosci.* **1,** 152–165.

Marc, R. E., and Lam, D. M. K. (1986). In preparation.

Marc, R. E., and Lui, W.-L. S. (1984). Horizontal cell synapses onto glycinergic interplexiform cells. *Nature (London)* 266–269.

Marc, R. E., and Liu, W.-L. S. (1985). [3H]Glycine-accumulating neurons in the human retina. *J. Comp. Neurol.* **232,** 241–260.

Marc, R. E., and Liu, W.-L. S. (1986). In preparation.

Marc, R. E., and Sperling, H. G. (1976a). Color receptor identities of goldfish cones. *Science* **191,** 487–489.

Marc, R. E., and Sperling, H. G. (1976b). The chromatic organization of the goldfish cone mosaic. *Vision Res.* **16,** 1211–1224.

Marc, R. E., and Summerall, R. D. (1980). Development of transmitter systems in the teleost retina. *Invest. Ophthalmol. Visual Sci. Suppl.* **21,** 247.

Marc, R. E., Stell, W. K., Bok, D., and Lam, D. M. K. (1978). GABA-ergic pathways in the goldfish retina. *J. Comp. Neurol.* **182,** 221–246.

Mariani, A. (1982). Biplexiform cells: Ganglion cells of the primate retine that contact photoreceptors. *Science* **216,** 1134–1136.

Masland, R. H. (1977). Maturation of function in the developing rabbit retina. *J. Comp. Neurol.* **175,** 275–286.

Masland, R. H., and Mills, J. W. (1979). Autoradiographic identification of acetylcholine in the rabbit retina. *J. Cell Biol.* **83,** 159–178.

Miller, R. F., and Dacheux, R. F. (1976). Synaptic organization and ionic basis of *on* and *off* channels in the mudpuppy retina: III. A model of ganglion cell receptive field organization based on chloride free experiments. *J. Gen. Physiol.* **67,** 661–678.

Morest, D. K. (1970). The pattern of neurogenesis in the retina of the rat. *Z. Anat. Entwicklungsgesch.* **131,** 45–67.

Morgan, W. W., and Kamp, C. W. (1980). A GABAergic influence on the light-induced increase in dopamine turnover in the dark-adapted rat retina *in vivo*. *J. Neurochem.* **34,** 1082–1086.

Morgan, W. W., and Kamp, C. W. (1982). Postnatal development of the light response of the dopaminergic neurons in the rat retina. *J. Neurochem.* **39,** 283–285.

Moscona, A. A. (1965). Recombination of dissociated cells and the development of cell aggregates. *In* "Cells and Tissues in Culture" (E. N. Willmar, ed.), pp. 489–529. Academic Press, New York.

Munz, F. W., and McFarland, W. A. (1975). Evolutionary adaptations of fishes to the photic environment. *In* "The Visual System in Vertebrates. Handbook of Sensory Physiology VIII/5" (F. Crescitelli, ed.), pp. 193–275. Springer-Verlag, Berlin and New York.

Murakami, M., and Shimoda, Y. (1975). Identification of amacrine and ganglion cells in the carp retina. *J. Physiol. (London)* **264,** 801–818.

Naka, K.-I. (1977). Functional organization of catfish retina. *J. Neurophysiol.* **40,** 26–43.

Naka, K.-I., and Witkovsky, P. (1972). Dogfish ganglion cell discharges resulting from extrinsic polarization of horizontal cells. *J. Physiol. (London)* **223,** 449–460.

Negishi, K., Kato, S., and Teranishi, T. (1983). Development of retinal monoamine neurotransmitters in larval goldfish: A histofluorescence study. *Dev. Brain Res.* **312,** 111–116.

Nelson, R. (1982). AII amacrine cells quicken time course of rod signals in the cat retina. *J. Neurophysiol.* **47,** 928–947.

Nguyen-Legros, J., Berger, B., Vigny, A., and Chantal, A. (1981). Presence of interplexiform dopaminergic neurons in the rat retina. *Brain Res. Bull.* **9,** 379–381.

Nguyen-Legros, J., Vigny, A., and Gay, M. (1983). Post-natal development of TH-like immunoreactivity in the rat retina. *Exp. Eye Res.* **37,** 23–32.

Nishimura, Y. (1980). Determination of the developmental pattern of retinal ganglion cells in chick embryos by golgi impregnation and other methods. *Anat. Embryol.* **158,** 239–247.

Patterson, P. H. (1981). Environmental determination of neurotransmitter functions in developing sympathetic neurons. *In* "Development and Chemical Specificity of Neurons. Progress in Brain Research" (M. Cuenod, G. W. Kreutzberg, and F. E. Bloom, eds.), Vol. 51, pp. 75–82. Elsevier, Amsterdam.

Pomeranz, B. (1972). Metamorphosis of frog vision: Changes in ganglion cell physiology and anatomy. *Exp. Neurol.* **34,** 187–199.

Pomeranz, B., and Chung, S. H. (1970). Dendritic-tree anatomy codes form vision physiology in tadpole retina. *Science* **170,** 983–984.

Pourcho, R. G. (1980). Uptake of [^{3}H]glycine and [^{3}H]GABA by amacrine cells in the cat retina. *Brain Res.* **198,** 333–346.

Pourcho, R. G. (1982). Dopaminergic amacrine cells in the cat retina. *Brain Res.* **252,** 101–109.

Proll, M. A., Kamp, C. W., and Morgan, W. W. (1982). Use of liquid chromatography with electrochemistry to measure effects of varying intensities of white light in DOPA accumulation in rat retinas. *Life Sci.* **30,** 11–19.

Rayborn, M. E., Sarthy, P. V., Hollyfield, J. G., and Lam, D. M. K. (1980). The emergence, localization and maturation of neurotransmitter systems during development of the retina in *Xenopus laevis.* II. Glycine. *J. Comp. Neurol.* **195,** 585–593.

Ruffolo, R. R., Jr., Eisenbarth, G. S., Thompson, J. M., and Nirenberg, M. (1978). Synapse turnover: A mechanism for acquiring synaptic specificity. *Proc. Natl. Acad. Sci. U.S.A.* **75,** 2281–2285.

Russoff, A., and Schnitzer, J. (1984). Horizontal cells of the mouse retina contain GAD-immunoreactivity during early developmental stages. *Invest. Ophthalmol. Visual Sci. Suppl.* **25,** 86.

Rutishauser, U., Thiery, J. P., Brackenbury, R., Sela, B. H., and Edelman, G. M. (1976). Mechanisms of adhesion among cells from neural tissue of the chick embryo. *Proc. Natl. Acad. Sci. U.S.A.* **73,** 577–581.

Rutishauser, U., Gall, W. E., and Edelman, G. M. (1978a). Adhesion among neural cells of the chick embryo. IV. Role of the cell surface molecule CAM in the formation of neurite bindles in cultures of spinal ganglia. *J. Cell Biol.* **79,** 382–393.

Rutishauser, U., Thiery, J. P., Brackenbury, R., and Edelman, G. M. (1978b). Adhesion among neural cells of the chick embryo. III. Relationship of the surface molecule CAM to cell adhesion and the development of histotypic patterns. *J. Cell Biol.* **79,** 371–381.

Sarthy, P. V., and Lam, D. M. K. (1979). The uptake and release of [^{3}H]dopamine in the goldfish retina. *J. Neurochem.* **32,** 1269–1277.

Sarthy, P. V., Rayborn, M. E., Hollyfield, J. G., and Lam, D. M. K. (1981). The emergence, localization and maturation of neurotransmitter systems during development of the retina in *Xenopus laevis.* III. Dopamine. *J. Comp. Neurol.* **195,** 595–602.

Scholes, J. H. (1975). Colour receptors and their synaptic connexions in the retina of a cyprinid fish. *Philos. Trans. R. Soc. London Ser. B* **270,** 61–118.

Schwartz, E. A. (1982). Calcium-independent release of GABA from isolated horizontal cells of the toad retina. *J. Physiol. (London)* **323,** 211–227.

Sharma, S. C., and Unger, F. (1980). Histogenesis of the goldfish retina. *J. Comp. Neurol.* **191,** 373–382.

Sheffield, J. B. (1982). Sorting behavior among cells from the 14-day embryonic chick neural retina. *Dev. Biol.* **89,** 41–47.

Sidman, R. L. (1961). Histogenesis of mouse retina studied with thymidine-^{3}H. *In* "Structure of the Eye" (G. K. Smelser, ed.), pp. 487–505. Academic Press, New York.

Slaughter, M. M., and Miller, R. F. (1984). 2-Amino-4-phosphonobutyric acid: A new pharmacologcal tool for retina research. *Science* **211,** 182–185.

Steinberg, M. S. (1962). On the mechanism of tissue reconstruction by dissociated cells. III. Free energy relationships and the reorganization of fused, heteronomic tissue fragments. *Proc. Natl. Acad. Sci. U.S.A.* **48,** 1769–1776.

Steinberg, M. S., and Poole, T. J. (1981). Strategies for specifying form and pattern: adhesion-guided multicellular assembly. *Philos. Trans. R. Soc. London Ser. B* **295,** 451–460.

Stell, W. K. (1967). The structure and relationships of horizontal cells and photoreceptor–bipolar synaptic complexes in goldfish retina. *Am. J. Anat.* **121,** 410–424.

Stell, W. K. (1972). The morphological organization of the vertebrate retina. *In* "Physiology of Photoreceptor Organs. Handbook of Sensory Physiology VII/2" (M. G. F. Fuortes, ed.), pp. 111–213. Springer-Verlag, Berlin and New York.

Stell, W. K., and Hárosi, F. (1976). Cone structure and visual pigment content in the retina of the goldfish, *Carassius auratus. Vision Res.* **16,** 647–657.

Stell, W. K., and Lightfoot, D. O. (1975). Color-specific interconnections of cones and horizontal cells in the retina of the goldfish. *J. Comp. Neurol.* **159,** 473–502.

Stell, W. K., Ishida, A. T., and Lightfoot, D. O. (1977). Structural basis for on- and off-center responses in retinal bipolar cells. *Science* **198,** 1269–1271.

Sterling, P. (1983). Microcircuitry of the cat retina. *Annu. Rev. Neurosci.* **6,** 145–185.

Stone, S., and Witkovsky, P. (1984). The actions of γ-aminobutyric acid, glycine and their antagonists upon horizontal cells of the *Xenopus* retina. *J. Physiol. (London)* **353,** 249–264.

Teranishi, T., Negishi, K., and Kato, S. (1983). Dopamine modulates S-potential amplitude and dye-coupling between external horizontal cells in the carp retina. *Nature (London)* **301,** 243–246.

Teranishi, T., Negishi, K., and Kato, S. (1984). Regulatory effect of dopamine on spatial properties of horizontal cells in carp retina. *J. Neurosci.* **4,** 1271–1280.

Terubayashi, H., Okamura, H., Fujisawa, H., Itoi, M., Yanaihara, N., and Ibata, Y. (1982). Postnatal development of vasoactive intestinal polypeptide immunoreactive amacrine cells in the rat retina. *Neurosci. Lett.* **33,** 259–264.

Thiery, J. P., Brackenbury, R., Rutishauser, U., and Edelman, G. M. (1977). Adhesion among neural cells of the chick embryo. II. Purification and characterization of a cell adhesion molecule from neural retina. *J. Biol. Chem.* **252,** 6841–6845.

Thompson, J. M. (1982). Increase in acetylcholine release from the chick embryo retina during development. *Dev. Brain Res.* **4,** 259–264.

Tornquist, K., Hansson, Ch., and Ehinger, B. (1983). Immunohistochemical and quantitative analysis of 5-hydroxytryptamine in the retina of some vertebrates. *Neurochem. Int.* **5,** 299–308.

Tucker, G. S., and Hollyfield, J. G. (1977). Modifications by light of synaptic density in the inner plexiform layer of the toad, *Xenopus laevis. Exp. Neurol.* **55,** 133–151.

Tumosa, N., Eckenstein, F., and Stell, W. K. (1984). Immunocytochemical localization of putative cholinergic neurons in the goldfish retina. *Neurosci. Lett.* **48,** 255–259.

Ventura, A. L. M., Klein, W. L., and DeMello, F. G. (1984). Differential ontogenesis of D1 and D2 dopaminergic receptors in the chick embryo retina. *Dev. Brain Res.* **12,** 217–223.

Voaden, M. J. (1976). Gamma-aminobutyric acid and glycine as retinal transmitters. *In* "Transmitters in the Visual Process" (S. L. Bonting, ed.), pp. 107–125. Pergamon, New York.

Vogel, Z., Maloney, G. J., Ling, A., and Daniels, M. P. (1977). Identification of synaptic acetylcholine receptor sites in retina with peroxidase-labeled α-bungarotoxin. *Proc. Natl. Acad. Sci. U.S.A.* **74,** 3268–3272.

Walls, G. L. (1935). The visual cells of lampreys. *Br. J. Ophthalmol.* **19,** 129–148.

Wässle, H., Boycott, B. B., and Piechl, L. (1978a). Receptor contacts of horizontal cells in the retina of the domestic cat. *Proc. R. Soc. London Ser. B* **203,** 247–267.

Wässle, H., Piechl, L., and Boycott, B. B. (1978b). Topography of horizontal cells in the retina of the domestic cat. *Proc. R. Soc. London Ser. B* **203,** 269–291.

West, R. W. (1978). Bipolar and horizontal cells of the gray squirrel retina: Golgi morphology and receptor connections. *Vision Res.* **18,** 129–136.

Witkovsky, P., and Powell, C. (1981). Synapse formation and modification between distal retinal neurons in larval and juvenile *Xenopus. Proc. R. Soc. London Ser. B* **211,** 373–398.

Witkovsky, P., and Stone, S. (1983). Rod and cone inputs to bipolar and horizontal cells of the *Xenopus* retina. *Vision Res.* **23,** 1251–1258.

Yazulla, S., and Kleinschmidt, J. (1983). Carrier-mediated release of GABA from retinal horizontal cells. *Brain Res.* **233,** 211–215.

DIFFERENTIAL ADHESION IN NEURONAL DEVELOPMENT

CHRISTOPHER C. GETCH AND
MALCOLM S. STEINBERG

Department of Biology
Princeton University
Princeton, New Jersey

I. Introduction

Neurogenesis involves a series of spatially and temporally coordinated events: the localized proliferation of cells in different regions, the navigation of cells from the positions in which they are generated to the places in which they will eventually reside, the differentiation of immature neurons, the formation of specific connections with other neurons and target cells, and the death of neurons that have become superfluous. Despite this complex development, the pattern of axonal growth remains remarkably consistent from embryo to embryo in a given species, indicating that this process is carefully regulated during development.

This book is about the retina. The function of the retina is to present to the brain an accurate representation of the external visual world so that the animal can respond to objects in its environment in an adaptive way. This is made possible by the ordered outgrowth of retinal axons, which extend their growth cones along the optic stalk, across the optic chiasm, and onto the surface of the optic tectum, where they make synapses with tectal neurons to form a retinotopic map. This map is a spatially ordered projection of the retina—and therefore of the animal's visual field—upon the surface of the brain. In order for the retinotectal map to be established, each migrating retinal axon must find its way to the correct tectal address. A large body of experimental data (reviewed in Fraser and Hunt, 1980; Constantine-Paton, 1982; Meyer, 1982; Fraser, 1985) has led to the conclusion (Fraser, 1985) that the information guiding retinal neurons to their appropriate destinations is principally adhesive in nature. The same conclusion has been reached for other neurogenetic systems also, both vertebrate and invertebrate. Indeed, neuronal assembly systems appear to be only a subset within a wider category of assembly systems that organize themselves according to the principle that "motile cells will naturally tend to group so as to maximize their adhesive interactions (minimize interfacial free energy)" (Steinberg and Poole, 1981).

In recent years much has been learned about both neuronal guidance and intercellular adhesion systems, and the implication of specific adhesion-mediating molecules in the control of neurogenetic events has now begun. It is our purpose here on the one hand to describe studies of neuronal assembly which have suggested an adhesion-based causality and on the other hand to examine the molecular systems that have been implicated in neuronal adhesion, together with evidence for their function in normal events of neurogenesis. We have necessarily been selective rather than comprehensive. Our aim in all of this is to present adhesion-based neuronal assembly as probably the most elaborated example of a wider class of morphogenetic phenomena we categorize as "adhesion-guided multicellular assembly systems."

Our presentation is organized with respect to different levels of scientific investigation. Initially, at the biological level, evidence is presented which suggests that axons respond through selective adhesion to guidance cues in the environment. At the biochemical level, various molecular mechanisms of adhesion are presented which have been implicated in cell–substrate and cell–cell adhesive interactions. Finally, we consider what kinds of specificity or selectivity are required to enable outgrowing neurites to make the choices they do. We conclude that while in some cases (particularly in insects) a specific neuronal association may be mediated by a corresponding, unique adhesion system, neuronal associations in many parts of the nervous system are mediated by shared adhesion systems. That is to say, "molecular specificity" is not the whole basis for neuronal specificity. Rather, confusion in associative decision-making is often avoided through temporal or spatial isolation. Even in the absence of temporospatial isolation, sufficient basis for an associative choice may sometimes be provided by differences in the amounts of a given ligand or by modifications of it which modulate its chemical affinities. Also, a given neuron can possess multiple adhesion systems and be involved in more than one associative decision at the same time.

Because insects have provided some of the most readily analyzed neurogenetic systems, and because neurogenesis in insects and vertebrates appears to follow similar principles, we deal at some length with studies of neuronal guidance in insect systems.

II. Neuronal Guidance Systems: Biological Studies

A. The Grasshopper Limb Bud

Clues concerning the nature of factors regulating the development of the peripheral nervous system have been derived from observations on the pioneer neurons of the grasshopper limb bud. It has been proposed that the morphology of the adult nerve in the limb bud is determined by a series of preaxogenesis neurons termed "stepping stones" (Bate, 1976b) or "guidepost" cells (Keshishian and Bentley, 1983) which guide the pioneering neurons from the distal part of the limb to the CNS. Observations on the development of the embryonic nerve reveal that the pathway is pioneered by a pair of afferent neurons, the tibial (Ti1) cells (Keshishian and Bentley, 1983). In their characteristic pattern of growth, the tibial Ti1 axons were found to grow anteriorly along the basal lamina of the ectodermal epithelium, eventually contacting the stomata of the F_2 neuron. Upon contacting this cell, the Ti1 growth cones turned 60–90° to project into the CNS (Keshishian and Bentley, 1983). When the developing chain of axons was injected with the dye Lucifer yellow, it was discovered that the cells had been selectively coupled through their axons, stomata, and filopodia, suggesting that these cells were in intercellular communication with one another (Keshishian and Bentley, 1983; Taghert *et al.*, 1982). Subsequent development of the embryonic nerve pioneered by the Ti1 axon was found to involve the selective fasciculation of later-developing neurons along this pathway. Axons of the femoral chordontal organ (FCO), located in the anterior ventral region of the mid-femur, were seen to project along the ventral ectoderm until they contacted the fibers of the embryonic nerve and then grow along the latter to the CNS (Keshishian and Bentley, 1983). In contrast to the pioneer neurons, these later-developing axons did not establish dye coupling with the cell at the point at which they joined the embryonic pathway. It was also reported that dye coupling was transient between the Ti1 axons and cells initially involved in establishing the nerve pathway. This coupling pattern seems to suggest a precise, temporally coordinated mechanism of cell recognition initially mediated through selective growth cone adhesion.

The importance of these guidepost cells in the proper development of the peripheral nerve pattern has been demonstrated in experiments in which the guidepost cells were lesioned with a UV microbeam. It was found that in limbs in which the guidepost cells were killed, pioneer growth cones failed to navigate their normal route, showed abnormal branching, and often failed to contact the CNS (Caudy and Bentley, 1983). It has been shown in the cricket, using a similar technique, that pioneer fibers are necessary for the proper guidance of later-developing axons. It was found that when the apical regions of the cercal rudiments of the cricket were lesioned with a UV laser before the pioneer fibers had elongated, the bundles of dorsal and ventral sensory axons were disorganized (Edwards *et al.*, 1981).

The studies on the grasshopper limb bud suggest a mechanism by which a stereotypic nerve pattern could be achieved by utilizing a series of guidepost cells spaced at intervals along the embryonic pathway. Yet in this system it is still unclear what initially guides the pioneer's axon toward the first guidepost cell, F_2. The observation that the axons almost always emerge from the proximal surface of the pioneer cells and almost always extend proximally for several cell diameters has led to the suggestion that this oriented growth could be due to the influence of internal organization of the Ti1 cells and to the presence of a weak, proximally oriented cue generated by a diffusible substance or a gradient of adhesiveness (Caudy and Bentley, 1983).

B. In Vitro Model Systems

The influence of cell-to-substratum adhesion on neuronal morphogenesis has been clearly demonstrated *in vitro*. Letourneau (1975a) has shown that cells from 4- to 8-day-old chicken embryos initiate more axonal growth, elongate faster, and show more axonal branching when cultured on a surface to which they are more adhesive, such as polyornithine, than on surfaces to which they are less adhesive, such as tissue culture plastic. Moreover, it has been shown that the pathway of axonal elongation can be influenced by the relative adhesiveness of the growth cone to the substratum. When sensory ganglia of chicken embryos were placed on a substratum bearing polyornithine- and palladium-shadowed areas in a gridlike pattern, the growth cones restricted their growth to that surface (polyornithine-coated) to which they adhere more strongly (Letourneau, 1975b). The same preference is shown for native laminin (a basal lamina protein discussed further in Section III,A,2) over UV-irradiated laminin (Hammarback and Letourneau, 1984).

In vitro experiments have shown that an adhesion gradient is capable of guiding the migration of cells. Mouse fibroblast cells, when placed on an artificial substratum upon which a gradient of palladium metal had been deposited, moved uniformly in the direction of increasing amounts of metal (greater adhesiveness) and migrated farther than cells placed on an evenly metallized surface (Carter, 1965, 1967; see also Harris, 1973). Given the results of these *in vitro* experiments, axons could in principle be guided *in vivo* by precisely defined pathways of relatively high adhesiveness or by broad adhesion gradients (Letourneau, 1975b).

C. *The Moth Wing*

Evidence which supports a role for adhesion gradients in the formation of a stereotypic neuronal pattern has been derived from studies of axonal behavior in the developing wings of a moth, *Manduca sexta*. During normal development, the sensory axons grow proximally along the basal lamina of the upper epithelial layer of the wing, toward the mesothoracic ganglion (Nardi, 1983). When the upper epithelial layer of the wing was altered, however, by exchanging epithelial grafts at various locations along the proximo-distal axis, the normal pattern of development was interrupted. It was found that when grafts were exchanged between different wing regions in the same pupa, axons encountering grafts from more proximal regions generally grew across them, while axons encountering grafts from more distal positions either failed to cross the grafts or bypassed them and joined a nearby nerve (Nardi, 1983). SEM analysis of the upper epithelial layer revealed a higher concentration of oriented fibrils and matrix particles in the proximal region than in the distal one, suggesting that a gradient of extracellular material to which the growth cones adhere could be responsible for guiding the axons proximally.

D. *The Grasshopper Antenna*

Further observations consistent with the action of adhesion gradients in guiding axonal migration come from studies on pioneer neuron development in the grasshopper antenna (Berlot *et al.*, 1983; Berlot and Goodman, 1984). In the course of development, axons arising from the dorsal pair neurons migrate proximally for 200 μm along the inside surface of the columnar epithelium without contacting special landmark cells. The axons eventually give up contact with the epithelial cells in order to fasciculate selectively with the axon of the base pioneer neuron which they follow to the CNS. By transmission electron microscopy (TEM), the growth cones of the dorsal pair axons were observed to migrate under the epithelial basement membrane with their filopodia in direct contact with the surface of the epithelial cells, as though the filopodia "are responding to an adhesive gradient along the epithelial face" (Berlot *et al.*, 1984).

E. The Grasshopper Central Nervous System

The central nervous system of the grasshopper embryo is particularly favorable for studying the generation of precise nerve patterns because in the early stages of its development there are only a few, large, easily accessible neurons whose cell lineages are known (Goodman and Bate, 1981; Goodman *et al.*, 1982). Cellular and immunological studies have revealed that nerve pathways in the grasshopper CNS are formed in a precise way (Raper *et al.*, 1983a) and that neuronal recognition through specific filopodial interactions could be important for mediating this development (Bastiani and Goodman, 1983; Bastiani *et al.*, 1984).

Neuroblast 7-4 is a neuronal precursor cell which gives rise to a number of ganglion mother cells, the first three of which divide to form neurons designated Q_1, Q_2, G, C, Q_5, and Q_6 (Raper *et al.*, 1983a). Of these progeny, Q_1 was found to initiate axonal outgrowth first and to pioneer a pathway across the T_2 ganglion in the posterior commissure to the contralateral neuropil (Raper *et al.*, 1983a). It is interesting to note that the Q_1 growth cone initially migrated in contact with the basement membrane at the T_2 ganglion, but later gave up this surface as it reached the ganglionic midline in order to contact the surface of its contralateral homolog growing in the opposite direction (Bate and Grunwald, 1981; Raper *et al.*, 1983a). This exchange could reflect a preference determined by the strength of adhesion of the axonal growth cone for one surface over the other. All of the progeny which subsequently initiated axonal outgrowth, Q_2, G, C, Q_5, and Q_6 (in this order), followed the Q_1 fiber despite the presence of other axon bundles in the posterior commissure (Raper *et al.*, 1983a). The two neurons G and C have been found to display a particularly interesting behavior once they arrive at the contralateral neuropil. G's growth cone, which was later joined by C's, remained stationary at this "choice" point for up to 10 hr, exhibiting a morphology quite different from that observed during growth. The shape of the growth cone at this choice point was found to be broad and complex, with filopodia extending in tufts from a few anteriorly directed lumps located proximal to the growth cone's distalmost tip (Raper *et al.*, 1983a). While the G and C growth cones were stationary in the contralateral neuropil, their filopodia were found in extensive contact with a particular fiber bundle and were observed to make only brief contacts with a number of nearby fascicles, suggesting that the growth cones have a high affinity for one particular fiber bundle (Bastiani *et al.*, 1984). This particular fiber bundle to which the filopodia selectively adhered was found to consist initially of several neurons, A_1, A_2, P_1, P_2, and P_3, which arise from unrelated neuronal precursor cells.

The A/P fascicle is formed from the selective fasciculation of these cells' axons which migrate in opposite directions through the dorsolateral neuropil. Observations on the migration of the A and P growth cones revealed that these, like the Q_1 axons, relinquish contact with the dorsal basement membrane in order to fasciculate with each other (Raper *et al.*, 1983b). Further analysis of the selective adhesion between the G and C growth cone filopodia and the A/P fascicle revealed that the G growth cone was always found associated with the P axon and not the A axon (Raper *et al.*, 1983b; Bastiani *et al.*, 1984), suggesting that the filopodia of G's growth cone can distinguish between subsets of axons in the A/P fascicle through relative strengths of adhesion. TEM serial section reconstructions also revealed that some of these filopodial contacts could represent highly specific interactions between these cells. In contrast to filopodia from other nearby neurons which only contacted the surface of the G growth cone, two filopodia from the P growth cone were found to insert into it and to induce the formation of coated pits and coated vesicles at the filopodial tips (Bastiani *et al.*, 1984). Similar growth cone insertions were found in studies on pioneer growth cones in the grasshopper CNS (Bastiani and Goodman, 1984). It has been suggested that these specific filopodial interactions could affect the adhesive properties of the migrating growth cones at each choice point through receptor-mediated endocytosis, causing the growth cones to fasciculate selectively with different axon bundles on their way to their appropriate targets (Bastiani and Goodman, 1984).

Further development of the G and C neurons results in the divergence of their growth cones, G's migrating anteriorly in contact with the P axons and C's posteriorly in association with the A axons (Raper *et al.*, 1983b). The shape of G's growth cone has been observed to change dramatically as it begins to migrate along the P axon, becoming long and tapered with fewer filopodia extending from only the tip of the growth cone (Raper *et al.*, 1983b). It is interesting to note here that the G growth cone responded to the P fiber bundle while the C growth cone, which had the opportunity to attach to the P fibers, chose instead to wait and follow the A fiber bundle. The selectivity of G and C growth cones for one pathway over another and the importance of the A/P fascicle for proper development of these axons have been further demonstrated in studies in which the A, P, or both axon bundles were deleted in the grasshopper embryo (Raper *et al.*, 1983c, 1984). Deletion of the A axons by cutting the lateral portion of the ganglionic connective had no effect on the migration of the G growth cone. Removal of the P axons or the entire A/P fascicle, however, prevented the migration of the G axon farther than the contralateral neuropil (Raper *et al.*, 1983c, 1984). Examination by TEM of the G growth cone in the contralateral neuropil revealed that in the absence of the P axons, it appeared abnormally branched in contact with or near more than 20 different axon fascicles. Yet none of the branches or filopodia of the growth cone, it was discovered, exhibited a particularly high affinity for any of these fiber bundles (Raper *et al.*, 1984). It is apparent from these results that there exists in the development of the grasshopper CNS a high degree of specificity but not a high degree of flexibility, since axons deprived of their normal pathways are unable to utilize other pathways. This contrasts with findings on chick motoneurons, which have been found not only to take aberrant pathways but also, in some cases, to reach their targets by unfamiliar routes (Lance-Jones and Landmesser, 1981b).

The high affinity expressed by the G growth cone for the axons and its inability to respond to other axons suggest that the surfaces of axons are specifically labeled such that each longitudinal axon fascicle, or even subsets of axons within a bundle, express different surface molecules and that the filopodia are only able to detect and respond to the appropriate labels (Raper *et al.*, 1983b; Goodman *et al.*, 1984). Evidence to support the existence of specific pathway labels has come from experiments utilizing monoclonal antibodies. A monoclonal antibody (mes-2) was isolated which detected a surface antigen expressed on only three neurons out of 1000 in the T_2 neuroganglion (Kotrla and Goodman, 1983). FETi and SETi, two of the neurons which possess this specific antigen, are derived from different precursor cells, yet have similar final pathways and target. During the course of development, the antigen cannot be detected while the growth cones of the two neurons migrate in different pathways; however, the antigen becomes detectable when the growth cone of FETi reaches the axon of SETi. Once both of these axons reach their target, the expression of the mes-2 antigen dramatically decreases (Kotrla and Goodman, 1983).

Observations derived from extensive studies on the development of the grasshopper CNS have led to the "labeled pathway" hypothesis (Raper *et al.*, 1983b), which states that pioneer neurons establish stereotypic pathways, that these pathways are differentially labeled on their cell surfaces, and that later growth cones are differentially determined in their ability to make specific choices of which labeled paths to follow. If this hypothesis is true, then a particular complex pathway might be generated by changing the relative adhesiveness of a growth cone for certain axon bundles at the "choice" points, thus linking together a series of smaller pathways (Raper *et al.*, 1983b).

F. Vertebrate Limb Motoneurons

A high degree of reproducibility has been demonstrated in the nerve pattern of vertebrate limbs. In studies on pathway selection by embryonic chick motor neurons prior to the normal period of cell death, it has been found that axons from a particular part of the spinal cord only project down certain nerves. Axons labeled in the third lumbosacral segment always project down the femorotibialis, adductor, and ischioflexorius nerves, whereas axons labeled in the first and second lumbosacral segments always project down the sartorius nerve (Lance-Jones and Landmesser, 1981a). Evidence supporting an active role for axons in generating this precise organization comes from experiments in which motor neurons were shifted away from their normal positions by anteroposterior reversal of segments of the lumbosacral spinal cord. Following such reversals, it was found that axons altered their direction of growth in the region of the plexus and proximal nerve trunk in order to enter their normal pathways. Moreover, it was noted that when axons were displaced too far from their normal points of entry into the limb, the axons projected to inappropriate targets (Lance-Jones and Landmesser, 1981b). These results suggest that axonal guidance cues for a specific pathway are localized in the limb bud or in the intervening tissue at the base of the limb (Landmesser, 1981).

The importance of initial pathway selection in determining the fate of a particular axon is also demonstrated in a set of experiments involving the dorsoventral rotation of distal limb segments in chick embryos. In earlier studies it had been found that although axons can take aberrant pathways and innervate inappropriate targets, they usually project correctly with respect to muscles derived from either the dorsal or ventral muscle mass (Landmesser, 1978a; Lance-Jones and Landmesser, 1981b; Whitelaw and Hollyday, 1983b). Yet Whitelaw and Hollyday (1983c) have found that when limbs were rotated at a level distal to that at which axons usually select ventral or dorsal pathways, the innervation of the rotated limb was in all cases opposite from normal. In this situation it appears that pathway selection rather than any dorsoventral muscle tissue recognition that might exist determines what muscle mass the axons will innervate.

In addition to a mechanism of axonal guidance by specific cues in the environment, results of experiments involving deletions of limb segments in the chick hindlimb indicate that competitive interactions within a pathway or at the site of termination are important for establishing normal connections. In partial limbs, missing either a thigh or both a calf and a foot, organization of the motor column was normal prior to the period of cell death. However, after the period of cell death, it was found that only the axons which normally serve the segment remained (Whitelaw and Hollyday, 1983a).

As is evident from all these experiments, the nature of the cues which guide the axons to their appropriate targets is quite specific. The fate of a motoneuron is determined by its affinities, most likely mediated through selective adhesive interactions, for a particular pathway which serves a given set of muscles. It is interesting, then, that despite this apparent specificity, axons have been observed to take aberrant or totally novel pathways to reach their correct muscles (Lance-Jones and Landmesser, 1981b). In order to explain these findings, Lance-Jones and Landmesser (1981b) proposed that when axons are introduced into sufficiently foreign regions they are guided in a "nonspecific" way down a stereotyped set of pathways which may simply represent regions in the developing limb where adhesive interactions favor axon elongation. However, when axons detect cues appropriate for them, they deviate from this inappropriate pathway toward their normal target. This proposal is consistent with the idea that axons can exhibit a certain degree of selectivity for a particular pathway through their adhesive interactions. Thus, it is not as though axons are rigidly specified, in this case, to recognize only one particular pathway; more likely, one pathway is favored over another because of the relative strengths of the different adhesive interactions.

G. *Vertebrate Central Nervous System*

Studies concerned with the growth of axon fibers in the spinal cord of newts and *Xenopus* have revealed that axons grow into a series of preexisting longitudinally oriented spaces (Norlander and Singer, 1978, 1982; Singer *et al.*, 1979). In *Xenopus*, it has been shown that these spaces not only appear in a particular sequence but also occur in reproducible positions along the circumference of the spinal cord (Norlander and Singer, 1982). Regeneration experiments with newts have shown that the fiber tracts determined by the primitive ependyma of the regenerating cord are similar to those generated by the primitive neuroepithelium in the developing embryo, thus implying that the formation of these spaces is not random but is determined by the neuroepithelium cells lining these pathways (Singer *et al.*, 1979).

Observations on the neuroepithelium cells that give rise to axon fiber pathways reveal that they surround the lumen of the neural tube. These cells, it has been discovered, send out radial processes to the periphery of the neural tube and have been found to be tightly bound to neighboring cells by desmosomes both at the periphery and at their apical ends near the central canal (Norlander and Singer, 1978; Singer *et al.*, 1979). The spaces that are formed between these radial processes constitute a section of the channel and when successively aligned, as in both the newt and *Xenopus*, give rise to long, longitudinally oriented pathways (Norlander and Singer, 1978, 1982; Singer *et al.*, 1979).

Fiber tracts first appear in these channels in the ventral part of the spinal cord and later develop in dorsolateral positions, suggesting that a gradient of "development'' exists in the ventrodorsal direction as well as the rostrocaudal one (Singer *et al.*, 1979). When these channels are invaded by growing axons, they become significantly enlarged radially, becoming more oval in shape (Norlander and Singer, 1978). Interestingly, it was also found that several membrane specializations, hemidesmosome-like structures, coated membranes and vesicles, and synapse-like configurations developed between the ingrowing axons and the neuroepithelial processes.

The first two membrane specializations, hemidesmosome-like structures and coated membranes and vesicles, were found predominantly in primitive regions at the channels which contain only a few recently acquired axons (Norlander and Singer, 1978). The third specialization, the synapse-like configurations which consisted of an aggregation of small vesicles near a dense presynaptic region, were distributed more widely, especially in rostral areas of the developing cord (Norlander and Singer, 1978; Singer *et al.*, 1979). As suggested by Norlander and Singer (1978), these ''synaptoids'' might be sites of nonsynaptic chemical release important for neurotropic interactions. Similarly, the coated vesicles could be carrying substances important for axon–pathway interactions. The hemidesmosome-like structures probably serve as temporary sites of attachment, facilitating axonal elongation (Norlander and Singer, 1978).

The presence of these membrane specializations seems to suggest not only that the channels serve as mechanical guides for elongating axons but also that some kind of ''recognition'' exists between these cell types. It is unclear from these results, however, whether these specializations reflect a highly specific interaction between a particular pathway and the ''appropriate'' set of axons or a more general developmental association that might occur between the neuroepithelial cells and any invading axons.

Further evidence to support the role of glial cells as mechanical guides in neuronal development has been derived from studies by Rakic (1972) on the development of the fetal monkey cortex. For new neural cells generated in the ventricular and interventricular zones to reach the cortical plate, it was discovered that they must migrate 3500 μm through an intermediate zone containing a complex mixture of closely packed cell processes and cell bodies that make up the developing cerebral wall. In order to accomplish this, these cells were found to migrate along elongated, radially oriented immature glial processes which spanned almost the entire width of the cerebral wall. Throughout their migration, the neural cells were observed to be in constant contact with only one particular fiber along its curving trajectory to the cortical plate and, in some instances, to have their leading processes actually wrapped around these fibers.

Rakic (1972) has suggested that these fibers, in addition to serving as simple mechanical guides, are important in the development of vertical cell columns in the adult neocortex. Electron microscopic analysis has revealed that several generations of neural cells migrate along the same radial fiber (Rakic, 1972). The close association of the migrating cells with one radial fiber and the finding that several generations follow the same fiber imply that a highly specific affinity could exist between developing neural cells and particular fibers. Yet due to the complexity required to produce and orchestrate a large number of unique cues between glial fibers and the neural cells which migrate along them, it seems more likely that the new neural cells simply have higher affinity for glial fibers than for surrounding cells and that they join the radial fiber closest to their site of origin in the ventricular and subventricular zone. Letourneau (1975b) has shown *in vitro* that elongating axons prefer a substratum of glial cells over a number of other surfaces.

Analysis of the pattern of development of ascending and descending axons in the embryonic chick spinal cord has provided evidence to suggest not only that axons serve as a suitable substratum to guide other axons but also that developmental timing can impose a topographic order upon fiber tracts of the spinal cord (Nornes *et al.*, 1980). By injecting [^{3}H]proline into the chick spinal cord, it was found that in early stages of development, axons extending in the descending direction are positioned in the most ventromedial marginal zone. In the same early stages of development, axons extending in the ascending direction were found to fasciculate along these descending fibers. In later stages, new axons emerging from neurons differentiating in a rostrocaudal direction were found to fasciculate upon "older" ascending fibers, i.e., from more rostral levels forming layers of age-related fibers and thus imparting a topographic order to the fiber tract.

Although it is argued that this laminar fiber organization is simply a consequence of the sequential development, it is also possible that this pattern is sharpened and maintained dynamically by axons exhibiting a continuing preference for fibers of the same age. Even if all these fibers fasciculate using a common adhesion mechanism, differences between age-related groups of fibers could arise from quantitative or qualitative differences in an adhesion molecule. Age-dependent qualitative differences could arise from structural changes in the cell adhesion molecule which stabilize bond formation. Evidence which suggests that axons can exhibit a selective affinity for a particular group of fibers has been obtained *in vitro*. When dissociated cells from rat retina and sympathetic ganglia were mixed and allowed to form aggregates, the sympathetic neurons were found in groups, frequently at the periphery of the aggregates, while the retina cells showed a tendency to group in small rosettes (Bray *et al.*, 1980). Moreover, the axons of each cell type in these aggregates were observed to form separate bundles of predominantly one kind of fiber.

The results of an experiment in which the acousticolateral placode was transplanted into the eye region of the developing frog, *Rana pipiens,* show that different sets of axons can respond to the same physical environment in different ways. When VIIIth nerve ganglion cells were transplanted to the evacuated eye position, their axons were observed to extend dorsally along the surface of the neural tube from a relatively ventrally placed ganglion and to penetrate the lateral region of the diencephalon, eventually reaching the periventricular cellular region of the frog's dorsal thalamus. This trajectory was similar to that within the medulla oblongata but unlike the trajectory of any axons normally found in the diencephalon (Constantine-Paton, 1983). Although this stereotyped behavior could be attributed to the possession by an axon of an internal program for generating a particular pattern, it more likely reflects the ability of axons to respond selectively to informational cues in the environment. Optic nerve axons have been found to grow in longitudinally oriented tracts (Constantine-Paton and Capranica, 1976; Constantine-Paton, 1978), while VIIIth nerve axons project inward toward the central regions of the neural tube. In the course of their development, the VIIIth nerve axons were also found not to alter their growth trajectory even though they grew at right angles across the retinal ganglion cell axons of the optic tract. This observation further suggests that axons are able to discriminate between potential pathways in the environment. It also indicates that not all sensory axons have an inherent affinity for other sensory fibers characteristically located in longitudinally oriented tracts in the dorsolateral quadrant of the neural tube.

H. The Vertebrate Retinotectal Projection

The axons of retinal neurons migrate along stereotypic pathways to the optic tectum, with which they establish precisely ordered connections. During the course of their development, the growth cones of these axons must choose between various pathways and upon reaching the tectum must make appropriate synaptic connections with a particular tectal region.

It was originally proposed by Sperry (1943, 1963) that the specific patterning of synaptic connections between retinal ganglion cell axons and tectal neurons arose through the formation of complementary biaxial gradients of cell surface molecules. Specific connections developed through the matching of complementary molecules on the pre- and postsynaptic cell surfaces. Support for the existence of stable affinity labels between retinal and tectal cells is provided by the results of experiments in which the optic nerve in goldfish was severed and part of the retina removed. Regenerating fibers from a ventral hemiretina selectively innervated the dorsal tectum and fibers from a nasal hemiretina bypassed rostral regions in the tectum in order to innervate more caudal regions. When a complete peripheral ring of retina was removed, leaving the central retina intact, the regenerating axons from this part were found to enter the parallel layer at the cortex but not to descend into the plexiform layer until reaching their normal termination sites in central regions of the tectum (Attardi and Sperry, 1963). The results of studies in which the tectal surface has been altered also suggest that affinity labels exist on the surface of the tectum. If a small rectangular piece of goldfish tectum was dissected out, rotated 180°, and then reimplanted in the same site, it was found that the retinal projection of the rotated tectal implant was also completely reversed (Yoon, 1973). In a more recent study, Gaze and Hope (1983) reported that if rectangular grafts were translocated without rotation between rostral and caudal regions of the goldfish optic tectum, the corresponding visual field was also translocated.

The results of *in vitro* experiments provide additional support for Sperry's hypothesis of neuronal specificity. Barbera (1975) has shown that dorsal retina cells preferentially initiate adhesions to ventral tectal halves and that ventral retina cells preferentially initiate adhesions to dorsal tectal halves. Interestingly, cells from the ventral retina were found to exhibit binding behavior which was dependent on the length of time these cells were allowed to recover after dissociation. Shortly after trypsinization in Ca^{2+}-free solution, these cells adhered preferentially to the ventral half of the tectum; however, 3–4 hr after dissociation, these cells adhered preferentially to dorsal tectal halves. In contrast, dorsal retinal cells adhered preferentially to ventral tectal halves regardless of the length of the recovery period, raising the possibility that two different cell adhesion molecules which differ in their sensitivity to trypsin in the absence of Ca^{2+} are located in different regions of the retina. It is important to note, however, that in this experiment all types of retinal cells were tested in the binding assays instead of only those that actually make contact with the tectum.

One type of experiment which attempts to solve this problem measured the adhesion of tectal membranes to neurites sprouting from strips of retinal tissue cut in either the anteroposterior or dorsoventral direction. Using this experimental system, it was discovered that tectal membranes bound in larger amounts to dorso ventral strips through the anterior part of the retina than through the posterior part (Hafter *et al.*, 1981). Bonhoeffer and Huf (1980, 1982) have shown that in a situation in which axons emerging from retinal explants are given a choice between two cell monolayers, the one consisting of retinal and the other of tectal cells, they grow preferentially over the tectal monolayer. In a more recent set of experiments using this same system, it was found that axons from the temporal region of the retina were able to distinguish between tectal cells originating from different positions within the tectum along the anteroposterior axis. Temporal axons, when exposed to monolayers of either anterior or posterior tectal cells, exhibited a strong preference for anterior tectal cells. This affinity decreased gradually from the anterior to the posterior tectum (Bonhoeffer and Huf, 1982).

Evidence which challenges the existence of rigid positional markers on the surface of the tectum is derived from experiments in which the relative size of the retinal or tectal populations was experimentally altered. In one such experiment, when the caudal half of the optic tectum was removed in goldfish, the axons of this missing region innervated the remaining rostral half of the tectum in appropriate retinotopic order (Gaze and Sharma, 1970). In the same set of experiments, removal of half of the tectum in conjunction with an optic nerve crush resulted in the compression of the entire visual field into the residual half tectum. Schmidt and co-workers (1978) similarly demonstrated that the site at which retinal ganglion cells can terminate on the tectum is not rigidly determined. After removal of the nasal half of the retina, the remaining axons projected to their appropriate region in the rostral half of the tectum, which is consistent with earlier findings of Attardi and Sperry (1963). However, over time, these fibers gradually spread caudally to cover the available tectal space (Schmidt *et al.*, 1978). The results of an experiment which suggests that an orderly retinotectal map can develop independently of tectal cues involved the rotation of all or part of the optic tectum. The array of optic axons invading the tectum was found to ignore the reversal of the target structure and to form a retinotectal projection normally oriented with respect to the whole animal but reversed with respect to the optic tectum (Chung and Cooke, 1978). Yet it was also reported that whenever the position of the diencephalon relative to the tectum was changed, the axial polarity of the retinotectal projection was also altered (Chung and Cooke, 1978), indicating that the diencephalon organizes the tectal components of the retinotectal mapping determinants. It was also observed that when the prospective preoptic nucleus and hypothalamus (both are components of the diencephalon, which normally lies anterior to the optic tectum) are transplanted to a position posterior to the optic tectum, the optic nerve deviates posteriorly, preserving a spatial relationship between the optic chiasm and these nuclei. An optic nerve deviated in this manner enters the tectum from the caudal rather than the rostral end. When this happens, the direction in which the histogenesis of the tectum proceeds (it is normally rostrocaudal) is correspondingly reversed (Chung and Cooke, 1978). These correlations between diencephalic position, optic nerve migration pathways, and orientation of the retinotectal projection on the optic tectum suggest causal relationships which have not yet, however, been firmly defined.

Detailed examination using small injections of horseradish peroxidase (HRP) has shown that fibers are topographically and chronotopically arranged in the optic nerve and tracts of both *Rana* (Reh *et al.*, 1983) and adult goldfish (Easter *et al.*, 1981; Bunt, 1982). In the development of the retinotectal projection of *Rana*, fibers are added to the periphery of the nerve, thus forming concentric rings of age-related fibers. Axons emerging from dorsal and ventral retina generally remain in the dorsal and ventral regions of the optic nerve, respectively. As a consequence of a different pattern of growth, however, the optic nerve in goldfish becomes layered with the youngest fibers present on the ventral side and the oldest fibers on the dorsal side of the nerve. Within each layer, the retinal fibers are also topographically ordered. Dorsal fibers are located in the central part of each layer, flanked by nasal fibers on one side and temporal fibers on the other. Ventral fibers occupy both ends of each layer (Bunt, 1982).

Interestingly, in *Rana* it was found that fibers of the same origin and age travel not as one but as two distinct groups located on either side of the dorsoventral midline. This division is more apparent with either temporal or nasal axons, approximately half of whose fibers were found to cross over to the opposite side of the nerve immediately central to their entry into the optic nerve head. As a result of these rearrangements, each concentric ring in the optic nerve is made up of two concentric semicircles, both of which contain age-related fibers from the full circumference of the retina. The optic nerve thus contains minor symmetric representations of the retinal surface on either side of the dorsoventral midline of the nerve (Reh *et al.*, 1983).

Immediately peripheral to the optic chiasm, the semicircles of age-related fibers in *Rana* become flattened, with the oldest groups of fibers located in the dorsal regions of the chiasm and the newest fiber bundles located below them adjacent to the glial limiting membrane on the ventral side of the brain (Reh *et al.*, 1983). Shortly after passing the optic chiasm, the fibers in each age group undergo a substantial rearrangement. Axons from dorsal retinal ganglion cells enter the lateral optic tract, while fibers from ventral retinal ganglion cells enter the medial tract. Nasal fibers are present in both the lateral and medial tracts in close association with dorsal and ventral fibers, respectively. Temporal retinal fibers do not enter either tract but grow between them (Fawcett and Gaze, 1982; Reh *et al.*, 1983). As the fibers continue to grow along the diencephalic tracts toward the tectum, age-related fibers remain bundled together (Reh *et al.*, 1983). Moreover, the target was found not to affect the organization at the optic nerve and tract, since retinotopic order was well preserved in tectumless frogs (Reh *et al.*, 1983).

It is evident from these results that "passive" factors such as developmental timing and mechanical guidance contribute to the organization of the optic nerve in *Rana* (Reh *et al.*, 1983) and in goldfish (Easter *et al.*, 1981; Bunt, 1982). In mice, a system of oriented spaces within the undifferentiated optic cup was detected, suggesting that mechanical guidance may contribute to the topographical patterning in this system as well (Silver and Sidman, 1980). As proposed by Reh and co-workers (1983), "passive" mechanisms could adequately account for the organization of axons in the optic nerve of *Rana* but these mechanisms could not account for the sorting-out of axons at the chiasm. Instead, it was suggested that the specific rearrangement of fibers at the chiasm reflects an active selection process by the growth cones of the axons for one pathway over another.

Observations from several studies suggest that axons can actively select specific pathways. When the medial and lateral tracts of the optic nerve in cichlid fish were cross-connected to the opposite central stump, later histological analysis revealed that each fiber bundle, instead of growing down the inappropriate channel, crossed back to its original pathway (Arora and Sperry, 1962). It was also reported that if the lateral optic bundle was dissected out and inserted deep into the channel at the medial bundle, the deflected fibers advanced into the foreign half of the tectum. Although this result seems to suggest that axons are not responding to specific cues present in the optic tract, it is possible that since they were inserted deep into the other fiber bundle they were deprived of access to their regular guidance cues. Thus they followed the only cues that were available in the medial optic tract. By removing specific parts of the retina and allowing fibers to regenerate back to the tectum, Attardi and Sperry (1963) demonstrated that fibers can distinguish between different pathways. With removal of the ventral half of the retina and complete sectioning of the nerve, fibers from the dorsal retina selectively entered the lateral tract to connect with the ventral tectum. Regenerating fibers from a ventral hemiretina consistently entered the medial tract and terminated on the dorsal tectum.

Studies in which compound eyes were constructed by the fusion of embryonic eye fragments also suggest that fibers can select appropriate pathways (Straznicky *et al.*, 1979, 1981; Fawcett and Gaze, 1982; Gaze and Fawcett, 1983). Fibers from compound ventral eyes (VV) were found to enter only the medial brachium even though both pathways were available. Axons of compound temporal eyes gave similar results in that they, too, followed a pathway to the tectum that was appropriate for their embryonic origin. Although it is apparent from these results that axons can actively select pathways, the results of studies by Gaze and Grant (1978) suggest that a complete visuotectal map can still be generated by axons without following their appropriate pathways. Regenerating fibers which grow in a compact band in an abnormal pathway along the far lateral edge of the diencephalon were found to form a complete visuotectal map 10–20 days after nerve sectioning. Even though it has been reported that the regenerating fibers do not follow the paths of degenerating optic fibers through the diencephalon, it is possible that fibers could be responding to degenerative materials left on the surface of the tectum.

The results of recent experiments in which only a small portion of one eye was replaced by similar wedge-shaped pieces from opposite regions of the other eye suggest that the retina contains information that is used in the assembly of the retinotectal map. Electrophysiological recording of the retinotectal projection from these compound eyes revealed that the axons from the transplanted wedge of retinal tissue projected to the region of the tectum which was appropriate for their original position within the donor eye rudiments (Cooke and Gaze, 1983; Willshaw *et al.*, 1983). These findings suggest, then, that different positional information is encoded in different parts of the retina (Willshaw *et al.*, 1983). Whether this information is expressed in terms of fiber–fiber interactions or fiber–tectal interactions, however, is still unclear.

The observation that the innervation of a single tectum by two eyes results in the formation of a pair of interrupted, interdigitated ("striped") projections (Levine and Jacobson, 1975; Constantine-Paton and Law, 1978) poses a challenge to any model of retinotectal "specificity" based solely upon concepts of positional correspondence between points on the retinal and tectal surfaces. It has become increasingly clear that explanation of this phenomenon as well as of regulative properties of the retinotectal matching system requires the operation of multiple determinants. Fraser (1980, 1985; Fraser and Hunt, 1980) has constructed a model, based largely upon adhesive interactions between neurites, which attempts to simulate both normal and experimental retinotectal mapping behavior with the minimal number of variables. His current model, which seems to achieve this goal, utilizes the following determinants: (1) C—a strong, position-independent adhesion between optic nerve fibers and the tectum; (2) R—a strong repulsion or competition between optic nerve fiber terminals; (3) DV—a weak, position-dependent adhesion between optic nerve fibers and the tectum, dorsoventrally graded on both; (4) AP—a somewhat weaker, position-dependent adhesion between optic nerve fibers and the tectum, anteroposteriorly graded on both; and (5) a marker, probably a "local label", that causes preferential association between fibers from neighboring retinal sites. Because treatments that abolish neural electrical activity prevent the formation of eye-specific bands, "the correlated activity of neighboring ganglion cells has become a prime candidate for this local label" (Fraser, 1985).

III. Molecular Mechanisms of Adhesion

The behavior of axons in a variety of biological systems such as those presented in the previous section suggests that their stereotypic patterning is dependent upon adhesive interactions with the substratum and with other axons. Investigations into the molecular basis of these interactions have revealed many details concerning the process of cell adhesion. A number of macromolecular substances isolated from cells or tissues have been found to affect both cell–substratum and cell–cell adhesion (for review see Garrod and Nicol, 1981; Hay, 1981; Edelman, 1983; Damsky *et al.*, 1984).

According to one view, each type of cell or tissue produces a unique cell-type-specific or tissue-specific adhesive "factor," qualitatively different from the "factors" produced by other cells or tissues (Moscona, 1962, 1968; Lilien and Moscona, 1967; Lilien, 1968; Garber and Moscona, 1972; Hausman and Moscona, 1976). Tissue affinities (Holtfreter, 1939) and the sorting-out of embryonic cells were ascribed by Moscona and associates to the action of such tissue-specific ligands, conferring upon cells and tissues a preference for "self"-

associations. Reasons to doubt this view were advanced quite early (Steinberg, 1963, 1964, 1970, 1978a,b, 1981; Garrod, 1981; Garrod and Nicol, 1981). As molecules that mediate cell attachments have come to be identified, it has become apparent that adhesion systems are not in general parceled out among cells and tissues in a one tissue : one ligand fashion. Not only can one and the same adhesion mechanism be found on the cells of a variety of tissues (e.g., Thiery *et al.,* 1982), but also cells of a given kind may possess more than one adhesion mechanism (Takeichi *et al.,* 1979; Thomas *et al.,* 1981a,b; Thomas and Steinberg, 1981; Thiery *et al.,* 1985). Although cell adhesion can be mediated through intercellular junctions such as gap junctions and desmosomes, this section will not address these structures but will focus instead on factors which have been implicated in the mediation of adhesions of neurons to each other or to environmental surfaces.

A. *Cell–Substratum Adhesion*

Neuronal growth cones show preferences in their adhesion to different surfaces *in vitro.* Growth cones from chick embryonic sensory ganglion cells were found to adhere much more strongly to polyornithine- or polylysine-coated surfaces and to the upper surfaces of explanted glial cells than to tissue culture plastic. Axons elongated more rapidly and branched more on polyornithine-coated dishes than on tissue culture plastic (Letourneau, 1975a). Such migratory growth cones, at the boundaries between areas of the culture dish treated to offer greater or lesser adhesiveness, selected the former over the latter (Letourneau, 1975b), leading to the conclusion that environmental adhesive anisotropies "are important determinants of the directions and pathways of axonal elongation."

1. Fibronectin

After the demonstration that one or more components of heart-cell-conditioned medium promotes outgrowth of ciliary ganglion neurites on a polyornithine-coated substratum (Helfand *et al.,* 1976), Hauschka and Ose (1979) reported that fiber outgrowth from chick embryonic retinal cells *in vitro* required a collagen substratum, a high molecular weight (>200) component of embryo extract, and serum. The serum, it was found, could be replaced by plasma fibronectin. The finding that fibronectin (reviewed in Hynes and Yamada, 1982) promotes retinal neurite outgrowth on a polycationic surface was confirmed and extended by Akers *et al.* (1981). Culp *et al.* (1980) reported that plasma fibronectin will bind to dishes and promote "normal" spreading and movement of both rat neuroblastoma and glioma cells. However, the most complete evidence for the role of fibronectin in guiding the migration of neural precursor cells has come from the study of the guidance of neural crest cell migration.

Using immunoperoxidase labeling, it was found that fibronectin is localized in the basement membranes surrounding neural crest cell populations both before and during migration but was absent when crest cells aggregated to form ganglion rudiments (Newgreen and Thiery, 1980). Similar findings were reported by Mayer *et al.* (1981). Greenberg *et al.* (1981) reported that cultured neural crest cells require serum or purified fibronectin to attach to collagen substrata and proposed that fibronectin mediates the attachment of migrating neural crest cells to collagen fibers which define the crest cells' migration pathways in the embryo. Newgreen *et al.* (1982) found, in addition, that fowl neural crest cells burrow into a fibronectin-rich matrix but generally remain on the surface of collagen–fiber gels and supported the hypothesis that fibronectin dictates neural crest migration pathways "by acting as a highly preferred physical substrate."

Both Newgreen *et al.* (1982) and Erickson and Turley (1983) reported that chondroitin sulfate (or an undersulfated form of it) can reduce crest cell responsiveness to fibronectin, and the former investigators further suggested that the utilization of certain fibronectin-containing pathways may be limited by the presence of proteoglycans containing undersulfated chondroitin sulfate.

Avian trunk neural crest cell adhesion to plasma fibronectin-bearing substrates or extracellular matrices was inhibited by antibodies to fibronectin or to a cell-binding fragment of it (Newgreen, 1982; Rovasio *et al.*, 1983), and these cells migrated on areas of substratum coated with fibronectin in preference to areas coated with laminin or type I collagen (Rovasio *et al.*, 1983). Interestingly, although fibronectin evidently defines the normal trunk neural crest migration pathways, *direction* is imparted to crest cell migration only by population pressure together with contact inhibition by other neural crest cells (Newgreen *et al.*, 1979; Rovasio *et al.*, 1983; see also Weston, 1963, 1970; Tosney, 1978).

2. Laminin

Avian trunk neural crest cells explanted onto glass coated with alternating tracks of adsorbed fibronectin and the basement membrane protein laminin (see Timpl *et al.*, 1979) were found by Newgreen (1984) to follow each kind of track equally well when each protein had been applied at a concentration high enough to produce the maximal response, in contrast to the finding of Rovasio *et al.* (1983). The biological significance of this affinity of neural crest cells for laminin is unclear, however, since laminin is evidently not normally present in their environment during their migration (Newgreen, 1984).

The first report that laminin promotes neurite outgrowth *in vitro* appears to be that of Baron van Evercoorin *et al.* (1982), who studied human fetal sensory ganglion cultures. Rogers *et al.* (1983) found that attachment to a plastic substratum and neurite extension by dissociated peripheral (embryonic chick dorsal root and sympathetic ganglion) and central (spinal cord and retina) neurons was promoted by substratum-attached laminin. Only peripheral and not central neurons initiated neurites on a fibronectin substratum. The authors speculate that "these responses to laminin and fibronectin *in vitro* may be related to the presence or absence of these glycoproteins in specific extracellular environments during specific developmental stages." Laminin deposited in extremely small amounts on a polyornithine-coated plastic surface strongly promoted the outgrowth of neurites from dissociated chick parasympathetic (ciliary ganglion) neurons in serum-free culture (Manthorpe *et al.*, 1983). Cultured neurons from a variety of peripheral as well as central nervous tissues responded in the same manner. The neurite-outgrowth-promoting activity of laminin was blocked by anti-laminin antibodies (Manthorpe *et al.*, 1983) and has been traced to certain isolated peptide fragments of the molecule (Wewer *et al.*, 1983). When this activity was reduced by local, patterned ultraviolet irradiation (Hammarback and Letourneau, 1984), neurites restricted their outgrowth to the unirradiated areas, demonstrating that mechanical boundaries or physical discontinuities are not required for neurite guidance on laminin pathways.

3. Other Neurite-Promoting Factors

Several studies have shown that heart-cell-conditioned medium (HCM) promotes neurite outgrowth (e.g., Helfand *et al.*, 1976; Ebendal, 1981). Experiments utilizing ganglion cells in cultures conditioned with heart explants have shown that axon development can be promoted by soluble factors released into the culture medium (Ebendal, 1981) and by materials which become attached to the culture substratum (Collins, 1978; Collins and Garrett, 1980; Adler and Varon, 1981). Embryonic chick heart explants have been observed to elicit neurite outgrowth from sympathetic spinal, ciliary, and Remak's ganglia placed in collagen gel cocultures (Ebendal, 1981). When heart explants were removed after only initial stimulation, less dense neurite outgrowth was observed. Furthermore, washing the cultures caused a cessation of extension, retraction, and even degeneration of the neurites. Similarly, Helfand and co-workers (1976), using ciliary ganglion plated on polyornithine-coated dishes, found that the effect of HCM on neurite growth was not due to the deposition of material firmly attached to the surface of the culture dish.

However, Collins (1978) has reported that ciliary or sympathetic ganglion neurons in fresh culture medium, when plated onto polyornithine-coated dishes pretreated with HCM and then thoroughly washed, exhibited rapid neurite outgrowth. These results suggest that a neurite-promoting effect of the HCM is mediated through factors bound to the substratum. Evidence suggesting that the surface-bound factors may promote neurite outgrowth by changing the relative adhesion between the cell and the substratum is derived from observations on HCM-induced changes at the cell surface. Before the addition of HCM, ciliary neurons generate membrane ruffles and extend filopodia around the entire periphery of the cell body. Yet following the addition of HCM, these seemingly random movements become localized, the cells forming highly active growth cones which subsequently form axons (Collins, 1978). Further support for the adhesion-mediating effect of this substratum-bound factor arises from experiments in which only narrow strips of the polyornithine-coated dishes were pretreated with HCM. Parasympathetic neurons, when placed on this specially prepared substratum, grew only on the areas conditioned by the embryonic heart cells, changing their direction of growth so as not to cross into the uncoated regions (Collins and Garrett, 1980). Attempts to characterize the HCM factor have shown that the neurite-inducing activity is trypsin sensitive but is not inactivated by the membrane-solubilizing detergent Triton X-100 or by the enzymes collagenase, RNase, or DNase (Collins, 1980). The factor is also not inactivated by antibodies directed against fibronectin. It has been suggested, however, that the trypsin sensitivity may be due not to degradation of the factor but to its release from the culture substratum, since almost all of the visible cell debris is also removed by the trypsin treatment (Collins, 1980).

A similar neurite-promoting polyornithine-binding factor (PNPF-1) released by cultured rat Schwannoma cells has been found to be a very large, acidic macromolecule that yields two major bands at 200 and 190 kDa after SDS–PAGE under reducing conditions. Antibodies to rat laminin bind strongly to the 200-kDa band but do not block the neurite-promoting activity of the parent material, although they do block the neurite-promoting activity of laminin itself (Manthorpe *et al.*, 1983; Varon *et al.*, 1984; Davis *et al.*, 1984). Manthorpe *et al.* (1983) conclude that laminin and PNPF-1 are different molecules. Lander *et al.* (1984) report that a similar factor from bovine corneal epithelium is a complex containing laminin in association with a heparan sulfate proteoglycan and a sulfated protein tentatively identified as entactin (Carlin *et al.*, 1981), a laminin-binding protein.

In addition to heart and Schwannoma cells, explanted ciliary ganglion has also been found to produce a neurite outgrowth-promoting factor (Adler and Varon, 1981). Autoradiographic techniques have been used to show that the explanted ciliary ganglia deposit a material on the polyornithine substratum around the explant prior to neurite outgrowth (Adler and Varon, 1981). Axons emerging later have been observed to confine their growth to the "conditioned" area of the substratum, avoiding the uncoated polyornithine regions. It was discovered that placing two explants so that their conditioned regions overlap can cause axons to pass from one explant to the other. This observation has led to the proposal that "avenues of preferred terrain" can develop between neighboring ganglia and that this type of mechanism might be used to guide axons *in vivo* (Adler and Varon, 1981).

B. *Cell–Cell Adhesion*

Studies of chick embryonic neural tissue by a number of groups have led to the implication of several different entities in neural cell adhesion: "ligatin" (Jakoi and Marchase, 1979), "cognin" (Hausman *et al.*, 1976), "ligand–agglutinin" (Rutz and Lilien, 1979), "adherons" (Schubert *et al.*, 1983), enzyme–substrate complexes (Balsamo and Lilien, 1975), N-CAM (Thiery *et al.*, 1977), L1 antigen (Rathjen and Schachner, 1984), Ng-CAM (Grumet *et al.*, 1984), and both calcium-dependent and calcium-independent adhesion systems (Takeichi *et al.*, 1979; Grunwald *et al.*, 1980; Magnani *et al.*, 1981; Thomas and Steinberg, 1981; Thomas *et al.*, 1981a,b; Brackenbury *et al.*, 1981). Further investigations of some of these factors have led to proposals for several different mechanisms of cell–cell adhesion. Whether other of these materials are in fact directly or indirectly involved in adhesion has yet to be firmly established.

1. Ligand and Agglutinin

"Ligand" and "agglutinin," stated by Rutz and Lilien (1979) to correspond, respectively, with "retinal aggregation-promoting material" and "ligator" of Balsamo and Lilien (1974a,b), were shown by Rutz and Lilien (1979) to cause agglutination of fixed retinal cells. "Ligand" was derived from tissue culture medium of intact retina and "agglutinin" was obtained from monolayers of retinal cells (Rutz and Lilien, 1979). The agglutination process mediated by these two factors was found to exhibit pH, temperature, and divalent cation (either Mg^{2+} or Ca^{2+}) dependence (Rutz and Lilien, 1979). These factors were also reported to possess tissue specificity. The order in which cells are exposed to these two factors was discovered to be important. Rutz and Lilien (1979) reported that "ligand" must be bound to the cell before "agglutinin" will act. The results of these studies led to the proposal that at least three functionally distinct activities interact to mediate retinal cell adhesion: a cell surface receptor, a cell surface-associated component ("ligand"), and "agglutinin" (Rutz and Lilien, 1979). Based on these requirements, these authors have suggested four possible adhesion mechanisms. Three of these mechanisms require that "ligand" be bound to the cell surface through specific receptors. "Agglutinin" would then mediate "ligand"–"ligand" interactions through either a catalytic or structural role. The fourth mechanism requires that "ligand" and "agglutinin" have their own cell surface receptors and interact directly to cause adhesion.

2. Cognin

Extraction of plasma membranes of 10-day-old chick neural retina with butanol yielded a 50- to 60-kDa glycoprotein, "cognin" (Hausman and Moscona, 1976), which markedly enhanced the aggregation of retina cells. McClay and Moscona (1974) also isolated a 50-kDa glycoprotein from culture medium after tryptic dissociation of neural retinas. The latter glycoprotein has been suggested to be a derivative of cognin, released from the cell surface as a result of trypsin treatment (Hausman and Moscona, 1976). The results of studies in which optic tectum, cerebrum, and liver were treated with retinal cognin suggest that this factor is tissue specific, since it did not enhance reaggregation of cells from these other tissues (Hausman and Moscona, 1976). Factors isolated from cerebrum and spinal cord by butanol extraction of plasma membranes display a similar tissue specificity, preferentially enhancing the reaggregation of that tissue from which they were derived (Hausman *et al.*, 1976). Further analysis of the cerebrum-aggregating factor revealed that it is, like cognin, a 50 to 60 kDa protein. The cell-aggregating activity of cognin has also been found to be age dependent. Aggregating activity was obtained only from retinas younger than the thirteenth day of development and only pre-day 13 cells responded to active preparations (Hausman and Moscona, 1976). These authors point out that this decrease in activity correlates with the completion of morphogenetic cell organization in the retina and believe that it is important for early retinal development. They suggest that, after day 13, it may be modified or replaced by another cell adhesion mechanism that stabilizes tissue architecture.

3. Ca^{2+}-Dependent and Ca^{2+}-Independent Adhesion Systems

Two different modes of adhesion, one Ca^{2+}-dependent (CD) and one Ca^{2+}-independent (CI), have been demonstrated in a variety of cell types (Takeichi *et al.*, 1979; Grunwald *et al.*, 1980; Magnani *et al.*, 1981; Thomas and Steinberg, 1981; Thomas *et al.*, 1981a,b; Brackenbury *et al.*, 1981). Analysis of these two categories of adhesion mechanisms has shown that the CI adhesion systems studied were active in the absence of proteolytic pretreatment, resistant to low levels of trypsin, and relatively temperature independent, while the CD adhesion could not be detected before proteolytic treatment, was expressed after treatment of cells with trypsin in the presence of Ca^{2+}, and was temperature dependent (Magnani *et al.*, 1981). Since these two modes of cell adhesion differ in their sensitivity to various concentrations of Ca^{2+} and trypsin, cells could be prepared which display one or the other, both, or neither of these categories of adhesion system (Magnani *et al.*, 1981). Cells prepared with low levels of trypsin in the absence of Ca^{2+} were found to express CI adhesion, while cells prepared with a high concentration of trypsin in the presence of Ca^{2+} displayed CD adhesion (Magnani *et al.*, 1981).

In experiments designed to test for cross-recognition between these two kinds of adhesion system, it was found that retina cells which were prepared to have one or the other adhesion system would only adhere to other retinal cells displaying the same system (Thomas *et al.*, 1981b). Moreover, this was found to be true for cross-adhesion between cells from different tissues. For example, cells from the optic tectum and heart expressing only CD adhesion were found to cross-adhere extensively with retinal cells also displaying only CD adhesion but not to retinal cells displaying only CI adhesion. Similar results were obtained for the CI adhesion systems of the same kinds of cells (Thomas *et al.*, 1981a). Anti-retinal Fabs which inhibit CD or CI aggregation of retinal cells were found also to inhibit the corresponding kinds of aggregation of optic tectum cells but not of heart cells (Thomas *et al.*, 1981a). These results suggest that there exist "families" of related but nonidentical CD and CI adhesion systems, present on the surfaces of cells of different kinds. The similarities are expressed in the ability of adhesion systems of different cell types to interact, while the differences are made apparent by the ability of antibodies to discriminate between them.

4. Ligatin

Ligatin, a filamentous 10-kDa cell surface protein which has been purified from embryonic chick neural retina (Jakoi and Marchase, 1979), has been found to inhibit the reaggregation of dissociated retina cells (Marchase *et al.*, 1981). Merrell and co-workers (1975) have also isolated a factor by treatment of retinal cell membranes with acetone, followed by extraction with lithium diiodosalicylate. This factor was reported to inhibit retina cell adhesion and has characteristics similar to ligatin. Both are extractable from retina plasma membranes, pass through filters with a nominal cutoff of 10 kDa, but are retained by filters with a nominal cutoff of 2 kDa, and are trypsin sensitive.

Insight into the manner by which ligatin acts to inhibit adhesion has come from experiments in which monolayers of 10-day-old chick neural retina were preincubated with ligatin. When labeled cells were added to these monolayer cultures, fewer cells were observed to be bound. Furthermore, upon adding ligatin-containing medium from the pretreated monolayers to the fresh monolayers, it was found that the medium was no longer capable of inhibiting adhesion, suggesting that the ligatin-mediated inhibitory effect was removed by binding to the retina cell surfaces of the first monolayers (Marchase *et al.*, 1981).

Studies on the rat ileum by Jakoi *et al.* (1976) have revealed that ligatin serves as a baseplate for the attachment of β-*N*-acetylglucosaminidase to the external surfaces of the plasma membranes of lumenal epithelial cells. A similar role as a baseplate has been proposed for retinal cells (Jakoi and Marchase, 1979; Marchase *et al.*, 1981); however, the associated proteins have not been identified. If ligatin does serve as a baseplate, then the observed inhibitory effect could be due to competition between ligatin and the associated proteins for binding sites (Jakoi and Marchase, 1979; Marchase *et al.*, 1981). Ligatin prepared from suckling rat ileum or mouse peritoneal macrophages was ineffective in inhibiting adhesion of embryonic chick retina cells, implying that it binds in a tissue- or species-specific manner (Marchase *et al.*, 1981).

5. Enzyme–Substrate Complexes

A potential mechanism based on the specific interactions between enzymes (glycosyltransferases) and their substrates on opposing cell surfaces has been proposed by Roseman (1970). Interaction between cells employing such a lock-and-key mechanism could provide very specific recognition and adhesion between cells. Cell–cell adhesion using this mechanism could be regulated by controlling the glycosyltransferase–substrate reaction which, when allowed to run to completion, would result in the detachment of the interacting cells (Roth *et al.*, 1971).

Balsamo and Lilien (1975) have isolated from a supernatant prepared from neural retina cells a factor that binds specifically to cells of the same type and promotes their aggregation. Analysis by paper chromatography reveals that enzymatic dissection causes the release of *N*-acetylgalactosamine residues from this retinal factor. Moreover, addition of *N*-acetylgalactosamine inhibits its binding to retinal cells (Balsamo and Lilien, 1975). These results suggest that carbohydrate residues present on the cell surface could be important for mediating cell–cell adhesion. It has also been suggested that -*N*-acetylgalactosamine residues are important in the formation of an adhesive bond between neural retina cells and cells of the optic tectum. Biochemical investigations into the possible mechanism of binding between these two cell types had previously led to the proposal that retinal and tectal cells adhere using two complementary molecules: one concentrated dorsally in both tissues and possessing a terminal -*N*-acetylgalactosamine residue, and the other, a protein concentrated ventrally, capable of binding this terminal residue (Marchase, 1977). One enzyme, UDPgalactose:GM_2galactosyltransferase, was found to be more concentrated ventrally than dorsally but only by 30% (Marchase, 1977).

Balsamo and Lilien (1982) have detected the presence of a different enzyme, -*N*-acetylgalactosaminyltransferase, on the surface of neural retina cells and have obtained results which suggest that ligatin may also be a part of an enzyme–substrate type of adhesion system. The transferase–acceptor complex was removed from the cell surface by trypsinization in the absence of Ca^{2+}. Subsequently this activity reappeared at the cell surface, where it later was converted to a denser form as measured by sedimentation on a sucrose gradient. Treatment of isolated membranes bearing this denser form of the complex with 40 m*M* Ca^{2+} or 0.5 m*M* glucose-1-phosphate results in a decrease in density which, it is suggested, is likely to be due to the extraction of ligatin and/or other components from the complex. As was mentioned earlier, ligatin was found to serve as a baseplate for the attachment of α-*N*-acetylglucosaminidase to the external surface of the rat ileal epithelial cell plasma membrane.

6. ADHERONS

Spherical particles 15 nm in diameter, designated "adherons," are released into culture medium by embryonic chick neural retina cells (Schubert *et al.*, 1983). Adherons were found to increase the rate of cell–cell aggregation, and when adsorbed onto petri dishes, were found to increase the initial rate of cell–substratum adhesion. Analysis of adherons revealed that they are composed of glycosaminoglycans, an unknown polysaccharide, and a large number of proteins. In order to further characterize these particles, antiserum was prepared in rabbits against purified neural retina adherons. Monovalent Fab′ fragments inhibited the spontaneous aggregation of neural retina cells but not the aggregation of chick skeletal muscle cells (Schubert *et al.*, 1983). Neural retina cells were found to adhere to petri dishes coated with particulate material from homologous tissues but not from pigmented epithelium or skeletal muscle myoblasts (Schubert *et al.*, 1983). Fab′ fragments directed against neural retinal adherons also inhibited cellular adhesion of neural retina cells to the petri dishes coated with neural retina adherons. Adherons could represent yet another factor utilized in an adhesion mechanism; however, there is evidence to suggest that adherons might be composed, at least in part, of adhesion factors already implicated from other studies. Schubert and co-workers (1983) have found that retinal adherons were able to block the activity of Fab′ fragments of antibodies prepared against the cell surface. This suggests that the neural retina particles share antigenic determinants with molecules on the cell surface. SDS–PAGE has revealed that the particles contain proteins with molecular weights of 20,000, 43,000, 50,000, and 140,000.

Both cognin and N-CAM, discussed elsewhere in this section, have been isolated from cells (Hausman and Moscona, 1976; Rutishauser *et al.*, 1982) and have molecular weights of 50,000 and 140,000, respectively. The activity of retinal adherons has been found to be age dependent, increasing from day 7 to day 12 and subsequently declining. A similar age dependence has been noted for cognin (Hausman and Moscona, 1976). Because adherons have also been isolated from skeletal muscle myoblasts (Schubert and LaCorbiere, 1980), the study of that system may provide insight into the mechanism of adheron-mediated adhesion in the neural retina. A skeletal muscle myoblast cell line, L6, and an adhesion-deficient variant, M3A, have been found which release adherons into the culture medium. It was reported by Schubert and LaCorbiere (1982) that adherons from both cell lines bound equally well to plastic surfaces and to cells in the absence of Ca^{2+}. However, in the presence of Ca^{2+}, L6 adherons induced cell–cell aggregation while the M3A adherons did not. Thus, in this system adherons appear to bind to the surface of cells in a Ca^{2+}-independent manner, but the surface particles promote intercellular adhesion in a Ca^{2+}-dependent manner (Schubert and LaCorbiere, 1982). Adherons of the neural retina do not require Ca^{2+} for binding and it has been proposed that adherons in this system might cause cell aggregation either by forming a direct bridge between cells or by binding to the cell surface, then becoming linked by an additional factor (Schubert *et al.*, 1983). It seems evident from these studies that further investigation into the protein components of adherons is needed in order to establish their precise roles in cell adhesion and their possible correspondence with other adhesion-promoting factors.

Although evidence summarized above indicates that each of the molecular systems described can influence or simulate retinal cell adhesion, the evidence falls short of demonstrating that each of these systems is a normal participant in cellular adhesive interactions. In the case of adherons, it is possible that only part of the particle may be a component of the binding system. Interpretation of the significance of factors which promote cell–cell aggregation or agglutination is particularly problematical. The fact that a cell membrane protein or macromolecular complex will agglutinate cells does not constitute evidence that it is an adhesion protein. Nor does the tissue-specificity of action of certain of these aggregation-enhancing "factors" strengthen the argument that they are adhesion proteins, since *the cells from which they are derived do not show a corresponding tissue specificity of adhesion.* Their adhesion systems do reveal a strong molecular specificity, but closely similar if not identical adhesion systems are present on cells from many tissues and have been shown to mediate cross-adhesions between their component cells (Thomas *et al.*, 1981a; Brackenbury *et al.*, 1981; Rutishauser *et al.*, 1983).

7. N-CAM

Considerable information has been developed concerning one neuronal cell adhesion molecule, N-CAM, and its developmental significance (Edelman, 1983; Rutishauser, 1984). Evidence to support the hypothesis that N-CAM is involved in the formation of cell–cell bonds (Rutishauser *et al.*, 1976) was initially provided by experiments utilizing rabbit antibodies prepared against whole retina cells and against polypeptides released from retinal tissue into the culture medium (Thiery *et al.*, 1977). Antibodies prepared against whole retina cells were found to inhibit the aggregation of cells from dissociated chick neural retinas. These same antibodies were also found to precipitate three active polypeptides from the culture supernatant. Antibodies prepared specifically against these polypeptides were found to inhibit cell adhesion and to precipitate a 140-kDa protein from a detergent extract of embryonic retina cell membranes.

Further chemical characterization of embryonic N-CAM has revealed that it is a large glycoprotein with an unusually high sialic acid content (Edelman and

Chung, 1982; Hoffman *et al.*, 1982). It can be extracted with detergent, indicating that it is an integral membrane protein (Hoffman *et al.*, 1982). Neuraminidase treatment of N-CAM isolated from the embryonic chick brain cell surface yielded two components with molecular weights of 140,000 and 170,000, which had the same amino-terminal sequence as N-CAM and gave nearly identical profiles on peptide maps (Cunningham *et al.*, 1983). Insight into the structure of N-CAM has been derived from the analysis of fragments of the molecule. Incubation of N-CAM at 37°C generates a fragment (F_1) which has a molecular weight of 65,000, lacks most of the sialic acid, and contains the amino-terminal of the polypeptide chain. Treatment of membranes with a protease releases a fragment (F_2) which has a molecular weight of 108,000 after neuraminidase treatment, has most of the sialic acid, and also contains the amino-terminal of the polypeptide chain (Cunningham *et al.*, 1983). These results suggest that the sialic acid region of the molecule is located in the middle of the polypeptide chain between the amino-terminal portion and the polypeptide cell attachment site. N-CAM isolated from embryonic plasma membranes and subjected to SDS–gel electrophoresis migrated as a broad, continuously stained band from molecular weight 200,000 to 250,000 (Hoffman *et al.*, 1982; Grumet *et al.*, 1984). This variability in the molecule has been attributed to its sialic acid portion. The sialic acid content of N-CAM has been found to decrease from 28 to 35% in the embryonic form to 10% in the adult form (Edelman *et al.*, 1982). The significance of this variability will be discussed later, when the binding properties of the molecule will be presented. In addition, amino acid sequence analysis and comparisons of proteolytic fragments of N-CAM suggest that all forms of the glycoprotein are derived from similar peptide chains (Hoffman *et al.*, 1982).

Rutishauser and co-workers (1982) have provided evidence which suggests that N-CAM mediates cell–cell adhesion through homophilic binding. When cells were preincubated with Fab′ fragments of antibodies against purified N-CAM, their ability to bind uncoated retina cells or soluble N-CAM was significantly reduced. More direct evidence for N-CAM–N-CAM binding was provided by the observations that ^{125}I-labeled N-CAM bound to N-CAM immobilized on Sepharose, that lipid vesicles containing N-CAM spontaneously formed large aggregates and that N-CAM appears to exist in solution as a heterodisperse oligomer (Rutishauser *et al.*, 1982). Furthermore, N-CAM was found to bind only to lipid vesicles and neurons which have N-CAM on their surfaces. Studies using neuraminidase treatment to remove 99% of the sialic acid from N-CAM have shown that the amino-terminal portion of the molecule contains at least part of the binding site, since desialylated N-CAM, when attached covalently to beads, bound specifically to cells (Cunningham *et al.*, 1983). It has also been reported that the 65-kDa polypeptide fragment appears to contain all the antigenic determinants of intact N-CAM that neutralize the adhesion-blocking ability of anti-retinal cell Fab′ fragments (Hoffman *et al.*, 1982).

Although the sialic acid region of N-CAM is apparently not directly important in binding, it was found that desialylated N-CAM bound to cells to a greater extent than did untreated N-CAM. A possible explanation for this observation is that removal of all or part of the negatively charged sialic-rich region of the N-CAM molecule would decrease the electrostatic repulsive forces between these regions of the molecule and would thus lead to an increase in the energy of adhesion (Edelman, 1983). The sialic acid content of N-CAM has been found to affect the rates of aggregation of reconstituted N-CAM vesicles and native brain vesicles. Vesicles containing N-CAM molecules which had been fully desialylated exhibited a fourfold increase in aggregation rate over vesicles containing the embryonic form of the molecule. Intermediate rates of aggregation were observed both with vesicles containing the adult form of N-CAM and with vesicles containing a partially desialylated embryonic form (Hoffman and Edelman, 1983).

Through the use of labeled anti-N-CAM antibodies, N-CAM was discovered to be uniformly distributed on both the axons and the cell bodies of neurons (Rutishauser *et al.*, 1978a). N-CAM has also been localized on a number of different embryonic tissues: retina, brain, optic nerve, spinal cord, and both sympathetic and dorsal root ganglia (Rutishauser *et al.*, 1978b). Furthermore, it has been reported (Thiery *et al.*, 1982) that N-CAM is found in the early embryo in regions which are involved in the induction of the primary axis, i.e., neural plate, neural tube, notochord, and somites, and in tissues in which later inductive events occur, i.e., neural crest cells, optic, otic and pharyngeal placodes, cardiac mesoderm, and mesonephric primordium. The distribution of N-CAM in this latter group of tissues has been found to be highly dynamic and transient (Thiery *et al.*, 1982). Neural crest cells, for example, are heavily stained with labeled anti-N-CAM antibodies before migrating from their dorsal position on the neural tube. During migration, however, the crest cells are devoid of N-CAM. At the site of arrest, when crest cells collected into ganglion rudiments, N-CAM was again detected in them (Thiery *et al.*, 1982).

Evidence to support a functional role for N-CAM in the development of the nervous system has been provided by studies both *in vivo* and *in vitro*. Rutishauser and co-workers (1978a) have shown *in vitro* that N-CAM is important in the formation of nerve bundles. Chick spinal root ganglia cultured with nerve growth factor were observed to form a halo of processes and fascicles. Addition of monovalent Fab′ fragments directed against N-CAM, however, caused a dramatic change in the morphology of the neurite halos, decreasing the thickness of the fascicles and increasing the number of single neurites. It was also reported that the anti-N-CAM antibody did not appear to affect growth cone function such as neurite elongation, mobility, and attachment to the substratum (Rutishauser *et al.*, 1978a). Two interesting results derived from these same studies were that when the substratum was made more adhesive by addition of lectin, neurite fascicles appeared even in the absence of anti-N-CAM and that thick fascicles not in contact with collagen or other cells were not affected by anti-N-CAM. Both of these observations suggest that nerve bundle morphology can be influenced by the relative adhesiveness of the axons to each other and to the substratum.

Anti-N-CAM antibodies have also been used to show that side-to-side interaction among neurites could influence the guidance of nerve bundles. In the absence of anti-N-CAM, focally applied nerve growth factor was observed to produce asymmetric outgrowth of neurites from the spinal ganglion explant (Rutishauser and Edelman, 1980). This asymmetry of growth, however, was found to be dramatically reduced by the addition of anti-N-CAM. The results of *in vitro* experiments on the histotypic development of retina cell aggregates have suggested that N-CAM is also important for the structural organization of tissues since treatment with anti-N-CAM was found to reduce the amount of sorting-out normally observed in the aggregate. Analysis of the aggregates by electron microscopy also revealed that both the cell bodies and neurites were less densely packed in the aggregate after treatment with the anti-N-CAM, suggesting that cell–cell, cell–neurite, and neurite–neurite interactions mediated by N-CAM are necessary for sorting-out of cell bodies and neurites into discrete regions (Rutishauser *et al.*, 1978b).

In vivo evidence to support a functional role for N-CAM has been derived from studies on the development of the cerebellum in normal and staggerer (*Sg*/*Sg*) mutant mice. In these mutants, Purkinje cells fail to form mature tertiary dendritic spines and thus do not synapse with the parallel fibers. The granule cells, which give rise to these fibers, subsequently die due to lack of functional connections. When the conversion of N-CAM from embryonic to adult form in mutant and wild-type mice was compared, it was found that this conversion was delayed in the mutant mouse. Embryonic-to-adult conversion in the mutant had not occurred by 21 days after birth, while in the normal mouse conversion was almost complete by this time (Edelman and Chuong, 1982). This result has led to the suggestion that one of the main roles of N-CAM conversion (i.e., loss of sialic acid) may be to terminate growth cone migration and to stabilize synapse formation at the proper time during development (Edelman, 1983).

8. L1 Antigen

L1 antigen is a glycoprotein which has been isolated from neural tissues in mice. Anti-L1 Fabs were found to inhibit the Ca^{2+}-independent aggregation of mouse cerebellar and neuroblastoma N2A cells (Schachner *et al.*, 1983; Rathjen and Schachner, 1984). Maximal inhibition of cerebellar cell aggregation was obtained with anti-L1 Fab and an Fab directed against an N-CAM-like antigen acting together, supporting the conclusion that their antigens are independent molecules (Schachner *et al.*, 1983; Faissner *et al.*, 1984). Unlike N-CAM, the

L1 antigen was found to yield the same banding pattern (a pair of prominent bands of 140 and 200 kDa) using SDS–PAGE in adult and embryonic stages of development. The two L1 glycoprotein bands obtained from N2A cells are of slightly higher molecular weight than the corresponding bands obtained from cerebellum (Schachner *et al.*, 1983). Digestion of N-CAM and L1 antigen with *Staphylococcus aureus* V8 protease generated two different peptide maps. Additional evidence showing that L1 antigen is different from N-CAM and thus seemingly represents another Ca^{2+}-independent adhesion molecule has been derived from immunohistochemical analysis of the distribution of L1 antigens and N-CAM in the mouse cerebellum. N-CAM was found to be uniformly distributed in all regions of the cerebellum, while L1 antigen was found to be restricted to neurons, primarily in the postmitotic inner granule layer, the developing molecular layer, and the inner part of the external granule layer (Rathjen and Rutishauser, 1984). Anti-L1 Fab has been found to inhibit the migration of mouse cerebellar corticogranular cells (Lindner *et al.*, 1983) but was without effect on a variety of other neuronal cell interactions (Schachner *et al.*, 1983).

9. Ng-CAM

Because antibodies to neuronal membranes inhibited neuron-glial cell adhesions but anti-N-CAM did not, Grumet *et al.* (1983) concluded that a specific neuron–glia adhesion mechanism must exist. Ng-CAM, a neuronal cell surface glycoprotein (MW 135,000), is thought to be the neuronal component of that adhesion system. Ng-CAM was found to neutralize the activity of Fab′ fragments that inhibited the binding of neuronal membrane vesicles to glial cells; in the same experiment, N-CAM was found to be ineffective at neutralizing this activity (Grumet *et al.*, 1984). A comparison between the peptide fragments of N-CAM and Ng-CAM showed that these molecules are structurally different. Simultaneous staining for Ng-CAM and N-CAM demonstrated that these molecules can coexist on neuronal cell surfaces. Although both Ng-CAM and N-CAM are Ca^{2+} independent, they mediate cell adhesion independently. In contrast to N-CAM, which links cells homophilically, Ng-CAM appears to link cells heterophilically since Ng-CAM molecules could not be detected on the surface of glial cells (Grumet *et al.*, 1984).

The appearance and distribution of Ng-CAM in the chick embryo have been studied by Thiery *et al.* (1985). The antigen is first seen in the ventral neural tube of the 3-day embryo at the surfaces of postmitotic motoneurons which send Ng-CAM-bearing fibers to the periphery. In the peripheral nervous system, the first neurons to show Ng-CAM were seen at 4 days of development in the cranial ganglia. Ultimately all peripheral neurons display Ng-CAM, as do the neurites of many CNS neurons.

IV. Neuronal Recognition and Differential Adhesion

"Cell recognition" is riveting as a pursuit but fugacious as a concept. One perhaps envisions a panorama of cells and tissues, each displaying its unique, proprietary banner: a molecular coat of arms conveying instant identification. The reality is far more intricate than this cartoon. One need only attempt an operational definition of "cell recognition" to sense the difficulty. This expression has been applied to many phenomena which have in common cells choosing. Recognition implies selectivity, which in turn implies differences that provide the basis for selection. Thus cell recognition is an amalgam of two components: there must be certain traits that serve as identifiers and these traits must be distributed differentially on the cellular or other surfaces to be discriminated. This does not mean, however, that each kind of cell or neuron must wave a unique banner. The grasshopper neurons recognized by monoclonal antibody mes-2 (Kotryla and Goodman, 1983) may come close to being such a case. Ng-CAM, also, may mediate neurite–glial associations quite specifically. More commonly, however, adhesion-mediating molecules are found in association with cells or tissues of many kinds. Fibronectin provides one clear example. Laminin is probably recognized by epithelial cells of all kinds in contact with a basal lamina. L-CAM (Gallin *et al.*, 1983) has been detected in all endodermal structures, in epidermal placodes, in extraembryonic ectoderm, and in some mesodermal structures such as Wolffian duct, the oviduct, and the kidney epithelium during organogenesis (Edelman *et al.*, 1983; Thiery *et al.*, 1984). N-CAM has been identified in the neural plate and tube, in early placodes, notochord, and somites, in kidney primordia, in limb buds, in cardiac mesoderm, and in striated muscle (Thiery *et al.*, 1982; Edelman *et al.*, 1983; Rutishauser *et al.*, 1983), where it has been implicated as functioning in initial neuromuscular associations (Rutishauser *et al.*, 1983).

All that is required of an adhesive recognition system is that it permit a desired choice to be made. If the choice is between only two alternatives, it is sufficient if one of the two possesses the relevant receptor or ligand *at the critical moment* and the other does not. In some cases it will even be sufficient if one of the two choices merely possesses more of the recognition marker than the other. There is no reason why a given marker may not be used in many places and at many times to direct choices in the same embryo, so long as adequate spatial or temporal resolution is maintained. In fact, this is a genetically economical strategy for providing many choices with a few well-programmed signals (Steinberg, 1978a,b). Temporal regulation has been documented for N-CAM, L-CAM, and Ng-CAM. In the case of N-CAM, it even appears that different degrees of sialylation of one and the same protein are used to modulate the intensity of neuronal interactions at different times during development (Rothbard *et al.*, 1982; Hoffman and Edelman, 1983).

Cells usually display multiple recognition banners simultaneously. Why? Consider the epithelial cell of the small intestine. Its luminal service is nonadhesive, a necessity to avoid obliterating the lumen. Subapically, a belt of tight junctions seals off the intercellular space from the gut lumen. Below that, the zonula adherens stabilizes the tight junction against mechanical disruption. Still farther below, spot desmosomes produce strong intercellular adhesions and gap junctions provide for passage of ions and small molecules from cell to cell (and incidentally must contribute to intercellular adhesion as well). On the basal surface, hemidesmosomes and receptors for fibronectin and laminin anchor the cell to the basal lamina. Each of these junctional structures serves a special purpose and each is reformed cooperatively and specifically after disruption, through a local recognition event either between two epithelial cells or between one such cell and the basal lamina. We have listed not fewer than six different kinds of recognition domains for these cells, and there may be more.

Epithelial cells are not neurons, to be sure, but neurons are cells and they also possess multiple kinds of recognition molecules to serve their own necessary functions. Retinal and other neurons simultaneously possess both CD and CI adhesion systems (see Section II,B,3). N-CAM and L-1 antigen, both calcium independent, coexist on postmitotic central neurons (Rathjen and Rutishauser, 1984). N-CAM and Ng-CAM, also calcium independent, also coexist on individual neurons (Grumet *et al.*, 1984). How many of these adhesion systems may coexist on an individual neuron has not yet been determined, but N-CAM and Ng-CAM clearly mediate different neuronal associations (interneuronal vs neuronal–glial) which occur simultaneously. How many different interactions a neuron may be involved in at one time is hinted at by the conclusion of Fraser (1985) that no fewer than five different sets of determinants are required to account for the documented behavior of retinal neurites in the process of organizing a retinotopic map on the optic tectum. In a similar context, it seems significant to us that the neurites of displaced motoneurons, which will alter their direction of locomotion in the vicinity of normal choice points in order to enter their normal pathways, can nevertheless follow inappropriate pathways to an incorrect destination when displaced from their normal points of entry into the limb (see Section II,F). This suggests to us that such neurites may possess major and minor recognition systems, the latter of which can act as effective guidance determinants only when the former are unable to act, as suggested by Fraser for the retinotectal system.

It has often been assumed that cell adhesion would boil down to recognition between complementary molecular locks and keys (see Sections II,B,1, 2, 5, and 6). This is evidently true for Ng-CAM, but N-CAM on one cell appears to interact with N-CAM on its neighbor (Rutishauser *et al.*, 1982). Cell recognition systems must have evolved, and it is particularly easy to see how a homophilic adhesion system might appear. Imagine that a multimer-forming protein with some other earlier function inside or outside the cell acquires, through genetic rearrangement, a signal sequence that causes it to be incorporated into the cell surface. Such molecules on the surfaces of two apposed cells could entwine in their accustomed way and constitute an instant self-recognition system. Specificity in this mechanism could arise not only from the specific molecular association between such fiber-forming polypeptides but also from their temporally coordinated expression on opposing cell surfaces. Creation of a new lock-and-key mechanism would require the coevolution of two complementary components, since a lock is of no use without its key. In a homophilic adhesion system, the lock *is* the key.

Finally, we anticipate that homologies among vertebrate adhesion systems will be found that will allow them to be classified into a not-very-large number of "families." This expectation is based upon our observations that retinal cells could adhere to heart cells specifically utilizing either the CD or the CI systems of each; that Fabs directed against each category of retinal adhesion system inhibited the aggregation of retinal cells via that adhesion system but not of heart cells utilizing the corresponding system; but that the same Fabs did inhibit the cross-adhesion between heart and retina cells utilizing either category of adhesion system (Thomas *et al.*, 1981a). Entirely similar results have been reported by Brackenbury *et al.* (1981) for the combination of retinal and liver cells: anti-L-CAM inhibits CI liver–liver and liver–retina adhesion but is without effect upon retina–retina adhesion. These observations call to mind the fact that embryonic cells of the most diverse kinds, most of which never encounter one another in the normal course of their existence, readily adhere to one another when placed together (e.g., Steinberg, 1970). In some cases, this may be due to their possession of identical adhesion systems. In the three cases cited above, however, the two types of cells cross-adhere via adhesion systems that are demonstrably nonidentical but nevertheless able to interact with each other. If these observations indeed presage the finding of *families* of adhesion systems possessing related cell recognition sites but differing in other molecular domains, cell (and neuronal) recognition will have banner days ahead.

Acknowledgment

Work from the authors' laboratory on this subject has been supported by U.S. PHS research Grant No. RO1 CA13605 from the National Cancer Institute to M.S.S.

References

Adler, R., and Varon, S. (1981). Neuritic guidance by polyornithine-attached materials of ganglionic origin. *Dev. Biol.* **81,** 1–11.

Akers, R. M., Mosher, D. F., and Lilien, J. E. (1981). Promotion of retinal neurite outgrowth by substratum-bound fibronectin. *Dev. Biol.* **86,** 179–188.

Arora, H. L., and Sperry, R. W. (1962). Optic nerve regeneration after surgical cross-union of medial and lateral optic tracts. *Am. Zool.* **2,** 389.

Attardi, D. G., and Sperry, R. W. (1963). Preferential selection of central pathways by regenerating optic fibers. *Exp. Neurol.* **7,** 46–64.

Balsamo, J., and Lilien, J. (1974a). Embryonic cell aggregation: Kinetics and specificity of binding of enhancing factors. *Proc. Natl. Acad. Sci. U.S.A.* **71,** 727–731.

Balsamo, J., and Lilien, J. (1974b). Functional identification of three components which mediate tissue-type specific embryonic cell adhesion. *Nature (London)* **25,** 522–524.

Balsamo, J., and Lilien, J. (1975). The binding of tissue-specific adhesive molecules to the cell surface. A molecular basis for specificity. *Biochemistry* **14,** 167–171.

Balsamo, J., and Lilien, J. (1982). An *N*-acetylgalactosaminyltransferase and its acceptor in embryonic chick neural retina exist in interconvertible particular forms depending on their cellular location. *J. Biol. Chem.* **257,** 349–354.

Barbera, A. J. (1975). Adhesive recognition between developing retinal cells and the optic tecta of the chick embryo. *Dev. Biol.* **46,** 167–191.

Baron van Evercoorin, A., Kleinman, H. K., Ohno, S., Marangos, P., Schwartz, J. P., and Dubois-Dalcq, M. E. (1982). Nerve growth factor, laminin and fibronectin promote neurite growth in human fetal sensory ganglia cultures. *J. Neurosci. Res.* **8,** 179–194.

Bastiani, M. J., and Goodman, C. S. (1984). Neuronal growth cones: Specific interactions mediated by filopodial insertion and induction of coated vesicles. *Proc. Natl. Acad. Sci. U.S.A.* **81,** 1849–1853.

Bastiani, M. J., Raper, J. A., and Goodman, C. S. (1984). Pathfinding by neuronal growth cones in grasshopper embryos. III. Selective affinity of the G growth cones for the P cells within the A/P fascicle. *J. Neurosci.* **4,** 2311–2328.

Bate, C. M. (1976). Pioneer neurons in an insect embryo. *Nature (London)* **260,** 54–56.

Bate, C. M., and Grunwald, E. B. (1981). Embryogenesis of an insect nervous system. II. A second class of neuron precursor cells and the origin of the intersegmental connectives. *J. Embryol. Exp. Morphol.* **61,** 317–330.

Berlot, J., and Goodman, C. S. (1984). Guidance of peripheral pioneer neurons in the grasshopper: An adhesive hierarchy of epithelial and neuronal surfaces. *Science* **223,** 493–496.

Berlot, J., Bastiani, M. J., and Goodman, C. S. (1983). Neuronal growth cones in the antennae of the grasshopper embryo: Guidance by epithelial gradients and adhesive hierarchies. *Soc. Neurosci. Abstr.* **9,** 1044.

Bonhoeffer, F., and Huf, J. (1980). Recognition of cell types by axonal growth cones *in vitro*. *Nature (London)* **288,** 162–164.

Bonhoeffer, F., and Huf, J. (1982). *In vitro* experiments on axon guidance demonstrating an anterior–posterior gradient on the tectum. *EMBO J.* **1,** 427–431.

Brackenbury, R., Rutishauser, U., and Edelman, G. M. (1981). Distinct calcium-independent and calcium-dependent adhesion systems of chicken embryo cells. *Proc. Natl. Acad. Sci. U.S.A.* **78,** 387–391.

Bray, D., Wood, P., and Bunge, R. P. (1980). Selective fasciculation of nerve fibers in culture. *Exp. Cell Res.* **130,** 241–250.

Bunt, S. M. (1982). Retinotopic and temporal organization of the optic nerve and tracts in the adult goldfish. *J. Comp. Neurol.* **206,** 209–226.

Carlin, B., Jaffe, R., Bender, B., and Chung, A. E. (1981). Entactin, a novel basal lamina-associated sulfated glycoprotein. *J. Biol. Chem.* **256,** 5209–5214.

Carter, S. B. (1965). Principles of cell motility. The direction of cell movement and cancer invasion. *Nature (London)* **208,** 1183–1187.

Carter, S. B. (1967). Haptotaxis and the mechanism of cell motility. *Nature (London)* **213,** 256–260.

Caudy, M., and Bentley, D. (1983). Guidance features for navigation by peripheral pioneer neurons. *Soc. Neurosci. Abstr.* **9,** 1043.

Chung, S.-H, and Cooke, J. (1978). Observations on the formation of the brain and of nerve connections following embryonic manipulation of the amphibian neural tube. *Proc. R. Soc. London Ser. B* **201,** 335–373.

Collins, F. (1978). Axon initiation by ciliary neurons in culture. *Dev. Biol.* **65,** 50–57.

Collins, F. (1980). Neurite outgrowth induced by the substrate-associated material from nonneuronal cells. *Dev. Biol.* **79,** 247–252.

Collins, F., and Garrett, J. E. (1980). Elongating nerve fibers are guided by a pathway of material released from embryonic nonneuronal cells. *Proc. Natl. Acad. Sci. U.S.A.* **77,** 6226–6228.

Constantine-Paton, M. (1978). Central projections of anuran optic nerves penetrating hindbrain or spinal cord regions of the neural tube. *Brain Res.* **158,** 31–43.

Constantine-Paton, M. (1982). The retinotectal hookup: The process of neural mapping. *In* "Developmental Order: Its Origin and Regulation" (S. Subtelny and P. Green, eds.), 40th Symposium, Society for Developmental Biology, pp. 317–349. Liss, New York.

Constantine-Paton, M. (1983). Trajectories of axons in ectopic VIIIth nerves. *Dev. Biol.* **97,** 239–244.

Constantine-Paton, M., and Capranica, R. R. (1976). Axonal guidance of developing optic nerves in the frog. I. Anatomy of the projection from transplanted eye primordia. *J. Comp. Neurol.* **170,** 17–31.

Constantine-Paton, M., and Law, M. I. (1978). Eye-specific bands in the tecta of three-eyed frogs. *Science* **202,** 639–641.

Cooke, J., and Gaze, R. M. (1983). The positional coding system in the early eye rudiment of *Xenopus laevis,* and its modification after grafting operations. *J. Embryol. Exp. Morphol.* **77,** 53–71.

Culp, L. A., Ansbacher, R., and Domen, C. (1980). Adhesion sites of neural tumor cells: Biochemical composition. *Biochemistry* **19,** 5899–5907.

Cunningham, B. H., Hoffman, S., Rutishauser, U., Hemperly, J. J., and Edelman, G. M. (1983). Molecular topography of the neural cell adhesion molecule N-CAM: Surface orientation and location of sialic acid-rich and binding regions. *Proc. Natl. Acad. Sci. U.S.A.* **80,** 3116–3120.

Damsky, C. H., Knudsen, K. A., and Buck, C. A. (1984). Integral membrane glycoproteins in cell–cell and cell–substratum adhesion. *In* "The Biology of Glycoproteins" (R. J. Ivatt, ed.), pp. 1–64. Plenum, New York.

Davis, G. E., Manthorpe, M., and Varon, S. (1984). Purification of rat Schwannoma neurite promoting factor. *Soc. Neurosci. Abstr.* **10,** 38.

Easter, S. S., Rusoff, A. C., and Kish, P. E. (1981). The growth and organization of the optic nerve and tract in juvenile and adult goldfish. *J. Neurosci.* **8,** 793–811.

Ebendal, T. (1981). Control of neurite extension by embryonic heart transplants. *J. Embryol. Exp. Morphol.* **61,** 289–301.

Edelman, G. M. (1983). Cell adhesion molecules. *Science* **219,** 450–457.

Edelman, G. M., and Chuong, C.-M. (1982). Embryonic to adult conversion of neural adhesion molecules in normal and staggerer mice. *Proc. Natl. Acad. Sci. U.S.A.* **79,** 7036–7040.

Edelman, G. M., Gallin, W. J., Delouvee, A., Cunningham, B. A., and Thiery, J.-P. (1983). Early epochal maps of two different cell adhesion molecules. *Proc. Natl. Acad. Sci. U.S.A.* **80,** 4384–4388.

Edwards, J. S., Chen, S. W., and Berns, M. W. (1981). Cercal sensory development following laser microlesions of embryonic apical cells in *Acheta domesticus. J. Neurosci.* **2,** 250–258.

Erickson, C. A., and Turley, E. A. (1983). Substrata formed by combinations of extracellular matrix components alter neural crest motility *in vitro. J. Cell Sci.* **61,** 299–323.

Faissner, A., Kruse, J., Goridis, C., Bock, E., and Schachner, M. (1984). The neural cell adhesion molecule L1 is distinct from the N-CAM related group of surface antigens BSP-2 and D2. *EMBO J.* **3,** 733–737.

Fawcett, J. W., and Gaze, R. M. (1982). The retinotectal fibre pathways from normal and compound eyes in *Xenopus. J. Embryol. Exp. Morphol.* **72,** 19–37.

Fraser, S. E. (1980). A differential adhesion approach to the patterning of nerve connections. *Dev. Biol.* **79,** 453–464.

Fraser, S. E. (1985). Cell interactions involved in neuronal patterning: An experimental and theoretical approach. *In* "The Molecular Basis of Neural Development" (G. M. Edelman, W. E. Gall, and M. Cowan eds.), pp. 481–507. Wiley, New York.

Fraser, S. E., and Hunt, R. K. (1980). Retinotectal specificity: Models and experiments in search of a mapping function. *Annu. Rev. Neurosci.* **3,** 319–352.

Garber, B. B., and Moscona, A. A. (1972). Reconstruction of brain tissue from cell suspensions. II. Specific enhancement of aggregation of embryonic cerebral cells by supernatant from homologous cell cultures. *Dev. Biol.* **27,** 235–243.

Garrod, D. R. (1981). Adhesive interactions of cells in development: specificity, selectivity and cellular adhesive potential. *In* "Fortschritte der Zoologie," Bd. 26, pp. 183–195. Fischer, Stuttgart.

Garrod, D. R., and Nicol, A. (1981). Cell behavior and molecular mechanisms of cell–cell adhesion. *Biol. Rev.* **56,** 199–242.

Gaze, R. M., and Fawcett, J. W. (1983). Pathways of *Xenopus* optic fibers regenerating from normal and compound eyes under various conditions. *J. Embryol. Exp. Morphol.* **73,** 17–38.

Gaze, R. M., and Grant, P. (1978). The diencephalic course of regenerating retinotectal fibers in *Xenopus* tadpoles. *J. Embryol. Exp. Morphol.* **44,** 201–216.

Gaze, R. M., and Hope, R. A. (1983). The visuotectal projection following translocation of grafts within an optic tectum in the goldfish. *J. Physiol. (London)* **344,** 257–275.

Gaze, R. M., and Sharma, S. C. (1970). Axial differences in the reinnervation of goldfish optic tectum by regenerating optic nerve fibres. *Exp. Brain Res.* **10,** 171–181.

Goodman, C. S., and Bate, C. M. (1981). Neuronal development in the grasshopper. *Trends Neurosci.* **4,** 208–214.

Goodman, C. S., Raper, J. A., Ho, R. K., and Chang, S. (1982). Pathfinding by neuronal growth cones in grasshopper embryos. *In* "Developmental Order: Its Origin and Regulation" (S. Subtelny and P. Green, eds.), 40th Symposium, Society for Developmental Biology, pp. 275–316. Liss, New York.

Greenberg, J. H., Seppa, S., Seppa, H., and Hewitt, A. T. (1981). Role of collagen and fibronectin in neural crest cell adhesion and migration. *Dev. Biol.* **87,** 259–266.

Grumet, M., Rutishauser, U., and Edelman, G. M. (1983). Neuron–glia adhesion is inhibited by antibodies to neural determinants. *Science* **222,** 60–62.

Grumet, M., Hoffman, S., and Edelman, G. M. (1984). Two antigenically related neuronal cell adhesion molecules of different specificities mediate neuron–neuron and neuron–glia adhesion. *Proc. Natl. Acad. Sci. U.S.A.* **81,** 267–271.

Grunwald, G. B., Geller, R. L., and Lilien, J. (1980). Enzymatic dissection of embryonic cell adhesive mechanisms. *J. Cell Biol.* **85,** 766–776.

Hafter, W., Claviez, M., and Schwarz, U. (1981). Preferential adhesion of tectal membranes to anterior embryonic chick retina neurites. *Nature (London)* **292,** 67–70.

Hammarback, J. A., and Letourneau, P. C. (1984). Guidance of neurite outgrowth by pathways of substratum-adsorbed laminin. *J. Cell Biol.* **99,** 72a.

Harris, A. (1973). Behavior of cultured cells on substrata of variable adhesiveness. *Exp. Cell Res.* **77,** 285–297.

Hauschka, S. D., and Ose, M. (1979). *In vitro* requirements for neural retina ganglion cell axon formation. *In Vitro* **15,** 204–205 (Abstr.).

Hausman, R. E., and Moscona, A. A. (1976). Isolation of retina specific cell-aggregating factor from membranes of embryonic neural retina tissue. *Proc. Natl. Acad. Sci. U.S.A.* **73,** 3594–3598.

Hausman, R. E., Knapp, L. W., and Moscona, A. A. (1976). Preparation of tissue-specific cell-aggregating factors from embryonic neural tissues. *J. Exp. Zool.* **198,** 417–422.

Hoffman, S., and Edelman, G. M. (1983). Kinetics of homophilic binding by embryonic and adult forms of the neural cell adhesion molecule. *Proc. Natl. Acad. Sci. U.S.A.* **80,** 5762–5766.

Hoffman, S., Sorkin, B. C., White, P. C., Brackenbury, R., Mailhammer, R., Rutishauser, V., Cunningham, B. A., and Edelman, G. M. (1982). Chemical characterization of a neural cell adhesion molecule purified from embryonic brain membranes. *J. Biol. Chem.* **257,** 7720–7729.

Holtfreter, J. (1939). Gewebeaffinitat, ein Mittel der embryonal Forbildung. *Arch. Exp. Zellforsch. Gewebezucht* **23,** 169–209.

Hynes, R. D., and Yamada, K. M. (1982). Fibronectins: Multifunctional modular glycoproteins. *J. Cell Biol.* **95,** 369–377.

Jakoi, E. R., and Marchase, R. B. (1979). Ligatin from embryonic chick neural retina. *J. Cell Biol.* **80,** 642–650.

Jakoi, E. R., Zampighi, G., and Robertson, J. D. (1976). Regular structures in unit membranes. II. Morphological and biochemical characterization of two water-soluble membrane proteins isolated from the suckling rat ileum. *J. Cell Biol.* **70,** 97–111.

Keshishian, H., and Bentley, D. (1983). Embryogenesis of peripheral nerve pathways in grasshopper legs. I. The initial nerve pathway to the CNS. *Dev. Biol.* **96,** 89–102.

Kotrla, K. S., and Goodman, C. S. (1983). Transient expression of cell surface antigen on two neurons that share a common final pathway and target in the grasshopper embryo. *Soc. Neurosci.* **9,** 1045 (Abstr.).

Lance-Jones, C., and Landmesser, L. (1981a). Pathway selection by embryonic chick lumbosacral motoneurons during normal development. *Proc. R. Soc. London Ser. B* **260,** 1–18.

Lance-Jones, C., and Landmesser, L. (1981b). Pathway selection by embryonic chick motoneurons in an experimentally altered environment. *Proc. R. Soc. London Ser. B* **214,** 19–52.

Lander, A. D., Fujii, D. K., Gospodarowicz, D., and Reichardt, L. F. (1984). ''Neurite outgrowth-promoting factors'' in conditioned media are complexes containing laminin. *Soc. Neurosci. Abstr.* **10,** 40.

Landmesser, L. (1978). The distribution of motoneurones supplying chick hind limb muscles. *J. Physiol. (London)* **284,** 371–389.

Letourneau, P. C. (1975a). Possible roles for cell to substratum adhesion in neuronal morphogenesis. *Dev. Biol.* **44,** 77–91.

Letourneau, P. C. (1975b). Cell-to-substratum adhesion and guidance of axonal elongation. *Dev. Biol.* **44,** 92–101.

Levine, R., and Jacobson, M. (1975). Discontinuous mapping of retina onto tectum innervated by both eyes. *Brain Res.* **98,** 172–176.

Lilien, J. E. (1968). Specific enhancement of cell aggregation *in vitro*. *Dev. Biol.* **17,** 657–678.

Lilien, J. E., and Moscona, A. A. (1967). Cell aggregation: Its enhancement by a supernatant from cultures of homologous cells. *Science* **157,** 70–72.

Lindner, J., Rathjen, F. G., and Schachner, M. (1983). L1 mono- and polyclonal antibodies modify cell migration in early postnatal mouse cerebellum. *Nature (London)* **305,** 427–430.

McClay, D. R., and Moscona, A. A. (1974). Purification of the specific cell aggregating factor from embryonic neural retina cells. *Exp. Cell Res.* **74,** 438–442.

Magnani, J. L., Thomas, W. A., and Steinberg, M. S. (1981). Two distinct adhesion mechanisms in embryonic neural retina cells. 1. A kinetic analysis. *Dev. Biol.* **81,** 96–105.

Manthorpe, M., Engvall, E., Ruoslahti, E., Longo, F. M., Davis, G. E., and Varon, S. (1983). Laminin promotes neuritic regeneration from cultured peripheral and central neurons. *J. Cell Biol.* **97,** 1882–1890.

Marchase, R. B. (1977). Biochemical investigations of retinotectal adhesive specificity. *J. Cell Biol.* **75,** 237–257.

Marchase, R. B., Harges, P., and Jakoi, E. (1981). Ligatin from embryonic chick neural retina inhibits retinal cell adhesion. *Dev. Biol.* **86,** 250–255.

Mayer, B. W., Jr., Hay, E. D., and Hynes, R. D. (1981). Immunocytochemical localization of fibronectin in embryonic chick trunk and area vasculosa. *Dev. Biol.* **82,** 267–286.

Merrell, R., Gottlieb, D. I., and Glaser, L. (1975). Embryonal cell surface recognition. *J. Biol. Chem.* **250,** 5655–5659.

Meyer, R. L. (1982). Ordering of retinotectal connections: A multivariate operational analysis. *Curr. Top. Dev. Biol.* **17,** 101–145.

Moscona, A. A. (1962). Analysis of cell recombinations in experimental synthesis of tissues *in vitro*. *J. Cell. Comp. Physiol.* **60** (Suppl. 1), 65–80.

Moscona, A. A. (1968). Cell aggregation: Properties of specific cell-ligands and their role in the formation of multicellular systems. *Dev. Biol.* **18,** 250–277.

Nardi, J. B. (1983). Neuronal pathfinding in developing wings of the moth *Manduca sexta*. *Dev. Biol.* **95,** 163–174.

Newgreen, D. F. (1982). Adhesion to extracellular materials by neural crest cells at the stage of initial migration. *Cell Tissue Res.* **227,** 297–317.

Newgreen, D. (1984). Spreading of explants of embryonic chick mesenchymes and epithelia on fibronectin and laminin. *Cell Tissue Res.* **236,** 265–277.

Newgreen, D. F., and Thiery, J.-P. (1980). Fibronectin in early avian embryos: Synthesis and distribution along the migration pathways of neural crest cells. *Cell Tissue Res.* **211,** 269–291.

Newgreen, D. F., Ritterman, M., and Peters, E. A. (1979). Morphology and behavior of neural crest cells of chick embryo *in vitro*. *Cell Tissue Res.* **203,** 115–140.

Newgreen, D. F., Gibbins, I. L., Sauter, J., Wallenfels, B., and Wutz, R. (1982). Ultrastructural and tissue-culture studies on the role of fibronectin, collagen and glycosaminoglycans in the migration of neural crest cells in the fowl embryo. *Cell Tissue Res.* **221,** 521–549.

Norlander, R. H., and Singer, M. (1978). The role of ependyma in regeneration of the spinal cord in the urodele amphibian tail. *J. Comp. Neurol.* **180,** 349–373.

Norlander, R. H., and Singer, M. (1982). Spaces precede axons in *Xenopus* embryonic spinal cord. *Exp. Neurol.* **75,** 221–228.

Nornes, H. O., Hart, H., and Carry, M. (1980). Pattern of development of ascending and descending fibers in embryonic spinal cord of chick: I. Role of positional information. *J. Comp. Neurol.* **192,** 119–132.

Rakic, P. (1972). Mode of cell migration to the superficial layers of fetal monkey neocortex. *J. Comp. Neurol.* **145,** 61–84.

Raper, J. A., Bastiani, M. J., and Goodman, C. S. (1983a). Pathfinding by neuronal growth cones in grasshopper embryos. I. Divergent choices made by the growth cones of sibling neurons. *J. Neurosci.* **3,** 20–30.

Raper, J. A., Bastiani, M. J., and Goodman, C. S. (1983b). Pathfinding by neuronal growth cones in grasshopper embryos. II. Selective fasciculation onto specific axonal pathways. *J. Neurosci.* **3,** 31–41.

Raper, J. A., Bastiani, M., and Goodman, C. S. (1983c). Ablation of a specific axonal pathway retards the extension of an identified growth cone in the CNS of the grasshopper embryo. *Soc. Neurosci. Abstr.* **9,** 1044.

Raper, J. A., Bastiani, M. J., and Goodman, C. S. (1984). Pathfinding by neuronal growth cones in grasshopper embryos. IV. The effects of ablating the A and P axons upon the behavior of the G growth cone. *J. Neurosci.* **4,** 2329–2345.

Rathjen, F. G., and Rutishauser, U. (1984). Comparison of two cell surface molecules involved in neural cell adhesion. *EMBO J.* **3,** 461–465.

Rathjen, F. G., and Schachner, M. (1984). Immunocytological and biochemical characterization of a new neuronal cell surface component (L1 antigen) which is involved in cell adhesion. *EMBO J.* **3,** 1–10.

Reh, T. A., Pitts, E., and Constantine-Paton, M. (1983). The organization of the fibers in the optic nerve of normal and tectum-less *Rana pipiens. J. Comp. Neurol.* **218,** 282–296.

Rogers, S. L., Letourneau, P. C., Palm, S. L., McCarthy, J., and Furcht, L. T. (1983). Neurite extension by peripheral and central nervous system neurons in response to substratum-bound fibronectin and laminin. *Dev. Biol.* **98,** 212–220.

Roseman, S. (1970). The synthesis of complex carbohydrates by multiglycosyltransferase systems and their potential function in intercellular adhesion. *Chem. Phys. Lipids* **5,** 270–297.

Roth, S., McGuire, E. J., and Roseman, S. (1971). Evidence for cell-surface glycosyltransferases. Their potential role in cellular recognition. *J. Cell Biol.* **51,** 536–547.

Rothbard, J. B., Brackenbury, R., Cunningham, B., and Edelman, G. M. (1982). Differences in the carbohydrate structure of neural cell adhesion molecule from adult and embryonic chicken brain. *J. Biol. Chem.* **257,** 11064–11069.

Rovasio, R. A., Delouvee, A., Yamada, K. M., Timpl, R., and Thiery, J. P. (1983). Neural crest cell migration: Requirements for exogenous fibronectin and high cell density. *J. Cell Biol.* **96,** 462–473.

Rutishauser, U. (1984). Developmental biology of a neural cell adhesion molecule. *Nature (London)* **310,** 549–554.

Rutishauser, U., and Edelman, G. (1980). Effects of fasciculation on the outgrowth of neurites from spinal ganglia in culture. *J. Cell Biol.* **87,** 370–378.

Rutishauser, U., Thiery, J.-P., Brackenbury, R., Sela, B.-A, and Edelman, G. M. (1976). Mechanisms of adhesion among cells from neural tissues of the chick embryo. *Proc. Natl. Acad. Sci. U.S.A.* **73,** 577–581.

Rutishauser, U., Gall, W. E., and Edelman, G. M. (1978a). Adhesion among neural cells of the chick embryo. IV. Role of the cell surface molecule CAM in the formation of neurite bundles in cultures of spinal ganglion. *J. Cell Biol.* **79,** 382–393.

Rutishauser, U., Thiery, J.-P., Brackenbury, R., and Edelman, G. M. (1978b). Adhesion among neural cells of the chick embryo. III. Relationship of the surface molecule CAM to cell adhesion and the development of histotypic patterns. *J. Cell Biol.* **79,** 371–381.

Rutishauser, U., Hoffman, S., and Edelman, G. M. (1982). Binding properties of a cell adhesion molecule from neural tissue. *Proc. Natl. Acad. Sci. U.S.A.* **79,** 685–689.

Rutishauser, U., Grumet, M., and Edelman, G. M. (1983). Neural cell adhesion molecule mediates initial interactions between spinal cord neurons and muscle cells in culture. *J. Cell Biol.* **97,** 145–152.

Rutz, R., and Lilien, J. (1979). Functional characterization of an adhesive component from the embryonic chick neural retina. *J. Cell Sci.* **36,** 323–342.

Schachner, M., Faissner, A., Kruse, J., Lindner, J., Meier, D. H., Rathjen, F. G., and Wernecke,

H. (1983). Cell-type specificity and developmental expression of neural cell-surface components involved in cell interactions and of structurally related molecules. *Cold Spring Harbor Symp. Quant. Biol.* **48,** 557–568.

Schmidt, J. T., Cicerone, C. M., and Easter, S. S. (1978). Expansion of the half retinal projection to the tectum in goldfish: An electrophysiological and anatomical study. *J. Comp. Neurol.* **177,** 257–278.

Schubert, D., and LaCorbiere, M. (1980). Role of a 16S glycoprotein complex in cellular adhesion. *Proc. Natl. Acad. Sci. U.S.A.* **77,** 4137–4141.

Schubert, D., and LaCorbiere, M. (1982). Properties of extracellular adhesion-mediating particles in myoblast clone and its adhesion-deficient variant. *J. Cell Biol.* **94,** 108–114.

Schubert, D., LaCorbiere, M., Klien, G., and Birdwell, C. (1983). A role for adherons in neural retina cell adhesion. *J. Cell Biol.* **96,** 990–998.

Silver, J., and Sidman, R. L. (1980). A mechanism for the guidance and topographical patterning of retinal ganglion cell axons. *J. Comp. Neurol.* **189,** 101–111.

Singer, M., Norlander, R. H., and Egar, M. (1979). Axonal guidance during embryogenesis and regeneration: The blueprint hypothesis of neuronal pathway patterning. *J. Comp. Neurol.* **185,** 1–22.

Sperry, R. W. (1943). Visuomotor coordination in the newt (*Triturus viridescens*) after regeneration of the optic nerves. *J. Comp. Neurol.* **79,** 33–35.

Sperry, R. W. (1963). Chemoaffinity in the orderly growth of nerve fiber patterns and connections. *Proc. Natl. Acad. Sci. U.S.A.* **50,** 703–710.

Steinberg, M. S. (1963). Reconstruction of tissues by dissociated cells. *Science* **141,** 401–408.

Steinberg, M. S. (1964). The problem of adhesive selectivity in cellular interactions. *In* "Cellular Membranes in Development" (M. Locke, ed.), pp. 321–366. Academic Press, New York.

Steinberg, M. S. (1970). Does differential adhesion govern self-assembly processes in histogenesis? Equilibrium configurations and the emergence of a hierarchy among populations of embryonic cells. *J. Exp. Zool.* **173,** 395–434.

Steinberg, M. S. (1978a). Cell–cell recognition in multicellular assembly: Levels of specificity. *In* "Cell–Cell Recognition" (A. S. G. Curtis, ed.), S. E. B. Symposium, No. 32, pp. 25–49. Cambridge Univ. Press, London and New York.

Steinberg, M. S. (1978b). Specific cell ligands and the differential adhesion hypothesis: How do they fit together? *In* "Specificity of Embryological Interactions" (D. Garrod, ed.), pp. 99–129. Chapman & Hall, London.

Steinberg, M. S. (1981). The adhesive specification of tissue self-organization. *In* "Morphogenesis and Pattern Formation" (T. G. Connelly, L. Brinkley, and B. Carlson, eds.), pp. 179–203. Raven, New York.

Steinberg, M. S., and Poole, T. J. (1981). Strategies for specifying form and pattern: Adhesion-guided multicellular assembly. *Philos. Trans. R. Soc. London Ser. B* **295,** 451–460.

Straznicky, C., Gaze, R. M., and Horder, T. J. (1979). Selection of appropriate medial branch of the optic tract by fibers of ventral retinal origin during development and in regeneration: An autoradiographic study in *Xenopus*. *J. Embryol. Exp. Morphol.* **50,** 253–267.

Straznicky, C., Gaze, R. M., and Keating, M. J. (1981). The development of the retinotectal projections from compound eyes in *Xenopus*. *J. Embryol. Exp. Morphol.* **62,** 13–35.

Taghert, P. H., Bastiani, M. S., Ho, R. K., and Goodman, C. S. (1982). Guidance of pioneer growth cones: Filopodial contacts and coupling revealed with an antibody to Lucifer yellow. *Dev. Biol.* **94,** 391–399.

Takeichi, M., Ozaki, H. S., Tokunaga, K., and Okada, T. S. (1979). Experimental manipulation of cell surface to affect cellular recognition mechanisms. *Dev. Biol.* **70,** 464–474.

Thiery, J.-P., Brackenbury, R., Rutishauser, U., and Edelman, G. M. (1977). Adhesion among neural cells of the chick embryo. *J. Biochem. Chem.* **252,** 6841–6845.

Thiery, J.-P., Duband, J. L., Rutishauser, U., and Edelman, G. M. (1982). Cell adhesion molecules in early chick embryogenesis. *Proc. Natl. Acad. Sci. U.S.A.* **79,** 6737–6741.

Thiery, J.-P., Delouvee, A., Grumet, M., and Edelman, G. M. (1985). Initial appearance and regional distribution of the neuron–glia cell adhesion molecule in the chick embryo. *J. Cell Biol.* **100,** 442–456.

Thomas, W. A., and Steinberg, M. S. (1981). Two distinct adhesion mechanisms in embryonic neural retina cells. II. An immunological analysis. *Dev. Biol.* **81,** 106–114.

Thomas, W. A., Edelman, B. A., Lobel, S. M., Breitbart, A. S., and Steinberg, M. S. (1981a). Two chick embryonic adhesion systems: Molecular vs. tissue specificity. *J. Supramol. Struct. Cell. Biochem.* **16,** 15–27.

Thomas, W. A., Thomson, J., Magnani, J. L., and Steinberg, M. S. (1981b). Two distinct adhesion mechanisms in embryonic neural retina cells. III. Functional specificity. *Dev. Biol.* **81,** 379–385.

Timpl, R., Rohde, H., Robey, P. G., Rennard, S. I., Foidart, J.-M., and Martin, G. R. (1979). Laminin—a glycoprotein from basement membranes. *J. Biol. Chem.* **254,** 9933–9937.

Tosney, K. W. (1978). The early migration of neural crest cells in the trunk region of the avian embryo: An electron microscopic study. *Dev. Biol.* **62,** 317–333.

Townes, P. S., and Holtfreter, J. (1955). Directed movements and selective adhesion of embryonic amphibian cells. *J. Exp. Zool.* **128,** 53–120.

Varon, S., Manthorpe, M., and Williams, L. (1984). Neuronotrophic and neurite-promoting factors and their clinical potentials. *Dev. Neurosci.* **6,** 73–100.

Weston, J. (1963). A radioautographic analysis of the migration and localisation of trunk neural crest cells in the chick. *Dev. Biol.* **6,** 279–310.

Weston, J. A. (1970). The migration and differentiation of neural crest cells. *Adv. Morphogen.* **8,** 41–114.

Wewer, U., Albrechtsen, R., Manthorpe, M., Varon, S., Engvall, E., and Ruoslahti, E. (1983). Human laminin isolated in a nearly intact, biologically active form from placenta by limited proteolysis. *J. Biol. Chem.* **258,** 12654–12660.

Whitelaw, V., and Hollyday, M. (1983a). Thigh and calf discrimination in the motor innervation of the chick hindlimb following deletions of limb segments. *J. Neurosci.* **3,** 1199–1215.

Whitelaw, V., and Hollyday, M. (1983b). Position-dependent motor innervation of the chick hindlimb following serial and parallel duplications of limb segments. *J. Neurosci.* **3,** 1216–1225.

Whitelaw, V., and Hollyday, M. (1983c). Neural pathway constraints in the motor innervation of the chick hindlimb following dorsoventral rotations of distal limb segments. *J. Neurosci.* **3,** 1226–1233.

Willshaw, D. J., Fawcett, J. W., and Gaze, R. M. (1983). The visuotectal projections made by *Xenopus* ''pie-slice'' compound eyes. *J. Embryol. Exp. Morphol.* **74,** 29–45.

Yoon, M. G. (1973). Retention of the original topographic polarity by the 180° rotated tectal reimplant in young adult goldfish. *J. Physiol. (London)* **233,** 575–588.

TROPHIC INTERACTIONS IN RETINAL DEVELOPMENT AND IN RETINAL DEGENERATIONS. *IN VIVO* AND *IN VITRO* STUDIES

RUBEN ADLER

The Michael M. Wynn Center for the
Study of Retinal Degenerations
The Wilmer Ophthalmological Institute
The Johns Hopkins University
School of Medicine
Baltimore, Maryland

I. Introduction

It seems fair to state, paraphrasing the Spanish philosopher Ortega y Gasset, that "a neuron is itself plus its microenvironment." Contemporary neurobiology is dominated by the concept that most neuronal phenotypic properties are regulated by molecular signals originating in the microenvironment surrounding the neurons. The regulation of neuronal survival and its counterpart, developmental neuronal death, occupy a place of preeminence among neuronal behaviors actively studied in recent years. Considerable effort is being invested in the search for "neuronotrophic factors," i.e., molecules capable of supporting neuronal survival. As a by-product, these investigations have led to the discovery of other molecular factors which, without affecting neuronal survival, are capable of controlling the expression of differentiated neuronal functions. Like most scientific fields undergoing rapid growth, trophic factor research is currently as confusing as it is interesting and challenging. But the notion that neuronal survival and differentiation are constantly regulated by microenvironmental molecular agents seems to remain unchallenged even while conceptual, technical, and semantic issues complicate the field and await clarification.

The discussion of neuronal regulation in a book devoted to cell biology of the retina is justified in more than one way. Information gathered in the study of other parts of the nervous system can illuminate the selection of experimental strategies for the investigation of similar problems in the retina. On the other hand the retina, although considerably more complex than most PNS organs, appears as one of the most suitable experimental systems for the investigation of trophic factors active on CNS neurons. Finally, several forms of blindness which result from the degeneration of retinal neurons might be caused by defects in the molecular mechanisms controlling the survival and differentiation of these neurons.

The search for molecular agents involved in neuronal regulation usually involves several complementary phases. The involvement of tissue or cell interactions in the control of a given neuronal property is usually disclosed through initial *in vivo* studies. An *in vitro* system is then developed in which these cell–cell interactions can be reproduced. This culture system is used as a bioassay in the search for and purification of *molecular factors* involved in the phenomenon. After these molecules are purified and antibodies against them are generated, research proceeds to the analysis of the *mechanism of action* of the regulatory agent, and to the investigation of its *physiological function* in the intact organism. As discussed in the following sections, the search for neuronotrophic factors active on retinal neurons has only reached the first phases of this sequence. For example, target-dependent neuronal survival and pigment epithelium-dependent photoreceptor maturation are well-known examples of cell–

cell interactions controlling the behavior of retina neurons. Although *in vitro* systems suitable for the investigation of some of these interactions at a molecular level have been developed in the last few years, they have not yet been exploited for this purpose. Consequently, this article will include a review of the accomplishment already made together with a discussion of some research strategies which deserve to be explored.

II. The Regulation of Neuronal Survival

As is the case throughout the nervous system, the occurrence of extensive cell death is necessary for normal retinal development (Glucksmann, 1940; Silver and Hughes, 1973). Developmental neuronal death occurs according to a stereotyped pattern, affecting defined numbers of cells in particular areas of the retina at specific times in development. Perhaps the most important concept derived from the study of this phenomenon is that a neuron, in order to survive, must have access to survival-promoting, regulatory macromolecules originating in the microenvironment (see below). These survival-promoting agents deserve additional attention in the context of pathological neuronal death which, in the mature retina, can give rise to blinding conditions such as retinitis pigmentosa or age-related maculopathy. It is an intriguing possibility that neuronal death might be controlled by similar mechanisms in normal development and in abnormal pathological conditions.

As mentioned above, normal developmental neuronal death is not exclusive to the retina, and the same is true for the occurrence of neuronal degenerative diseases of unknown etiology. Alzheimer's disease, Parkinson's disease, amyotrophic lateral sclerosis, Huntington's chorea, muscular dystrophy, and familial dysautonomia are examples from other regions of the nervous system (Appel, 1981; Price *et al.*, 1982; Barbeau, 1980; Johnson *et al.*, 1979). The general principles emerging from studies in various regions of the nervous system will be considered first, as a preamble to the discussion of neuronal death in the retina.

A. *Developmental Neuronal Death: Issues, Concepts, and Speculations*

The investigation of the developmental neuronal death phenomenon has generated some concepts which, although based predominantly on circumstantial evidence, have gained general acceptance. Some of these principles will be presented here in general terms, but the reader is referred to some recent reviews for more details (Berg, 1982; Cowan, 1973; Glucksman, 1951; Hamburger and Oppenheim, 1982; Jacobson, 1978; Oppenheim, 1981; Silver, 1978; Varon and Adler, 1980).

1. Neuronal Survival Requires Specific Regulatory Factors

Developmental neuronal death, which in most neural centers affects about 50% of the neurons which are generated, occurs predominantly at the time of synaptogenesis between neurons and their postsynaptic targets and seems to affect those neurons which fail to make appropriate contact with target cells. If the latter are surgically removed before synaptogenesis there is marked exacerbation of neuronal death and, correspondingly, death is reduced if additional target cells are made available. *In vivo* experiments (reviewed by Landmesser and Pilar, 1978; Hamburger and Oppenheim, 1982) have disclosed the following features of neurons which undergo developmental death: (1) they develop normally until the time of synaptogenesis, (2) they are not predetermined to die, (3) they reach the target cell region with their axons, but (4) they appear to compete with other neurons for a limited number of synaptic sites available in target cells. These findings led to the hypothesis that neurons require specific target-derived, survival-supporting molecular signals which are not freely diffusible but, rather, become available only to neurons which succeed in establishing stable contacts with postsynaptic cells (i.e., Landmesser and Pilar, 1978; Varon and Adler, 1980). As discussed below, this concept has received indirect support from *in vivo* and *in vitro* studies with the nerve growth factor (NGF) and other "trophic" molecules.

2. The Acquisition of Neuronal Dependency

Theoretically, neuronal target independence during early ("presynaptic") life could reflect either the availability of alternative sources of survival-supporting influences or an authentic lack of trophic dependency. Although the evidence is not conclusive, several *in vitro* studies appear to indicate that young neurons are indeed capable of surviving without special trophic support (Coughlin and Black, 1978; Coughlin *et al.*, 1978; Manthorpe *et al.*, 1981a; Adler and Varon, 1982). If this were the case, then, neurons would appear to undergo a real switch from target independence to target dependence at a specific time in development, a possibility which raises some challenging questions. Are there inductive signals responsible for the acquisition of trophic dependency at the time of synaptogenesis? Are target cells simultaneously programmed to produce the survival-promoting molecules necessary to satisfy this neuronal requirement? Could new trophic dependencies be "induced" in adult neurons under abnormal circumstances? Is it likely that some neuronal degenerations might be caused by the appearance of new trophic requirements? If these regulatory signals do actually exist, their identification may well represent one of the most important challenges faced by contemporary neurobiology, given the importance of the neuronal death phenomenon in morphogenesis, in the adjustment of neuronal numbers, and in the genesis of pathological conditions. We must humbly acknowledge our complete ignorance in this area.

3. Multiple Sources of Survival-Promoting Support?

Availability of target-derived survival-supporting factors is, at least during a particular phase of development, the leading mechanism controlling selective neuronal survival. Studies to be summarized below have shown the existence of alternative sources of survival-promoting factors, although they have failed to disclose their role in normal development. Theoretically, these factors could operate either in conjunction with or instead of target-derived factors (Varon and Adler, 1980). They could be particularly important during target-independent phases of the neuron's life cycle, or in those pathological situations when neurons become disconnected from their postsynaptic partners.

a. Presynaptic Influences. These can have profound effects on neuronal survival as illustrated by extensive neuronal death observed in the deafferented optic tectum (Filogamo, 1950; Kelly and Cowan, 1972) and other visual centers (reviewed by Cowan, 1973). In other organs, such as the ciliary ganglion, presynaptic innervation does not seem to affect neuronal survival, although it may have effects on neuronal maturation (Landmesser and Pilar, 1978). It has also been well documented that at least some neurotransmitter activities can be regulated by presynaptic influences (i.e., Black, 1978).

b. Glia and Other "Satellite" Cells. Glial-conditioned media have been shown to support the survival of both PNS and CNS neurons, thus demonstrating that glial cells can release survival-supporting factors *in vitro* (Banker, 1980; Barde *et al.*, 1978; Burnham *et al.*, 1972; Ebendal and Jacobson, 1975; Lindsay *et al.*, 1982; Varon *et al.*, 1981). However, the involvement of glial cells in the regulation of neuronal survival *in vivo* is yet to be demonstrated. Investigation of this issue has been stimulated in recent years by studies indicating that some glial cells might act as a "permissive terrain" for the elongation of regenerating axons (reviewed in Aguayo *et al.*, 1983). These investigations, however, have not addressed directly the possible involvement of glial-derived, survival-promoting factors in the regenerative response. The role of other "satellite" cells such as the retinal pigment epithelium is discussed below (Section V).

c. Blood-Borne Factors. Blood-borne factors would probably reach most neurons in a population and would therefore fail to satisfy the conditions of restricted availability necessary to determine selective neuronal survival in embryonic life. However, it has been known since the early years of NGF research that injections of this trophic agent resulted in stimulation of neuronal survival and maturation, whereas injection of NGF antibodies caused destruction of NGF-dependent neurons (review in Levi-Montalcini and Angeletti, 1968). More recently, studies in which pregnant animals were immunized by injection of heterologous NGF showed that transplacental transfer of anti-NGF antibodies had devastating effects on the survival and development of NGF-dependent neurons in the fetus (Johnson *et al.*, 1980). The systemic circulation, then, should not be ignored as a possible route through which trophic factors (or their antibodies) could reach neurons. In fact, circulating trophic factors or their antibodies could potentially be involved in the genesis of neuronal degenerations in the adult.

4. Is Neuronal Death a Universal Phenomenon?

Although there are no conclusive reports of neuronal populations which are *not* affected by developmental neuronal death, this phenomenon should not be automatically considered ubiquitous (Hamburger and Oppenheim, 1982). Practically all the instances of developmental neuronal death which have been studied in detail (e.g., spinal cord motor neurons, retina ganglion cells, ciliary and sensory ganglia) involve neurons which (1) are big and, therefore, easy to identify, (2) exist in discrete nuclei or layers, (3) project their axons to remote regions of the CNS or to the periphery, and (4) have target territories amenable to surgical extirpation in early embryonic life. It is also noteworthy that time-consuming quantitative analysis is necessary to detect neuronal death even in populations of large neurons in which the phenomenon is very extensive. There is great need for careful, systematic investigation of the phenomenon in other neuronal subpopulations, even if they are experimentally less attractive. Small ‘‘local’’ neurons, for example, are frequently intermingled with other neurons, are difficult to identify, and cannot be easily separated from their postsynaptic targets with which they coexist within a same organ. These technical difficulties, although real and important, cannot justify extrapolating to the entire nervous system a phenomenon that has only been investigated in cases particularly well suited for experimental analysis.

B. Developmental Neuronal Death in the Retina

An initial description of cell death during eye development was published by Glucksmann in 1940 (see also reviews by Glucksmann, 1951; Cunningham,

1982). Since that time, and until the second half of the 1970s, very little attention was paid to the issue. The number of papers devoted to its study increased in recent years, in parallel with the recognition of its importance in retinal development and pathology. Several mammalian species have been included in these recent studies.

1. Retina Ganglion Cells

a. Chick Embryos. In birds, retinal ganglion cells make connections with postsynaptic neurons located in the contralateral optic tectum. At least 20% of the cells present in the ganglion cell layer of chick embryos die during normal development between days 12 and 16, the period when retinotectal synapses are developing (Hughes and McLoon, 1979). If the primordial tectum is destroyed at 4 days of development (i.e., before the arrival of retinal axons) retinal development continues normally until day 11 (Hughes and LaVelle, 1972, 1975). This finding shows that retina ganglion cells (as other neurons) are target independent in early development. The operated embryos, however, lose about 72% of their ganglion cells after day 11, indicating that neuronal death results from target deprivation at the time when retinotectal connections would normally become organized. Rager and Rager (1978) reached similar conclusions through electron microscopical quantitation of axonal profiles in the optic nerve.

b. Rodents. Both retinotectal synaptogenesis and ganglion cell degeneration occur largely postnatally in rodents. In the mouse retina, for example, degenerating cells are abundant in the ganglion cell layer between embryonic day 16 and postnatal day 5, with a peak in number at birth (Hume *et al.*, 1983). A thorough study of cell death in the developing mouse retina was recently published by Young (1984). Cell death was found to be extensive during the first 2 weeks after birth, and to affect most cell types present in the retina. Peaks of degeneration (as judged by frequency of pycnotic cells) occurred at days 2–5 (ganglion cells), 3–8 (amacrine cells), and 8–11 (bipolar and Müller cells). Interestingly, cell death was also abundant among photoreceptors, particularly those which are "born" on the inner side of the outer plexiform layer and fail to migrate outward toward the outer nuclear layer. The author estimates that a minimum of 7.8% of the amacrine cells and 2.4% of the photoreceptors degenerate in the mouse retina. In the rat, substantial ganglion cell death can be observed during the first 10 postnatal days, with a peak around days 6–7 (Cunningham *et al.*, 1982; Potts *et al.*, 1982; Drehler *et al.*, 1983). Potts *et al.* (1982) have estimated that at least 35% of the ganglion cells degenerate during this period. Ganglion cell death is also particularly abundant in hamsters during early postnatal life, with a peak around day 5 (Figs. 1 and 2). As many as 49% of the cells originally present in the ganglion cell layer are affected (Sengelaub and Finley, 1982). There seems to

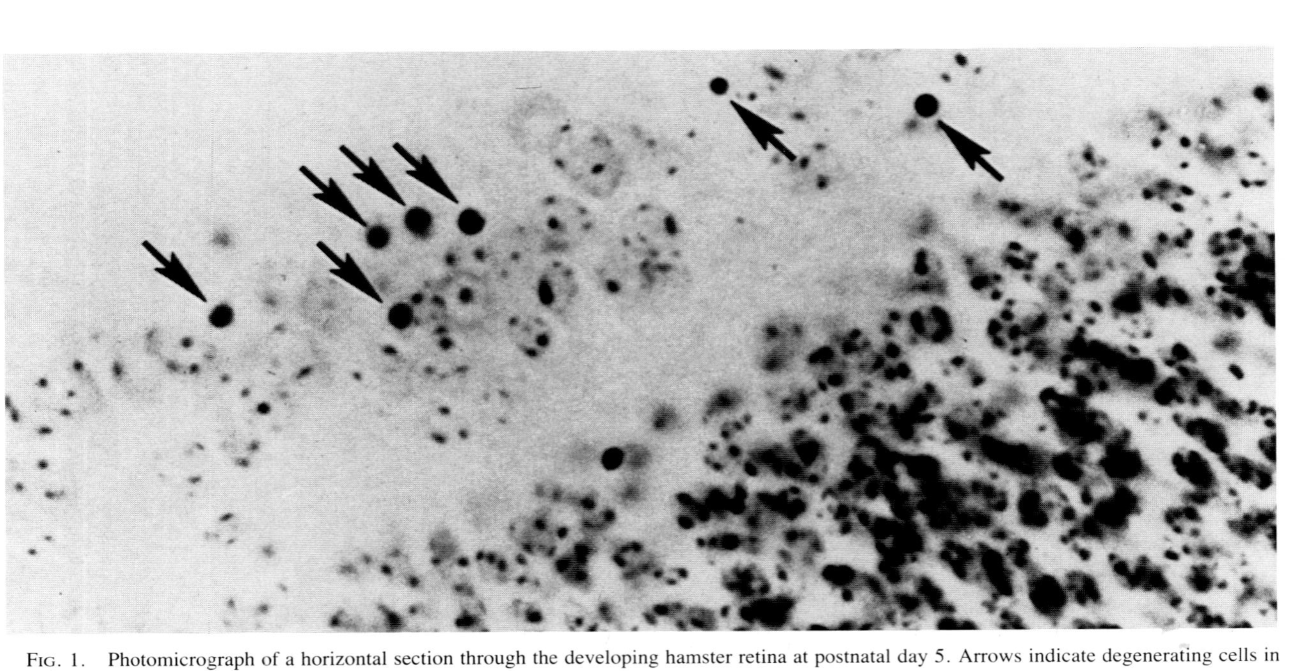

FIG. 1. Photomicrograph of a horizontal section through the developing hamster retina at postnatal day 5. Arrows indicate degenerating cells in the ganglion cell layer. Cresylcresyl violet stain, ×400. (From Sengelaub and Finlay, 1982, with permission.)

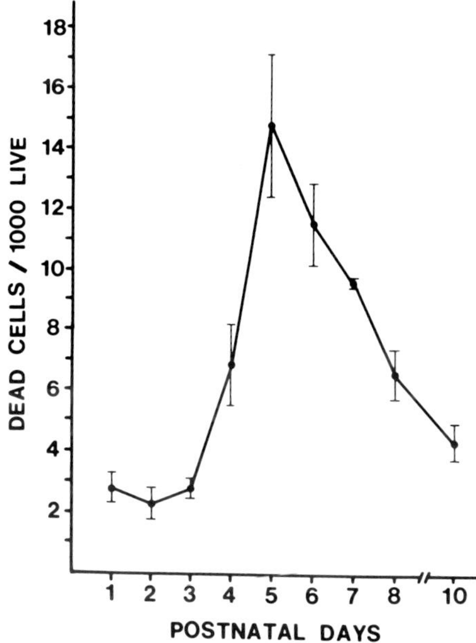

FIG. 2. Ratio of degenerating cells to normal cells in the ganglion cell layer of the hamster retina over the first 10 postnatal days. Points represent averaged values ± standard error of the mean. (From Sengelaub and Finlay, 1982, with permission.)

be a temporal correlation between natural neuronal death and retinotectal synaptogenesis in both rat and hamster, suggesting that competition for synaptic sites may be responsible for neuronal death (Fig. 2). Supporting this explanation, removal of one eye in the newborn hamster was found to produce a substantial decrease in retina ganglion cell death in the contralateral eye, apparently through reduced competition for synaptic sites (Sengelaub and Finley, 1981; Insausti *et al.*, 1984).

2. Are "Local Interneurons" and Photoreceptors also Affected by Developmental Death?

The difficulties found in investigating developmental neuronal death in populations of "local interneurons" have been mentioned above. The retina is no exception in this context. Papers devoted to the study of retina ganglion cell death in the chick (Hughes and LaVelle, 1975; Hughes and McLoon, 1979), rat (Kuwabara and Weidman, 1974; Cunningham *et al.*, 1982), and mouse (Hume *et al.*, 1983) contain, at best, passing references to the presence of pycnotic cells in the inner nuclear layer of the retina. Practically no information is available regarding the timing and extent of this phenomenon, its response to experimental manipulation, the cell types which are involved, etc. There is even less information regarding neuronal degeneration in the outer nuclear layer of the retina. Hume *et al.* (1983), for example, stated that in the mouse, "Virtually no pyknotic nuclei are found in the outer nuclear layer at any time during development." This study covered only the period between 16 days of embryonic life and 20 days of postnatal life. However, Young (1984) found evidence of abundant cell death in all layers of the retina, including the layer of photoreceptor cells.

Glucksmann (1940), on the other hand, mentioned the occurrence of a wave of cell degeneration in the amphibian outer nuclear layer at the onset of its development. According to Kuwabara and Weidman (1974), moreover, during rat retinal embryogenesis, "Degenerating cells in the inner layers disappear when the development of the nerve fiber layer is completed on the 16th day. They appear again in the outer layers when the differentiation of the photoreceptor cells begins at one week postnatal." There is obviously great need to investigate these issues systematically and in detail. If retinal interneurons and photoreceptors do undergo a period of developmental neuronal death, the search for factors limiting the magnitude of the phenomenon by promoting neuronal survival would be encouraged. If they do not, the universality of the developmental neuronal death phenomenon will be in question, and it will be pertinent to inquire which are the mechanisms through which the adjustment of neuronal numbers is accomplished in the retina. Also, a comparison of neuronal populations which do or do not undergo developmental death should provide useful insight about the regulation of neuronal survival.

3. Special Features of Neuronal Death in the Retina

a. Spatial Gradients. Many developmental phenomena occur in the retina according to one or more spatiotemporal gradients (e.g., from fundus to periphery, from nasal to temporal). Similar gradients have been observed in the progression of pathological photoreceptor degeneration in man and in animal mutants (LaVail, 1981; Merin and Auerback, 1976). Interestingly, developmental ganglion cell death also occurs along defined spatiotemporal gradients which, however, appear to differ in different animal species. In the chick, ganglion cell death advances from the temporal to nasal (Hughes and McLoon, 1979) and from the central to peripheral retina (Rager and Rager, 1978). Degeneration progresses nasotemporally in the rat (Cunningham *et al.*, 1982; see Table I) and along a central to peripheral path in the hamster (Sengelaub and Finley, 1982) and in mouse photoreceptors (Young, 1984). The existence of stereotyped patterns of progression in neuronal death both in normal development and in degen-

TABLE I

NASAL–TEMPORAL AND CAUDAL–ROSTRAL DIFFERENCES IN NEURONAL DEGENERATION[a]

	Postnatal day	Region	Degenerating profiles per 1000 normals	Nasal/temporal
Ganglion cell layer	6	Dorsal nasal	13.1	1.5
	6	Dorsal temporal	8.5	
	6	Ventral nasal	11.6	1.8
	6	Ventral temporal	6.3	
	7	Dorsal nasal	21.3	2.2
	7	Dorsal temporal	9.9	
	7	Ventral nasal	20.0	2.7
	7	Ventral temporal	7.5	
				Caudal/rostral
Superior colliculus (superficial layers)	6	Caudal	31.0	1.6
	6	Rostral	19.3	
	6	Caudal	41.8	2.0
	6	Rostral	21.0	
	7	Caudal	10.9	1.7
	7	Rostral	6.3	
	7	Caudal	22.2	1.5
	7	Rostral	14.6	

[a]From Cunningham *et al.* (1982), with permission.

erative diseases may perhaps be related to the cellular mechanisms underlying neuronal death. The question, however, has not been experimentally analyzed to any significant extent.

b. Reciprocity of Survival-Supporting Interactions between Retina and Brain Centers. Retinal ganglion cells, which require survival-supporting stimuli from postsynaptic brain neurons, are in turn capable of stimulating the survival and differentiation of the brain neurons with which they make connections. Since the pioneer studies of Filogamo (1950), it has been known that early removal of the chick embryo eye results in extensive cell death in the optic lobe (see also Kelly and Cowan, 1972; Mathers and Ostrach, 1979). Biochemical changes are also conspicuous in the cholinergic system in the deafferented tectum (Marchisio, 1969; Ciani *et al.*, 1978). Although causal relationships have not been established, attention has been called to a remarkable correlation in the incidence and localization of developmental neuronal death in regions of retina and brain which are normally interconnected (Cunningham *et al.*, 1982; Sengelaub and Finley, 1982). Peaks of neuronal degeneration seem to occur simultaneously in the caudal superior colliculus and the nasal ganglion cell layer as well as in the rostral colliculus and the temporal ganglion cell layer. Based on these observations, Cunningham *et al.* (1982) have suggested a possible reciprocal control of neuronal survival in "matching" regions of the retina and the tectum.

c. Glia, Macrophages, and the Removal of Neuronal Debris. The prevalence of neuronal death during retinal development raises the issue of the fate of the degenerating cells. Müller cells have been reported to phagocytize debris generated during developmental neuronal death in the rat (Kuwabara and Weidman, 1974), and chick (Rager and Rager, 1978; Hughes and McLoon, 1979). Hughes and McLoon (1979) also indicate that "Other cells of unknown origin were seen to sequester debris" (p. 591), while Young (1984) states that in the mouse retina "some of the phagocytes may be derived from the invading vascular tissue." Using antibodies against the macrophage-specific antigen F4-80, Hume *et al.* (1983) showed that macrophages from the general circulation invade the developing mouse retina at the time of neuronal death. Surprisingly, these macrophages remain in the retina after phagocytizing degenerating neurons and seem to differentiate as microglia. The developmental neuronal death phenomenon thus brings to light a complex and still poorly understood association between glia and macrophages in the retina, in which glial cells appear assuming a phagocytic function and macrophages seem to differentiate into glia. It would be of great interest to investigate the involvement of these phagocytic elements in pathological conditions characterized by extensive neuronal degeneration in the retina.

III. The Search for "Trophic" and Other Regulatory Factors

A. Conceptual Considerations

A very substantial effort has been invested in recent years in the investigation of molecules capable of supporting neuronal survival (for recent reviews, see Adler, 1982b, 1986a; Berg, 1982; Black and Patterson, 1980; Coughlin, 1984; Greene and Shooter, 1980; Harper and Thoenen, 1980; Varon and Adler, 1980, 1981, among others). As mentioned above, a very important by-product of this research was the realization that survival was not the only neuronal behavior subject to regulation by microenvironmental molecular factors. For example, NGF, a molecule indispensable for the survival of sympathetic and dorsal root ganglionic neurons, is also capable of stimulating specialized neuronal activities such as neurotransmitter synthesis and neurite growth.

These different effects of NGF could theoretically be mediated by entirely

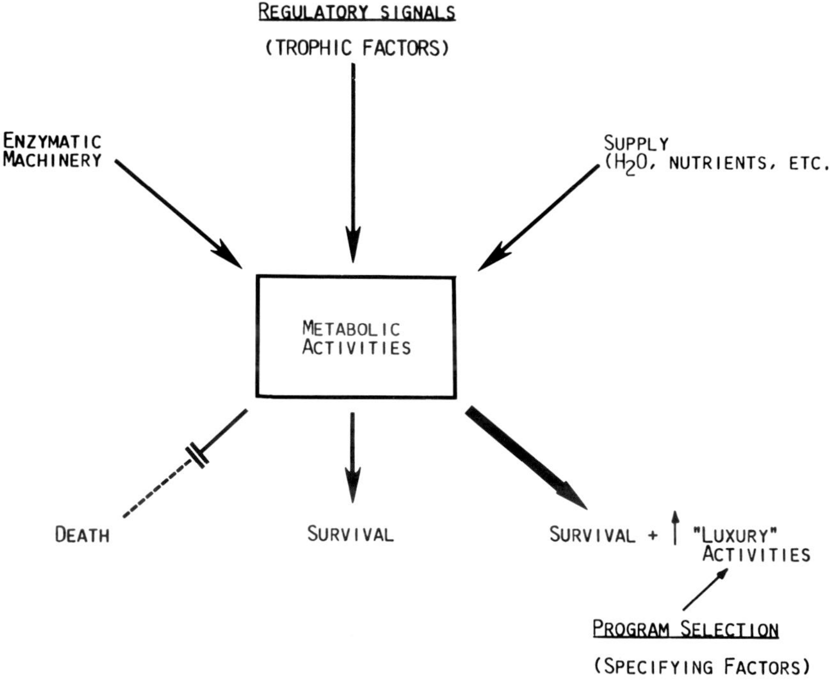

FIG. 3. Hypothetical scheme indicating the regulation of cell survival and differentiated function, as initially proposed by Varon (1977). The cell's metabolic activities not only require the necessary enzymes and an adequate supply of nutrients, but also are dependent upon special regulatory signals (trophic factors). These factors would control the occurrence of metabolic reactions necessary for cell survival, as well as for the execution of specialized cell functions (''luxury'' activities). The *nature* of these activities, however, would be regulated by a different set of ''specifying'' factors.

different molecular mechanisms. Varon (1977; see also Varon and Adler, 1980, 1981) proposed an alternative model to explain these different regulatory activities. According to this concept, the main function of a ''trophic'' factor such as NGF is to regulate metabolic activities essential for neuronal survival (Fig. 3). Through this stimulation of ''housekeeping'' metabolic activities, trophic factors could also have an *indirect,* but noticeable effect on the expression of specialized phenotypic activities by the neurons. In other words, a higher level of metabolic activity in the cell not only would satisfy the requirements for cell survival, but would also generate a surplus which would become available for other cellular activities. The model also postulates the existence of a different set of factors (''specifying'' factors; see Fig. 3). which would dictate the *selection* of the phenotypic activities expressed by a neuron at any particular point in time. It is implicit in this definition that specifying factors do not affect neuronal survival or have any generalized stimulatory effect upon neuronal metabolism.

B. *Culture Systems as Bioassays for the Study of Neuronotrophic Factors*

A tissue culture system must have some special features in order to be useful as a bioassay for the study of neuronotrophic activities (for recent reviews, see Varon *et al.*, 1982; Adler, 1986a). The cultures should allow accurate cell identification as well as direct, quantitative assessment of the cellular responses under investigation. Adequate PNS neuronal culture systems have been more readily available in recent years than their CNS counterparts (see Adler, 1986a). An important complicating factor in CNS cultures has been the coexistence of different neuronal types within one population, thus allowing cultured CNS neurons to interact not only with glial cells but also with other neurons from the same organ. These endogenous interactions have the potential of masking possible neuronal responses to exogenous trophic agents. Moreover, identification of different neuronal types is frequently very difficult in the absence of special cytochemical or autoradiographic techniques. It has only been recently that cell separation and cell identification techniques necessary to overcome these limitations have become available (see below).

C. *The Ciliary Ganglion as a Model: Lessons from Recent Studies*

Recent investigations of trophic and specifying factors active on peripheral nervous system neurons have provided important information about neuronal regulation in general and neuronal survival in particular. Since these studies illustrate so well the advantages and pitfalls of different research strategies, they should be taken into consideration before undertaking equivalent studies with more complex systems such as the neural retina.

Given the amount of data accumulated in recent years, however, even minimum coverage of the field would require considerably more space than is available in this article. An attempt will be made here to solve this dilemma by highlighting some of the most important lessons learned in work using the chick embryo ciliary ganglion—perhaps the peripheral organ most extensively used for these investigations in recent years. Factors active on ciliary ganglia have been reviewed in detail by Varon and Adler (1980, 1981), Berg (1982), Ebendal *et al.* (1982), Coughlin (1984), and Manthorpe and Varon (1984), among others.

1. Ciliary Ganglion Development *in Vivo*

To be successful, the *in vitro* investigation of the regulation of neuronal survival by trophic factors must be supported by a detailed description of the *in vivo* development of the neurons under investigation. In the case of the ciliary ganglion, this information was available through a series of excellent studies by Landmesser and Pilar (reviewed in Landmesser and Pilar, 1978). These *in vivo* investigations established (1) the duration of the target-independent phase of neuronal survival from the appearance of the ganglion [embryonic day (ED) 5] until the time of synaptogenesis between ciliary axons and intraocular target cells in the ciliary body, the choroid, and the iris (ED-8); (2) the magnitude (50% of the neurons) and duration (ED-8 and ED-14) of the neuronal death phase; (3) the correlation between neuronal survival and the formation of synaptic contacts with intraocular target cells; (4) the extension of the death phenomenon to 100% of the ciliary neurons caused by early removal of the eye; (5) the morphology of neuronal degeneration as it occurs both in the presence and in the absence of the embryonic eye; and (6) the role of competition for synaptic sites in neuronal degeneration. Work by Narayan and Narayan (1978) demonstrated, moreover, that increasing target cell availability by grafting a supplementary eye resulted in a reduction in ciliary ganglionic neural death. These *in vivo* studies established a baseline that proved essential for the design and interpretation of the *in vitro* studies described below.

2. Neuronal Death *in Vitro:* A Bioassay for Ciliary Neuronotrophic Factors

Helfand *et al.* (1976) prepared cell suspensions from ciliary ganglia dissected from 8-day chick embryos (the time of developmental neuronal death onset). Nonneuronal cells (Schwann cells and fibroblasts) survived normally in monolayer cultures from these cell suspensions, but neurons failed to survive for even 24 hr in the absence of special supplements. Helfand *et al.* also identified heart-conditioned medium as a source of survival-supporting factors for ciliary neurons. A useful feature of the ciliary cultures was that neuronal death could be recognized after only 24 hr in culture. Simultaneous survival of nonneuronal cells, moreover, indicated that neuronal death was due to the lack of a specific survival-supporting agent for neurons rather than to a general inadequacy of culture conditions.

This culture system provided the basis for a bioassay extensively used in the investigation of target-derived trophic factors for ciliary neurons. In this assay, the number of neurons surviving in 24-hr cultures is measured for different dilutions of a putative source of trophic support (Adler *et al.*, 1979b). A trophic unit is defined as the concentration of trophic support in 1 ml of medium which supports the survival of 50% of the maximum number of neurons which can be rescued in the cultures. The main disadvantage of this bioassay is that it requires visual counting of the number of neurons surviving in the cultures—a time-consuming and tiresome activity. However, it is likely that recently developed computer-assisted image analysis techniques will help in overcoming this drawback. Positive features of the bioassay include its accuracy and its short duration (24 hr).

3. CNTF, a Target-Derived Neuronotrophic Factor

The bioassay described above allowed detailed investigation of CNTF, a "ciliary neuronotrophic factor" present in the embryo which is capable of supporting the survival of ciliary neurons. Although direct evidence supporting the role of CNTF in target-dependent neuronal survival *in vivo* has not yet been generated, abundant circumstantial evidence suggests that this factor probably has a physiological role in this phenomenon. CNTF is highly concentrated in the intraocular tissues normally innervated by the ciliary ganglion (Adler *et al.*, 1979b); these intraocular target tissues are rich in CNTF at the time in development (days 8–14) when the fate of the ciliary neurons is being determined (Landa *et al.*, 1980). Interestingly, ciliary neurons become target-dependent *in vivo* and CNTF-dependent *in vitro* at the same time (Manthorpe *et al.*, 1981a; Adler and Varon, 1982). The factor, which is also active on other neurons (review in Manthorpe and Varon, 1984), behaves as a protein with a molecular weight of approximately 20,000 and which has been recently purified (Manthorpe *et al.*, 1980; Barbin *et al.*, 1984). Other laboratories have also attempted to purify survival-supporting molecules active on ciliary neurons, utilizing other bioassays and different sources including heart tissue from beef and other species (i.e., Bonyhady *et al.*, 1980; Ebendal *et al.*, 1982; Nishi and Berg, 1981). Chick embryo eye tissue has more recently been reported to contain a second factor which stimulates choline acetyltransferase activity in cultured ciliary neurons (Nishi and Berg, 1981).

4. PNPF, a Neurite-Promoting Factor for Ciliary Neurons

An important by-product of the investigation of trophic factors active on ciliary neurons was the discovery of a different molecule which, while lacking survival-supporting properties, has a dramatic effect on the production of nerve fibers by ciliary neurons and, as will be described later, also by other neuronal types. Collins (1978) first described in heart-conditioned medium an activity which bound to polyornithine but not to collagen or tissue culture plastic, and which was essential for the development of nerve fibers by ciliary neurons. Adler and Varon (1980) were able to segregate this "polyornithine-binding neurite-promoting factor" (or "PNPF") from the trophic activity also present in the conditioned medium. These investigators also developed a bioassay which allowed the measurement of PNPF activity in different sources. In this bioassay, survival of ciliary neurons was independently supported by a constant amount of CNTF added to the system, and PNPF activity was measured by counting the percentage of neurite-bearing neurons. PNPF behaves as a high molecular weight glycoprotein (Manthorpe *et al.*, 1981b). The activity is retained by XM-100 ultrafiltration membranes, is trypsin sensitive, and binds to concanavalin A or wheat germ agglutinin affinity columns from which it can be eluted with the specific sugars (Adler *et al.*, 1983). The biological activity of the factor can be inhibited by concanavalin A but, surprisingly, is not affected by wheat germ agglutinin. This neurite-promoting factor can be found in medium conditioned over a variety of cell types (Adler *et al.*, 1981). Recent studies have shown that the spatial distribution of PNPF (Collins and Garret, 1980) or related molecules produced by the ciliary ganglion itself (Adler and Varon, 1981a) can act as a guidance mechanism controlling the directionality of growing ciliary neurites. As discussed in some detail below, PNPF can also stimulate the production of nerve fibers by retina neurons (Adler, 1982a).

IV. The Search for Molecules Controlling Survival and Differentiation of Retinal Neurons: *In Vitro* Studies

Perhaps few regions of the central nervous system have been utilized as extensively as the neural retina for *in vitro* studies. Retinal histogenesis in explant cultures has been described, among others, by Dorris (1938), Hild and Callas (1967), Stefanelli *et al.* (1967), LaVail and Hild (1971), and Tamai *et al.* (1978). Since the pioneer studies by Moscona (1965), retinal cells have also been frequently used for reaggregation cultures (see also Steinberg, 1963; Suburo and Adler, 1977). Complex monolayer cultures, containing both multicellular clumps and a confluent layer of glial cells, have been used, among others, by Barr-Nea and Barishak (1970), Crisanti-Combes *et al.* (1977), Itoh (1976), Kaplowitz and Moscona (1976), Okada (1980), Pritchard *et al.* (1978), Puro *et al.* (1977), and Thompson and Pelto (1982). The application of retinal cultures to the study of neuronotrophic influences is only in its beginning. The information so far generated by these studies is suggestive of the existence of survival-promoting factors active on retinal neurons, and recent technical improvements offer promise for fast progress in the near future. However, the available information is still very fragmentary and not completely conclusive, and these deficits will necessarily be reflected in the following section, which discusses the current status of the field and the advantages and disadvantages of the techniques available for these studies.

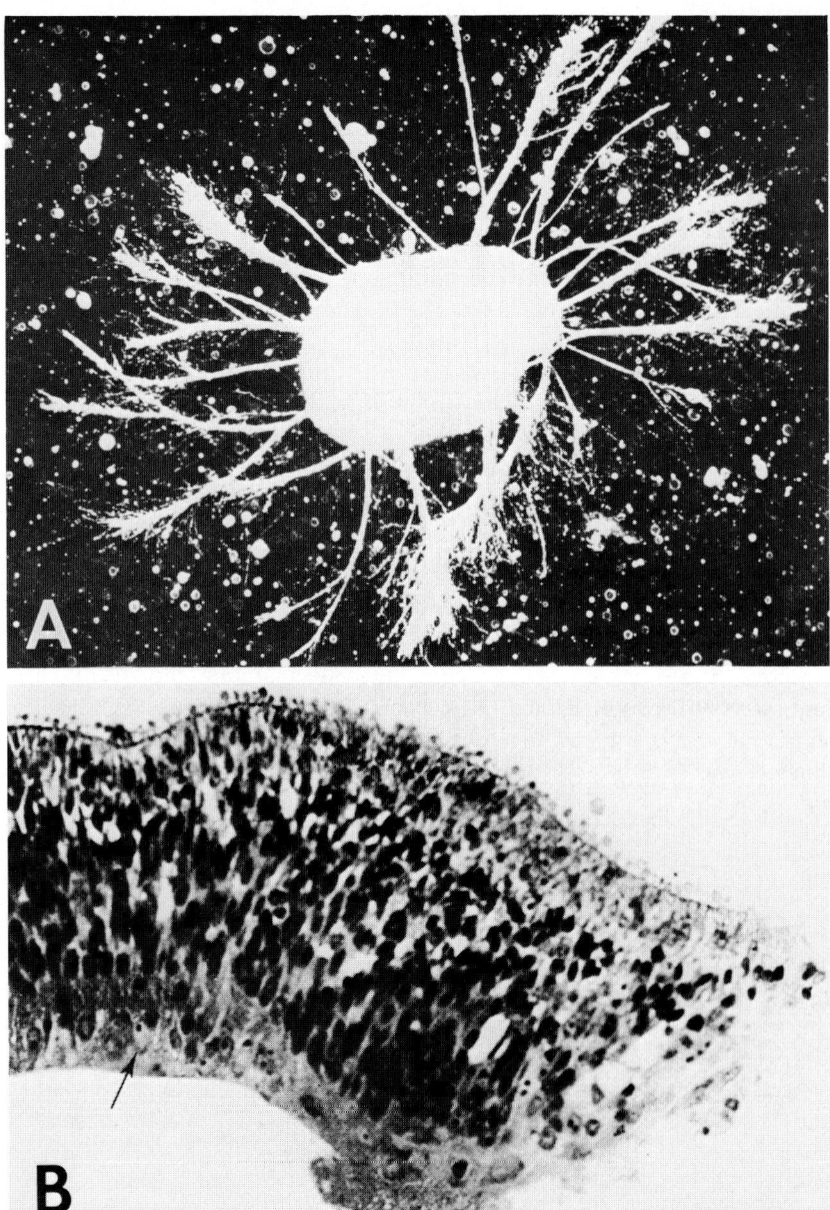

FIG. 4. (A) Retinal explant stimulated with optic lobe extract for 4 days. The size and density of fascicles are much greater than in the corresponding untreated controls. (B) Methylene blue-stained plastic section of a neural retina explant cultured on collagen gels. Cross section of the entire thickness of the explant cultured with optic lobe added. (Reproduced from Carri and Ebendal, 1983, with permission.)

A. *Studies Using Retinal Explants*

Explant cultures occupy a place of historical preeminence in neuronotrophic factor research, since they were extensively used as a bioassay during NGF purification. Probably because of that initial success, explant cultures are still extensively used in trophic factor research in spite of their drawbacks (see discussion in Adler, 1986a). A critical limitation of these cultures is that they do not allow direct evaluation of neuronal survival without time-consuming histological procedures (Fig. 4). In an attempt to circumvent this problem, neurite outgrowth and the activity of neurotransmitter enzymes are frequently used as indicators of neuronal viability. These, however, are *indirect* parameters which should be interpreted with caution. An increase in neurite outgrowth, for example, might be due not only to improved neuronal survival (a trophic effect), but also to specific stimulation of neurite elongation (a specifying effect), to changes in the proliferation and/or properties of glial cells (secondary effects), etc. (Adler and Varon, 1981b; Letourneau, 1982). In spite of these limitations, however, explant cultures have been widely used to investigate trophic influences active on retinal cells from different species.

Goldfish. Goldfish occupy a site of preeminence in studies of optic nerve regeneration (see article by Bernice Grafstein in Part II). Not surprisingly, therefore, many *in vitro* studies have been devoted to the search for factors capable of stimulating survival of and neurite outgrowth by goldfish retinal neurons. Although monolayer culture procedures for goldfish retina are available (Schwartz and Agranoff, 1981), most studies of trophic influences have been carried out using explants. Retinal tissue is usually obtained a few days after a "conditioning" optic nerve crush, which markedly increases the production of nerve fibers by retinal cells. A variety of agents have been found which stimulate neuritic outgrowth from these explants. Neurite-stimulating treatments have included (1) retinal coculture with tectum and other brain regions (Mizrachi and Schwartz, 1982), (2) brain extracts (Schwartz *et al.*, 1982a; Johnson and Turner, 1982), (3) conditioned medium from neuronal and glial cell lines (Schwartz *et al.*, 1982b), and (4) fetal calf serum (Schwartz *et al.*, 1982a; Johnson and Turner, 1982). The molecules responsible for these activities are yet to be identified, although it is known that the activity in brain extracts is heat labile and nondialyzable (Schwartz *et al.*, 1982a). The finding by Benowitz and Green (1979) that the goldfish brain is rich in NGF led Turner and co-workers to investigate the effects of this factor and its antiserum on axonal growth from retina neurons *in vivo* and *in vitro*. These studies demonstrated that NGF stimulates and anti-NGF prevents neurite outgrowth from retinal explants (Turner *et al.*, 1981, 1982). The neurite-promoting activity detected in brain extracts by Schwartz *et al.* (1982a), however, seems to be different from NGF.

Rodents. Rat and mouse retinal explants have also been used to investigate neurite-promoting activities for retinal neurons. Turner *et al.* (1983) found that rat fetal retinal explants, unlike those from goldfish, do not respond to NGF. However, neurite outgrowth could be stimulated in a dose-dependent fashion by extracts from pig brain. The activity responsible for this stimulation has been partially purified. Fetal mouse retinal explants have been used in Crain's laboratory to study development and specific connectivity of retinal neurons (Smalheiser *et al.*, 1981a,b). These workers observed neurite "stabilization" in retinas cocultured with the appropriate target tissue (optic tectum). This finding was considered indicative of a trophic effect of tectal cells upon retinal neurites, but the molecular bases of this interaction have not been investigated.

Chick. Carri and Ebendal (1983) used 6-day chick embryo neural retinal explants on collagen gels as a bioassay to study regulation of neurite outgrowth (Fig. 4). Optic lobe or forebrain extracts caused a sixfold increase in neurite length. Histological analysis suggested that these neurites originated in retinal ganglion cells. It is not clear, however, whether increased neurite production was due to improved ganglion cell survival or to direct stimulation of nerve fiber outgrowth. The neurite-promoting activity was retained by pressure dialysis membranes with a nominal cutoff of 100,000 Da. NGF was not active on this preparation.

Kato *et al.* (1983) cultured retinal explants from 8-day-old chick embryos on polyornithine substrata. An extract from chick gizzard was found to stimulate neurite outgrowth as well as neurotransmitter-related activities. Neurite stimulation could also be obtained after pretreating the polyornithine substratum with the extract, suggesting that the activity was similar to "PNPF" (Section III,C).

B. Reaggregation Cultures

Reaggregation cultures are particularly well suited for the investigation of contact-mediated cell–cell interactions, since they allow the generation of chimeric tissues containing controlled proportions of different cell types (Moscona, 1965; Steinberg, 1963; Seeds, 1983). This technique has been useful to analyze retinotectal interactions controlling development of the cholinergic system in these visual organs. It was known from *in vivo* experiments that the cholinergic enzyme choline acetyltransferase (CAT) fails to develop normally in the optic tectum deprived of retinal afferents (Marchisio, 1969). To investigate this interaction *in vitro,* CAT activity in aggregates of neural retina or optic lobe cells was compared vis-à-vis the activity in chimeric aggregates containing a mixture of both cell types. The experiments showed that retinotectal interactions taking place in the chimeric aggregates led to a significant increase in CAT activity (Adler and Teitelman, 1974). CAT stimulation in the combined aggregates was specific for this enzyme, was tissue and developmental stage dependent, and required defined proportions of each cell type (Adler *et al.*, 1976). These studies, which were subsequently reproduced by Ramirez and Seeds (1977), emphasized the importance of specific cell–cell interactions in the regulation of neuronal development.

A different application of neural retina aggregates was developed by Akers *et al.* (1981). In order to study responses of retina cells to substratum-bound materials, these authors used aggregates as "explants" for culturing on different substrata. Aggregates were preferred to the usual tissue fragments because they allowed more uniform sampling of the retina. The main finding of this study was a stimulatory effect of fibronectin on neurite outgrowth by explanted aggregates. It is interesting that fibronectin failed to stimulate neurite formation by dissociated retinal neurons grown as monolayers rather than as reaggregates (Manthorpe *et al.*, 1983; Rogers *et al.*, 1983; Adler and Hewitt, 1983; Adler *et al.*, 1985). It is not unlikely that cell–cell interactions within the aggregates might play a role in allowing neurons to respond to fibronectin with increased neurite production.

C. Monolayer Cultures

The retinal monolayer cultures available until recently were extremely complex and unsuitable for trophic factor investigation. As shown in Fig. 5A, in these complex cultures neurons coexist with multicellular clumps and with a confluent monolayer of nonneuronal, "flat" cells. This close physical contact with other cells renders the neurons potentially subject to endogenous interactions which might hinder the detection of exogenous trophic influences (see Section III,B). Moreover, the analysis of neuronal behaviors at the cellular level is also impaired by high cell density and multicellular clumps. Although inappropriate as a test system to detect trophic molecules, these complex monolayers are not completely devoid of applications in trophic factor research. For example, it is possible to remove mechanically both the neurons and the multicellular clumps, thus generating a highly purified population of flat cells (Adler *et al.*, 1982). Extensive circumstantial evidence suggests that these "flat" cells are glial, although there is still some controversy about their nature (Moscona and Linser, 1983; Li and Sheffield, 1984). The potential usefulness of these populations as a source of factors active on retinal neurons is obvious (see Hyndman and Adler, 1982c; and Section II,A,3,b). A note of caution is necessary, however, since upon prolonged cultivation retinal "flat" cells undergo a process of transdifferentiation, leading to the appearance of lens-like structures and pigmented epithelial cells in the cultures (Okada, 1980).

New techniques have become available in recent years which allow the generation of glia-free, purified retinal neuronal monolayers. Simultaneously, pro-

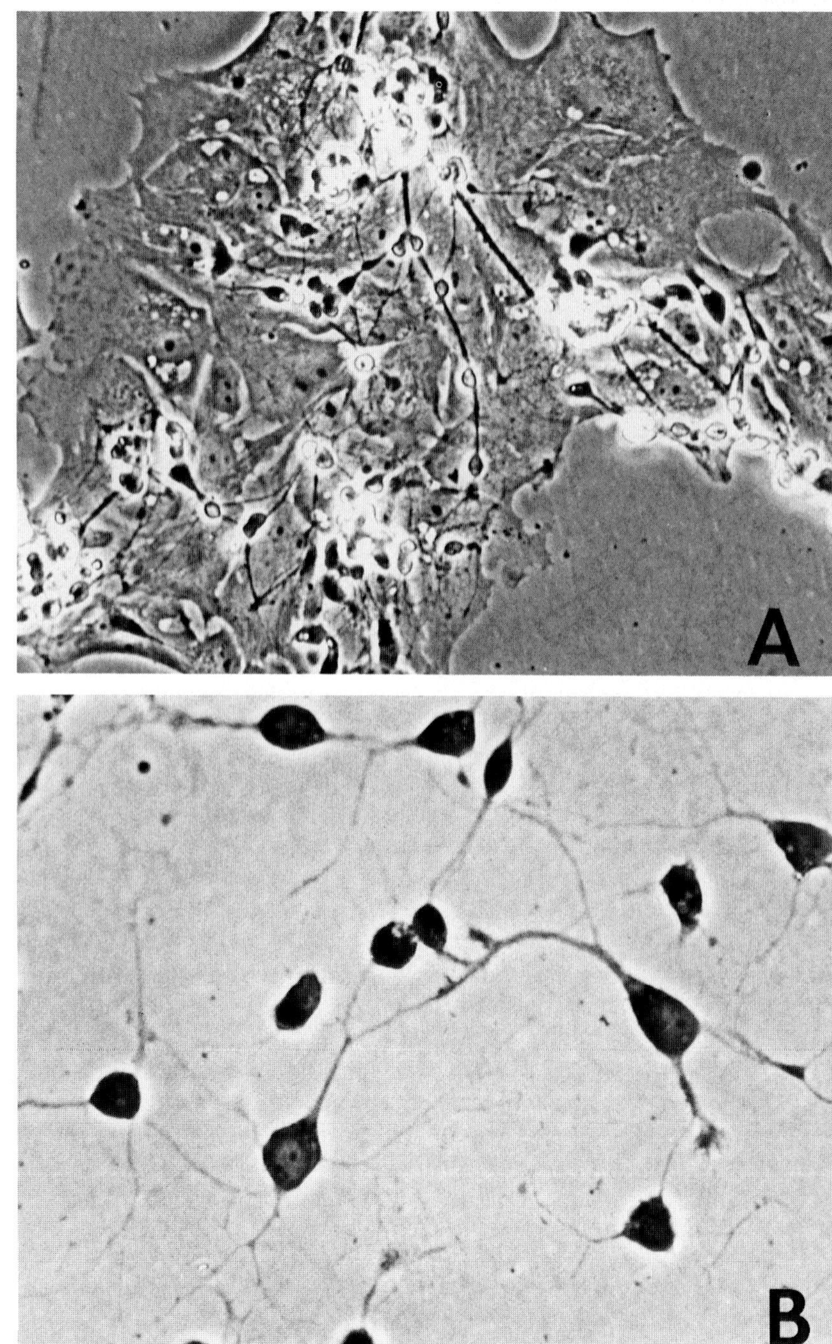
A
B

cedures have been developed which facilitate the identification of some neuronal subpopulations in these cultures. It seems appropriate to consider these technical developments in this section, since their availability is having a significant impact upon the application of monolayer culture techniques to the investigation of trophic factors for retinal neurons.

1. Cell Separation and Cell Identification Techniques for Retinal Neurons

a. Cell Separation. It is of interest that relatively small changes in culture conditions allow retinal cell suspensions to develop into a glia-free, purified neuronal monolayer (Fig. 5B) rather than into a complex clump- and glia-containing culture (Fig. 5A). The adhesiveness of the substratum to which cultured cells attach is one of the crucial factors responsible for these different behaviors (Adler *et al.*, 1979a, 1982). "Clumped" cultures are generated when a poorly adhesive substratum permits cell reaggregation to occur before cell–substratum attachment takes place. In the presence of fetal calf serum (which is also required), retinal cell clumps give rise to an expanding population of "flat" glial cells, characteristic of the complex cultures described above. That cell–cell interactions taking place within the clumps are essential for the generation of glial cells is indicated by the fact that highly adhesive substrata, which prevent cell clumping, do not permit glial cell development even if fetal calf serum is present in the medium. The resulting sparse populations contain neurons without any detectable glial contamination (Fig. 5B). These purified neurons can also be cultured in serum-free medium containing the N1 supplement described by Bottenstein and Sato (1979). Similar results have recently been described by Betz and Muller (1982).

Other recent studies have attempted to separate ganglion cells from other cell types present in retinal cell suspensions using two different approaches. Sarthy and colleagues (1983) achieved 50- to 100-fold enrichment of retinal ganglion cells by centrifugation in a 5–10% metrizamide gradient. Ganglion cells, which had been prelabeled by retrograde transport of the fluorescent dye True Blue (Fig. 6), could be maintained *in vitro* for up to 48 hr. Armson and Bennett (1983) also reported purification of True Blue-labeled rat ganglion cells. Their approach, however, was based on the use of fluorescence-activated cell sorting

FIG. 5. Chick embryo neural retina cells in monolayer culture. (A) Complex cultures containing a confluent monolayer of "flat cells" (probably glial in nature), as well as multicellular clumps and process-bearing neurons. These cultures are generated by seeding chick embryo retinal cell suspensions at high density on substrata of low adhesiveness and in the presence of fetal calf serum. (B) Glia-free, purified retinal neuronal monolayers are generated by seeding the same cell suspensions at lower density on highly adhesive substrata. These cultures can be supported by either serum-containing or serum-free media.

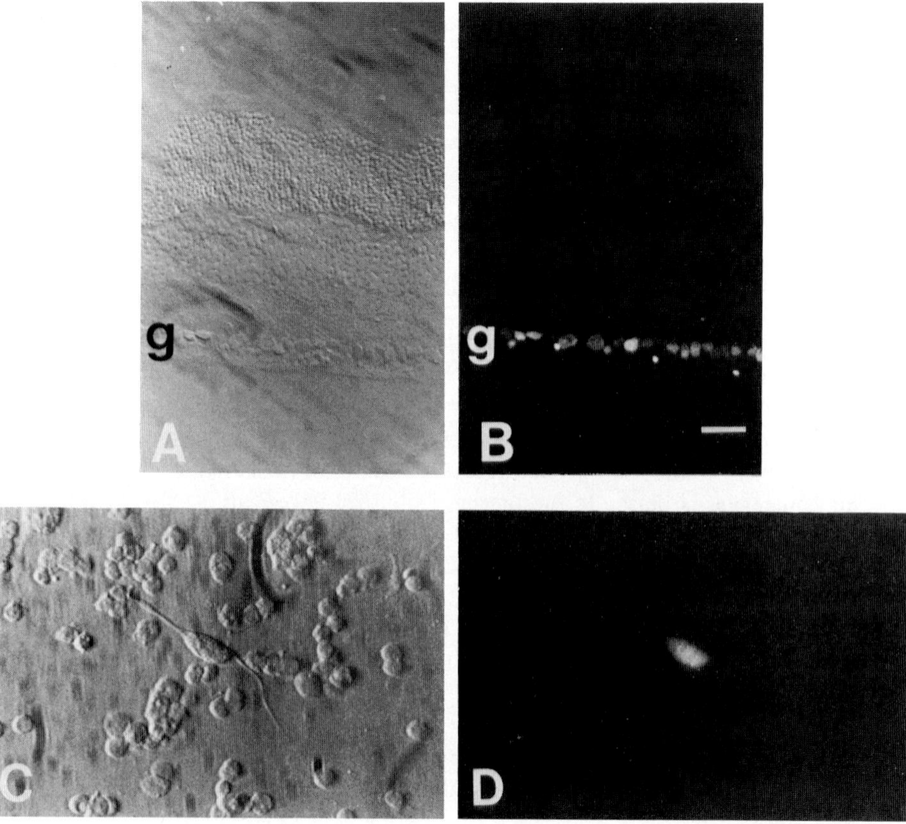

FIG. 6. Localization of True Blue (TB)-labeled cells in the retina and in culture. Newborn rat pups were injected with 1 μl of 10% TB solution at the optic chiasm. After 2 weeks of survival, retinas were dissected out, fixed in 4% paraformaldehyde, and processed for fluorescence microscopy. Alternatively, labeled retinas were dissociated by trypsinization and plated on polyornithine dishes coated with optic tectum conditioned medium. (A,B) Transverse sections of the retina were viewed with Hoffman modulation optics (A) and epifluorescence (B). The fluorescent cell bodies are localized to the ganglion cell layer. (C,D) Cultured cells seen with Hoffman optics (C) or epifluorescence (D). Note a labeled retinal ganglion cell, g. (Reproduced from Sarthy *et al.*, 1983, with permission.)

machines which segregate cells on the basis of size and fluorescent intensity. These authors reported a 150-fold increase in retina ganglion cell concentration; survival of these cells was apparently enhanced by coculturing with target tissues (see Section IV,C,2).

b. Cell Identification. Molecular properties characteristic of different retina neurons might assist in their identification *in vitro*. Some of the properties which have been studied *in vivo* and are suitable as *in vitro* ''markers'' are (1) autoradiographic determination of high-affinity uptake mechanisms for neurotransmitters (e.g., Hollyfield *et al.*, 1979), (2) immunocytochemical localization of neurotransmitter enzymes and neuroactive peptides (i.e., Brecha *et al.*, 1979), (3) lectin cytochemistry (Blanks and Johnson, 1983, 1984), and (4) use of cell-specific monoclonal antibodies (Barnstable, 1980). As recently discussed, it might be very misleading to rely on only one of these ''markers'' to establish the identity of a cultured cell (Adler, 1986a). Some molecular properties, for example, can be expressed *in vitro* by cell types different from those which normally display those features *in vivo*. Moreover, even in the *in vivo* situation, molecules once considered to be specific ''markers'' for one cell type were later found to be present in other, apparently unrelated cell classes (Adler, 1986a). A relevant example is the glial fibrillary acidic protein (GFAP), which was initially thought to be restricted to astrocytes but has recently been found also in peripheral glia (Barber and Lindsay, 1982) as well as in lens cells (Hatfield *et al.*, 1984). Cell identification, therefore, should be based on the accumulation of data regarding a variety of cell properties, rather than on the mere detection of an isolated marker.

The identification of neuronal from glial cells in retinal cultures has been facilitated by immunocytochemical detection of tetanus toxin receptors in neurons (Adler *et al.*, 1982), using the procedure described by Mirsky *et al.* (1978). Autoradiographic and immunocytochemical studies of neurotransmitter-related properties have shown the presence of different neuronal subpopulations in the cultures, without allowing a definitive correlation between those cultured neurons and their *in vivo* counterpart. For example, a subpopulation of neurons which become heavily labeled with radioactive GABA through a high-affinity uptake mechanism can be identified autoradiographically in retinal cultures (Hyndman and Adler, 1982a). It is not clear, however, whether these cells are amacrine, horizontal, and/or ganglion neurons, since all these cells have been reported to have high-affinity GABA uptake mechanisms in the chick retina *in vivo* (Baughman *et al.*, 1977). Moreover, a recent study has shown that as many as 50% of the neurons present in retinal cultures have high-affinity uptake mechanisms for three different putative neurotransmitter molecules (Pessin and Adler, 1985), thus emphasizing the unreliability of this parameter for neuronal identification *in vitro*.

Positive identification of cultured retinal neurons has been achieved so far only for ganglion cells and photoreceptors. Retina ganglion cells have been identified in cultures from both the chick (Nurcombe and Bennett, 1981) and rat (Sarthy *et al.*, 1983; McCaffery *et al.*, 1982). In those studies, brain regions innervated by ganglion cells were injected with fluorescent dyes or with horseradish peroxidase. Ganglion cell axons take up these molecules, and the cells thus labeled can then be identified *in vivo* as well as in monolayer cultures (Fig. 6). The intracellular markers are apparently devoid of deleterious effects, although this issue probably deserves closer scrutiny. The identification of photoreceptor cells in chick embryo retinal cultures on the basis of cytochemical and electron microscopical criteria is described below in some detail (section IV,C,3; see Figs. 7 and 8).

2. Neuronal Responses to Putative Regulatory Factors

a. Optic Lobe Extracts. As mentioned above, the optic lobe contains the postsynaptic elements with which retina ganglion cells make synapsis. Using glial-free retinal neuronal monolayers, Hyndman and Adler (1982b) reported the appearance in optic lobe-treated cultures of a neuronal type showing a characteristically long neurite. This response to optic lobe extracts was not accompanied by changes in the total number of neurons or the overall neuritic development of the cultures. The optic-lobe dependent neurons (which could perhaps be retinal ganglion cells) represented no more than 1% of the cultured neurons. This low frequency impaired further analysis of the chemical properties of the optic lobe activity (see also Adler, 1986a). Nurcombe and Bennett (1981) reported that coculturing retinal suspensions with fragments from chick optic lobe or rat superior colliculus stimulated ganglion cell survival. The cells were identified by retrograde transport of horseradish peroxidase, as described in the preceding section. Labeled retinal ganglion cells were reported to persist for 16 hr in the presence but not in the absence of optic lobe fragments in the cultures. Morphological analysis was complicated by the presence of tissue fragments in the cocultures, which showed practically no neuritic development.

b. Substratum-Bound Materials. Adler (1982a) reported that chick embryo retinal neurons are responsive to the neurite-promoting factor PNPF (see Section III,C). After 6 hr of culture, only 7% of the retinal neurons seeded on a fetal calf serum-coated polyornithine substratum showed neurite development. This limited neuritic development was in marked contrast with the behavior observed when the cells were seeded on conditioned medium-treated substrata, where as many as 45% of thc cclls showed neurites after 6 hr of culture. Increasing numbers of cells grown on serum-treated surfaces showed some degree of neurite development upon longer cultivation, but the differences with cultures on PNPF substrata remained evident. The stimulatory effect of PNPF on neurite development could be blocked by retreatment of the PNPF-coated surfaces with the lectin concanavalin A, suggesting the involvement of sugar moieties in the biological activity of the agent. Recent studies have shown that substratum-bound extracellular matrix molecules have rather specific effects on retinal neurons. These studies are discussed in detail in the article by A. T. Hewitt in Part II.

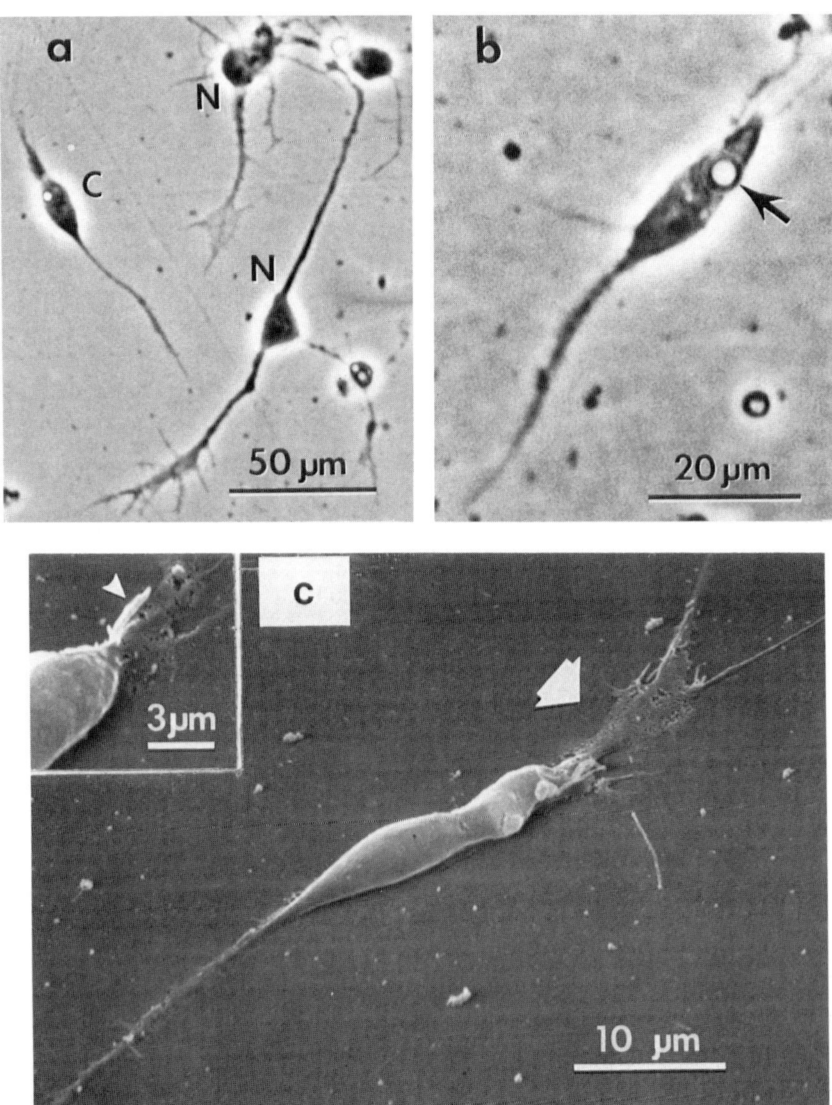

FIG. 7. Retinal cells expressing characteristic photoreceptor phenotypic properties in cell culture. (a) Phase-contrast image of a cone photoreceptor, C, and two multipolar retinal neurons, N. (b) Enlarged phase-contrast microscopy image of a cone-like cell illustrating the characteristic lipid droplet (arrow). (c) Scanning electron micrograph of a cone cell in culture. Note its apical membranous expansion and a small, outer segment-like process (inset). (Reproduced from Adler *et al.*, 1984.)

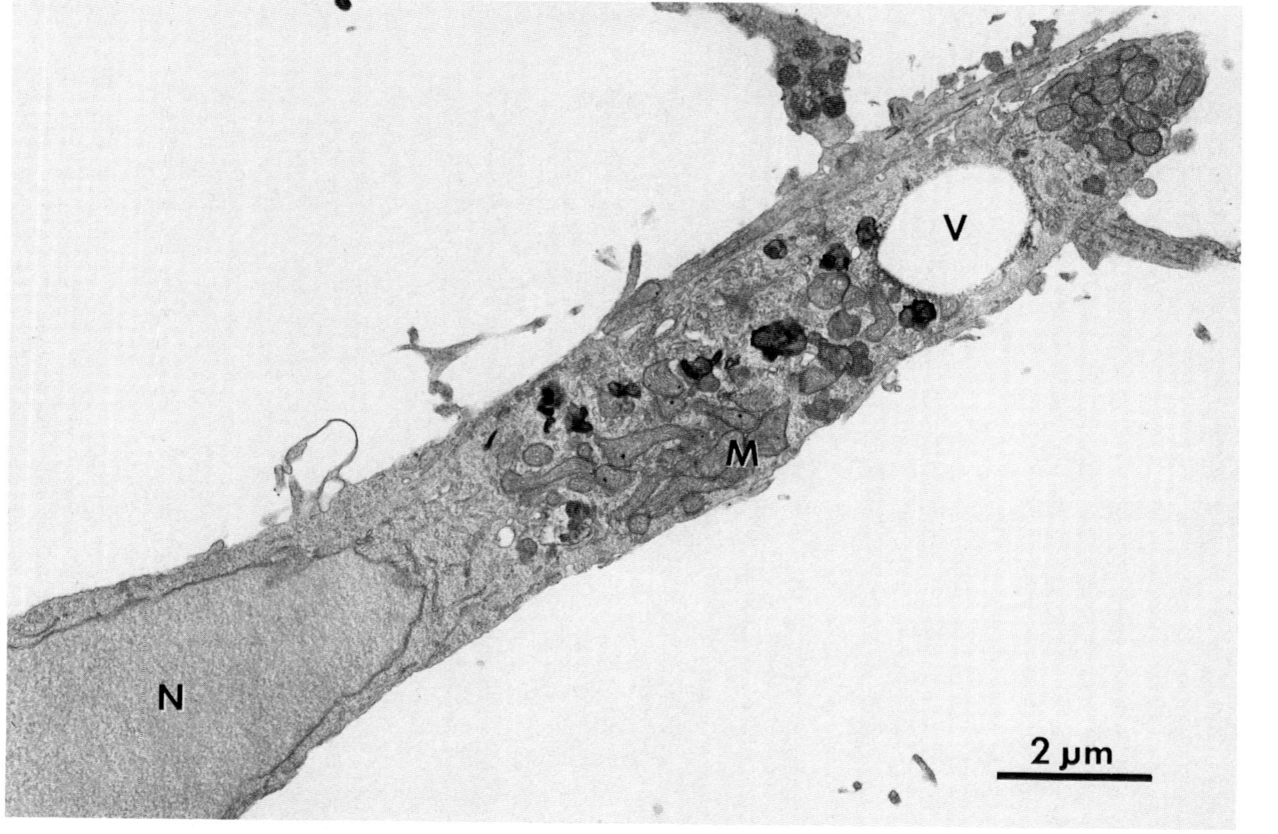

FIG. 8. Transmission electron micrograph of a cultured cone cell illustrating the polarized position of the nucleus, N, and mitochondria, M, as well as a large vacuole, V, corresponding to the lipid droplet. Other typical constituents of the photoreceptor inner segment, such as the Golgi apparatus and the paraboloid, could be seen in other sections. (Reproduced from Adler *et al.*, 1984.)

3. A Monolayer Culture System for the Study of Photoreceptor Survival and Differentiation

The inner layer of the optic vesicle, from which the neural retina is derived, consists of a homogeneous population of neuroepithelial "matrix" cells which are closely apposed to the future pigment epithelium. As they become postmitotic, a majority of these cells migrate toward the vitreal surface of the retina, giving rise to its ganglion and inner nuclear layers. Some retinal neuroepithelial cells, however, retain a close proximity with the pigment epithelium throughout their lives; these are the cells which differentiate as photoreceptors. This physical association suggests that the pigment epithelium may be the source of signals inducing and regulating the differentiation of photoreceptor cells. In fact, a study by Hollyfield and Witkovski (1974) demonstrated that the development of photoreceptor outer segments only occurs in the presence of the pigment epithelium.

In spite of this and other examples of photoreceptor-pigment epithelium interdependence (see articles by Clark, Part II and Besharse, Part I) specific molecules involved in these interactions have yet to be identified. One of the factors impairing their investigation has been the unavailability of monolayer cultures allowing photoreceptor development. Our laboratory, however, has recently reported a chick embryo neural retina monolayer culture system in which photoreceptors survive and express differentiated properties (Adler *et al.*, 1984). The cultures are started at embryonic day 8, when photoreceptors are already postmitotic but have not yet expressed any overt differentiation. After 6 days *in vitro,* monopolar cells displaying phenotypic properties characteristic of chick cones are the most abundant cell type in these cultures (Figs. 7 and 8). Those features include (1) a highly polarized structure; (2) a single, short, usually unbranched neurite; (3) a polarized nucleus located near the origin of the neurite; (4) inner segment elements including abundant free ribosomes, a polarized Golgi apparatus, a cluster of polarized mitochondria, a big, membrane-bound, pigment-containing vacuole (lipid droplet) and, at least in some cases, a well-developed paraboloid; (5) the presence of a complex of apical differentiations including abundant microvilli and a cilium; and (6) the staining of the apical region of the cell, including the inner segment, with peanut lectin. This staining reproduces the selective distribution of peanut lectin receptors characteristic of the embryonic chick retina *in vivo* (Blanks and Johnson, 1983, 1984). A small, finger-like process present in the apical region of the cultured cones (Fig. 7) appears to represent the first indication of outer segment development. This identity is suggested by a recent immunocytochemical study showing the accumulation of opsin-like immunoreactive materials in this process (Adler, 1985). Thus, these monolayer cultures appear as an attractive experimental system for the investigation of molecular mechanisms controlling outer segment differentiation (for a review see Adler, 1986b). Particular attention is being paid in our laboratory to the possible role of pigment epithelium-derived molecules, including the interphotoreceptor matrix.

V. Concluding Remarks

A. Studying Trophic Interactions in the Retina: The State of the Art

It seems clear that the investigation of the role of trophic factors in retinal development and pathology is going through a transition phase. The conceptual foundation on which these investigations must be based has already been laid down. We are clearly aware of the questions which should be investigated, and, fortunately, the necessary techniques to accomplish this task have undergone an almost revolutionary improvement in the last few years. Little is known about the cellular and molecular bases of trophic interactions in the retina, but much can be learned in this area using the methodology currently available. Retinal cells can now be cultured in chemically defined media, on chemically defined substrata. The cultured cells can be studied with immunocytochemical, autoradiographic, and ultrastructural techniques. They can be identified on the basis of objective criteria which reduce the impact of subjective (and potentially arbitrary) decisions by the observer. More and more cell separation techniques are becoming available, and recently developed cell sorting machines seem to offer great promise to make cultures of pure neuronal subpopulations a reality. And, last but not least, biochemical techniques are now available which allow the purification and characterization of even small amounts of bioactive molecules.

B. Trophic Interactions in Retinal Degenerations: A Working Hypothesis

It may be appropriate to close this article by discussing the impact that modern concepts and research strategies in the trophic factor field are likely to have on our understanding of retinal degenerations. As is the case for *developmental*

FIG. 9. This figure illustrates some hypothetical mechanisms capable of producing neuronal degenerations. Survival of the neuron under consideration requires is dependent upon the availability of a "trophic" factor, TF. The basic premise of this model is that the binding of TF molecules to receptors located at the neuronal surface, C, triggers a series of metabolic reactions, D, which are crucial for neuronal survival. TF is normally supplied to the neuron by a "factor-producing cell" which has the necessary machineries for synthesis (A) and delivery (B) of these molecules. Neuronal degenerations can be caused by primary defects in the neuron itself or in the factor-producing cell.

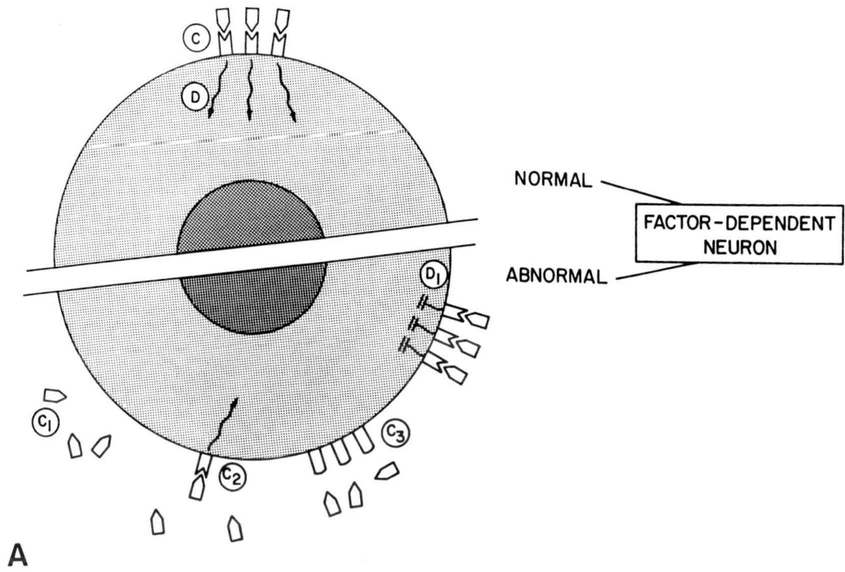

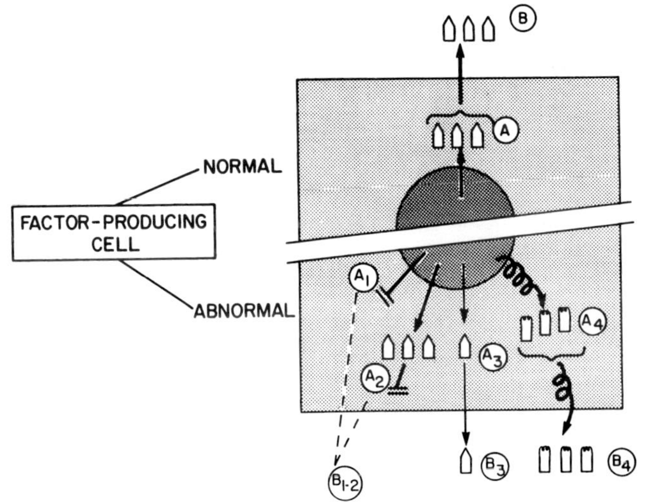

Neuronal defects include absence, C_1, scarcity, C_2, or abnormal properties, C_3, of cell surface receptors, as well as deficiencies in postreceptor mechanisms, D_1. In turn, the factor-producing cell can be responsible for the unavailability of TF molecules, B_{1-2}, through defective synthesis, A_1, or release, A_2. Alternatively, TF molecules can be produced in insufficient amounts, A_3 and B_3, or can show qualitative changes that make them inactive, A_4 and B_4. The relevance of this model for our understanding of retinal degenerations is discussed in the text.

neuronal death, these degenerative diseases are examples of *selective* neuronal death: defined numbers of neurons undergoing degeneration in a particular location of the brain at a defined time in the life of an organism. A variety of possible pathogenetic mechanisms have been proposed for neuronal degenerations in the retina and other regions of the nervous system. A hereditary component is present in most of these diseases, suggesting the importance of genetic factors. Autoimmune mechanisms must also be taken into consideration either as primary agents responsible for the onset of the disease, or as secondary responses involved in the spreading of the degenerative phenomenon. Other possible causal factors include infection with slow viruses and the action of neuronotoxic molecules. As summarized in Fig. 9, these etiological factors could lead to neuronal degeneration through alterations in a trophic mechanism by (1) increasing the demand for a trophic factor beyond normal levels and beyond the levels available in the neuronal microenvironment, (2) blocking the capacity of the neurons to respond to a trophic factor, (3) creating a dependency for trophic factors different from those required under normal circumstances, (4) affecting those cells which normally act as sources of trophic factors (postsynaptic target cells, glial cells, etc.), and (5) reducing the availability or level of activity of a trophic factor by direct modification of the trophic molecules.

Some observations in animal mutants affected by retinal degenerations are relevant in this context. For example, the onset of photoreceptor degeneration in the *rd* mouse coincides with a defect in synaptogenesis between photoreceptors and one of the components of the postsynaptic triad (Blanks *et al.*, 1974). Although it is not clear whether this defect in synaptogenesis is the cause or the consequence of photoreceptor degeneration, it is worth investigating whether postsynaptic elements associated with photoreceptors play a survival-supporting role similar to that shown by "target" cells in other parts of the nervous system. In other animal mutants, photoreceptor degenerations are accompanied by alterations in "satellite" cells. Müller cells, for example, show significant alterations at the time of photoreceptor degeneration in the *pcd* mouse (Blanks *et al.*, 1982). In the RCS rat, moreover, the primary defect leading to photoreceptor degeneration is found in the retinal pigment epithelium, which fails to phagocytize those portions of the photoreceptor outer segments which are "shed" on a daily basis (LaVail, 1981). This last example, although apparently not related to a "trophic" mechanism, illustrates how an abnormal gene leading to photoreceptor death may act primarily in cells other than the photoreceptors themselves.

C. A Note of Caution

This article has presented a mixture, hopefully in well-balanced proportions, of solid experimental data, extrapolations, working hypotheses, and speculations. It is hoped that the limits between these complementary but nonetheless different approaches to reality have been clearly delineated in the text. I would like to reiterate once again the hypothetical nature of many of the mechanisms proposed in this review. They have been presented here with the expectation that they will be challenged, but also with the belief that they deserve to be investigated. There seems to be enough solid evidence to support the concept that trophic interactions play at least some role in retinal development and perhaps also in retinal pathology. The task of disclosing the details of that involvement must be considered a challenge that the scientific community interested in the retina cannot postpone undertaking.

Acknowledgments

Studies from the author's laboratory included in this article were supported by Grant EY04859 and Core Grant EY07047. R. A. is a William and Mary Greve Scholar from Research to Prevent Blindness. The author is indebted to Drs. N. Carri, T. Cunningham, B. Finlay, and V. Sarthy for allowing reproduction of the figures and tables. Thanks are also due to Dr. J. Lindsey for critical reading of the manuscript and to Mrs. Doris Golembieski for typing it.

References

Adler, R. (1982a). Regulation of neurite growth in purified retina cultures. Effects of PNPF, a substratum-bound neurite-promoting factor. *J. Neurosci. Res.* **8,** 165–177.

Adler, R. (1982b). Tropic and neurite-promoting factors in eye development. *In* "The Structure of the Eye" (J. G. Hollyfield, ed.), pp. 215–228. Elsevier, Amsterdam.

Adler, R. (1985). Photoreceptor development and polarization in low density retinal cell cultures. *J. Cell Biol.* **101,** 222 (a).

Adler, R. (1986a). *In vitro* techniques for the investigation of trophic factors active on CNS neurons. *In* "Handbook of Nervous System and Muscle Factors" (J. R. Perez-Polo, ed.). C. R. C. Press, Boca Raton, Florida, in press.

Adler, R. (1986b). The differentiation of retinal neurons and photoreceptors *in vitro*. *Prog. Ret. Res.*, in preparation.

Adler, R., and Hewitt, A. T. (1983). Responses of cultured neural retinal cells to substratum-bound extracellular matrix molecules. *J. Cell Biol.* **97,** 97a.

Adler, R., Jerdan, J., and Hewitt, A. T. (1984). Responses of cultured neural retinal cells to aminin and other substratum-bound extracellular matrix molecules. *Dev. Biol.* **112,** 100–114.

Adler, R., and Teitelman, G. (1974). Aggregates formed by mixtures of embryonic neural cells: Activity of enzymes of the cholinergic system. *Dev. Biol.* **39,** 317–321.

Adler, R., and Varon, S. (1980). Cholinergic neuronotrophic factors. V. Segregation of survival and neurite-promoting activities in heart-conditioned media. *Brain Res.* **188,** 437–448.

Adler, R., and Varon, S. (1981a). Neuritic guidance by polyornithine-attached materials of ganglionic origin. *Dev. Biol.* **81,** 1–11.

Adler, R., and Varon, S. (1981b). Neuritic guidance by nonneuronal cells of ganglionic origin. *Dev. Biol.* **86,** 69–80.

Adler, R., and Varon, S. (1982). Neuronal survival in intact ciliary ganglia *in vivo* and *in vitro:* CNTF as a target surrogate. *Dev. Biol.* **92,** 470–475.

Adler, R., Teitelman, G., and Suburo, A. M. (1976). Cell interactions and the regulation of cholinergic enzymes during neural differentiation "*in vitro.*" *Dev. Biol.* **50,** 48–58.

Adler, R., Manthorpe, M., and Varon, S. (1979a). Separation of neuronal and nonneuronal cells in monolayer cultures from chick embryo optic lobe. *Dev. Biol.* **69,** 424–435.

Adler, R., Landa, K. B., Manthorpe, M., and Varon, S. (1979b). Cholinergic neuronotrophic factors. II. Selective intraocular distribution of soluble trophic activity for ciliary ganglionic neurons. *Science* **204,** 1434–1436.

Adler, R., Manthorpe, M., Skaper, S. D., and Varon, S. (1981). Polyornithine-attached neurite-promoting factors (PNPFs): Culture sources and responsive neurons. *Brain Res.* **206,** 129–144.

Adler, R., Magistretti, P. J., Hyndman, A. G., and Shoemaker, W. J. (1982). Purification and cytochemical identification of neuronal and nonneuronal cells in chick embryo retina cultures. *Dev. Neurosci.* **5,** 27–39.

Adler, R., Manthorpe, M., and Varon, S. (1983). Lectin reactivity of PNPF, a polyornithine-binding neurite-promoting factor. *Dev. Brain Res.* **6,** 69–75.

Adler, R., Lindsey, J. D., and Elsner, C. L. (1984). Expression of cone-like properties by chick embryo neural retina cells in glial-free monolayer cultures. *J. Cell Biol.* **99,** 1173–1178.

Aguayo, A. J., Benfey, M., and David, S. (1983). A potential for axonal regeneration in neurons of the adult mammalian nervous system. *In* "Nervous System Regeneration" (B. Haber, J. R. Perez-Polo, G. A. Hashim, and A. M. Stella, eds.), pp. 327–340. Liss, New York.

Akers, R. M., Mosher, D. F., and Lilien, J. E. (1981). Promotion of retinal neurite outgrowth by substratum-bound fibronectin. *Dev. Biol.* **86,** 179–188.

Appel, S. (1981). A unifying hypothesis for the course of amyotrophic lateral sclerosis, Parkinsonism and Alzheimer's disease. *Ann. Neurol.* **10,** 499–505.

Armson, P. F., and Bennett, M. R. (1983). Neonatal retinal ganglion cell cultures of high purity: Effect of superior colliculus on their survival. *Neurosci. Lett.* **38,** 181–186.

Banker, G. A. (1980). Trophic interactions between astroglial cells and hippocampal neurons in cultures. *Science* **109,** 809–810.

Barbeau, A. (1980). Biochemical aging in Parkinson's disease. *In* "Aging of the Brain and Dementia (Aging Vol. 13)" (L. Amaducci, A. N. Davison, and P. Antuono, eds.), pp. 275–285. Raven, New York.

Barber, P. C., and Lindsay, R. M. (1982). Schwann cells of the olfactory nerves contain glial fibrillary acidic protein and resemble astrocytes. *Neuroscience* **7,** 3077.

Barbin, G., Manthorpe, M., and Varon, S. (1984). Purification of the chick eye ciliary neuronotrophic factor. *J. Neurochem.* **43,** 1468–1478.

Barde, Y.-A., Lindsay, R. M., Monard, D., and Thoenen, H. (1978). New factor released by cultured glioma cells supporting survival and growth of sensory neurones. *Nature (London)* **274,** 818.

Barnstable, L. J. (1980). Monoclonal antibodies which recognize different cell types in the rat retina. *Nature (London)* **286,** 231.

Barr-Nea, L., and Barishak, R. Y. (1970). Tissue culture studies of the embryonal chicken retina. *Invest. Opthalmol.* **9,** 447–457.

Baughman, R. W., Schwarz, T. L., and Baker, C. R. (1977). Uptake systems for γ-aminobutyric acid in the chicken retina. *Neurosci. Abstr.* **3,** 553.

Benowitz, L. I., and Green, L. A. (1979). Nerve growth factor in the goldfish brain: Biological assay studies using pheochromocytoma cells. *Brain Res.* **162,** 164–168.

Berg, D. K. (1982). Cell death in neuronal development. *In* "Neuronal Development" (N. C. Spitzer, ed.), pp. 297–331. Plenum, New York.

Betz, H., and Muller, U. (1982). Culture of chick embryo neural retina in serum-free medium. *Exp. Cell Res.* **138,** 297–302.

Black, I. B. (1978). Regulation of autonomic development. *Annu. Rev. Neurosci.* **1,** 183–214.

Black, I. B., and Patterson, P. H. (1980). Developmental regulation of neurotransmitter phenotype. *Curr. Top. Dev. Biol.* **15,** 27–39.

Blanks, J. C., and Johnson, L. V. (1983). Selective lectin binding of the developing mouse retina. *J. Comp. Neurol.* **221,** 31.

Blanks, J. C., and Johnson, L. V. (1984). Specific binding of peanut lectin to a class of retinal photoreceptor cells. *Invest. Ophthalmol. Visual Sci.* **25,** 546–557.

Blanks, J. C., Adinolfi, A. M., and Lolley, R. N. (1974). Photoreceptor degeneration and synaptogenesis in retinal degeneration (*rd*) mice. *J. Comp. Neurol.* **156,** 95–106.

Blanks, J. C., Mullen, R. J., and LaVail, M. M. (1982). Retinal degeneration in the *pcd* cerebellar mutant mouse. II. Electron microscope analysis. *J. Comp. Neurol.* **212,** 231–246.

Bonyhady, R. E., Hendry, I. A., Hill, C. E., and McLennan, I. S. (1980). Characterization of a cardiac muscle factor required for the survival of cultured parasympathetic neurons. *Neurosci. Lett.* **18,** 197–201.

Bottenstein, J. E., and Sato, G. (1979). Growth of a rat neuroblastoma cell line in serum-free supplemented medium. *Proc. Natl. Acad. Sci. U.S.A.* **76,** 514–517.

Brecha, N., Karten, H. J., and Laverack, C. (1979). Enkephalin-containing amacrine cells in the avian retina: Immunohistochemical localization. *Proc. Natl. Acad. Sci. U.S.A.* **76,** 3010–3014.

Burnham, P., Raiborn, C., and Varon, S. (1972). Replacement of nerve-growth factor by ganglionic non-neuronal cells for the survival *in vitro* of dissociated ganglionic neurons. *Proc. Natl. Acad. Sci. U.S.A.* **69,** 3556–3560.

Carri, N. G., and Ebendal, T. (1983). Organotypic cultures of neural retina: Neurite outgrowth stimulated by brain extracts. *Dev. Brain Res.* **6,** 219–229.

Ciani, F., Contestabile, A., and Villani, L. (1978). Acetylcholinesterase activity in the normal and retino-deprived optic tectum of the quail. *Histochemistry* **59,** 81–95.

Collins, F. (1978). Induction of neurite outgrowth by a conditioned-medium factor bound to the culture substratum. *Proc. Natl. Acad. Sci. U.S.A.* **75,** 5210–5213.

Collins, F., and Garrett, J. E., Jr. (1980). Elongating nerve fibers are guided by a pathway of material released from embryonic nonneuronal cells. *Proc. Natl. Acad. Sci. U.S.A.* **77,** 6226–6228.

Coughlin, M. D. (1984). Growth factors regulating autonomic nerve development. *Adv. Cell. Neurobiol.* **5,** 53–112.

Coughlin, M. D., and Black, I. B. (1978). NGF-independent development of mouse embryo sympathetic neurons in cell culture. *Soc. Neurosci. Abstr.* **4,** 601.

Coughlin, M. D., Dibner, M. D., Boyer, D. M., and Black, I. B. (1978). Factors regulating development of an embryonic mouse sympathetic ganglion. *Dev. Biol.* **66,** 513–528.

Cowan, W. M. (1973). Neuronal death as a regulative mechanism in the control of cell number in the nervous system. *In* ''Development and Aging in the Nervous System'' (M. Rockstein, ed.), pp. 19–41. Academic Press, New York.

CrisantiCombes, P., Privat, A., Pessac, B., and Calothy, G. (1977). Differentiation of chick embryo neuroretina cells in monolayer cultures. An ultrastructural study. *Cell Tissue Res.* **185,** 159–173.

Cunningham, T. J. (1982). Naturally occurring neuron death and its regulation by developing neural pathways. *Int. Rev. Cytol.* **74,** 163–186.

Cunningham, T. J., Mohler, I. M., and Giordano, D. L. (1982). Naturally occurring neuron death in the ganglion cell layer of the neonatal rat: Morphology and evidence for regional correspondence with neuron death in superior colliculus. *Dev. Brain Res.* **2,** 203–215.

Dorris, F. (1938). Differentiation of the chick eye *in vitro*. *J. Exp. Zool.* **78,** 385–415.

Dreher, B., Potts, R. A., and Bennett, M. R. (1983). Evidence that the early postnatal reduction in the number of rat retinal ganglion cells is due to a wave of ganglion cell death. *Neurosci. Lett.* **36,** 255–260.

Ebendal, T., and Jacobson, C.-O. (1975). Human glial cells stimulating outgrowth of axons in cultured chick embryo ganglia. *Zoon* **3,** 169–172.

Ebendal, T., Hedlund, K.-O., and Norrgren, G. (1982). Nerve growth factors in chick tissues. *J. Neurosci. Res.* **8,** 153–164.

Filogamo, G. (1950). Consequenze della demolizione dell'abbozzo dell'occhio sullo sviluppo del lobo ottico nell'embrione di pollo. *Riv. Biol.* **42,** 73–79.

Glucksmann, A. (1940). Development and differentiation of the tadpole eye. *Br. J. Ophthalmol.* **24,** 153–178.

Glucksman, A. (1951). Cell death in normal vertebrate ontogeny. *Biol. Rev.* **26,** 59–86.

Greene, L. A., and Shooter, E. M. (1980). The nerve growth factor: Biochemistry, synthesis, and mechanism of action. *Annu. Rev. Neurosci.* **3,** 353–402.

Hamburger, V., and Oppenheim, R. W. (1982). Naturally occurring neuronal death in vertebrates. *Neurosci. Comment.* **1,** 39–55.

Harper, G. P., and Thoenen, H. (1980). Nerve growth factor: Biological significance, measurement and distribution. *J. Neurochem.* **34,** 5–16.

Hatfield, J. S., Skoff, R. P., Maisel, H., and Eng, L. (1984). Glial fibrillary acidic protein is localized in the lens epithelium. *J. Cell Biol.* **98,** 1895–1898.

Helfand, S. L., Smith, G. A., and Wesells, N. K. (1976). Survival and development in culture of dissociated parasympathetic neurons from ciliary ganglia. *Dev. Biol.* **50,** 541–547.

Hild, W., and Callas, G. (1967). The behavior of retinal tissue in vitro, light and electron microscopic observations. *Z. Zellforsch* **80,** 1–21.

Hollyfield, J. G., and Witkovsky, P. (1974). Pigmented retinal epithelium involvement in photoreceptor development and function. *J. Exp. Zool.* **189,** 357–378.

Hollyfield, J. G., Rayborn, M. E., Sarthy, P. V., and Lam, D. M. K. (1979). The emergence, localization and maturation of neurotransmitter systems during development of the retina in *Xenopus laevis. J. Comp. Neurol.* **188,** 587–598.

Hughes, W. F., and Lavelle, A., (1972). Influence of tectal ablation on the morphogenesis of retinal ganglion cells in the chick. *Anat. Rec.* **172,** 333.

Hughes, W. F., and Lavelle, A. (1975). The effects of early tectal lesions on development in the retinal ganglion cell layer of chick embryos. *J. Comp. Neurol.* **163,** 265–284.

Hughes, W. F., and McLoon, S. C. (1979). Ganglionic cell death during normal retinal development in the chick: Comparisons with cell death induced by early target field destruction. *Exp. Neurol.* **66,** 587–601.

Hume, D. A., Perry, V. H., and Gordon, S. (1983). Immunohistochemical localization of a macrophage-specific antigen in developing mouse retina: Phagocytosis of dying neurons and differentiation of microglial cells to form a regular array in the plexiform layers. *J. Cell Biol.* **97,** 253–257.

Hyndman, A. G., and Adler, R. (1982a). GABA uptake and release in purified neuronal and nonneuronal cultures from chick embryo retina. *Dev. Brain Res.* **3,** 167–180.

Hyndman, A. G., and Adler, R. (1982b). Neural retina development *in vitro.* Effects of tissue extracts on survival and neuritic development in purified neuronal cultures. *Dev. Neurosci.* **5,** 40–53.

Hyndman, A. G., and Adler, R. (1982c). Analysis of glutamate uptake and monosodium glutamate toxicity in neural retina monolayer cultures. *Dev. Brain Res.* **3,** 167–180.

Insausti, R., Blakemore, C., and Cowan, W. M. (1984). Ganglion cell death during development of ipsilateral retino-collicular projection in golden hamster. *Nature (London)* **308,** 362–365.

Itoh, Y. (1976). Enhancement of differentiation of lens and pigment cells by ascorbic acid in cultures of neural retinal cells of chick embryos. *Dev. Biol.* **54,** 157–162.

Jacobson, M. (1978). "Developmental Neurobiology." Plenum, New York.

Johnson, J. E., and Turner, J. E. (1982). Growth from regenerating goldfish retinal cultures in the absence of serum or hormonal supplements: Tissue extract effects. *J. Neurosci. Res.* **8,** 315–329.

Johnson, R. T., Katzman, R., Shooter, E., McGeer, E., Silberberg, D., and Price, D. (1979). "Report of the Panel on Inflammatory, Demyelinating, and Degenerative Diseases to the National Advisory Neurological and Communicative Disorders and Stroke Council." U.S. Department of Health, Education, and Welfare. *NIH Publ.* **79,** 1916.

Johnson, E. M., Gonn, P. D., Brandeis, L. D., and Pearson, J. (1980). Dorsal root ganglion neurons are destroyed by exposure *in utero* to maternal antibody to nerve growth factor. *Science* **210,** 916–918.

Kaplowitz, P. B., and Moscona, A. A. (1976). Lectin-mediated stimulation of DNA synthesis in cultures of embryonic neural retina cells. *Exp. Cell Res.* **100,** 177–189.

Kato, S., Negishi, K., Hayashi, Y., and Miki, N. (1983). Enhancement of neurite outgrowth and aspartate–glutamate uptake systems in retinal explants cultured with chick gizzard extract. *J. Neurochem.* **40,** 929–938.

Kelly, J. P., and Cowan, W. M. (1972). Studies on the development of the chick optic tectum. III. Effects of early eye removal. *Brain Res.* **42,** 263–288.

Kuwabara, T., and Weidman, T. A. (1974). Development of the prenatal rat retina. *Invest. Ophthalmol. Visual Sci.* **3,** 725–739.

Landa, K., Adler, R., Manthorpe, M., and Varon, S. (1980). Cholinergic neuronotrophic factors. III. Developmental increase of trophic activity for chick embryo ciliary ganglion neurons in their intraocular target tissues. *Dev. Biol.* **74,** 401–408.

Landmesser, L., and Pilar, G. (1978). Interactions between neurons and their targets during *in vivo* synaptogenesis. *Fed. Proc. Fed. Am. Soc. Exp. Biol.* **37,** 2016–2022.

LaVail, M. M. (1981). Analysis of neurological mutants with inherited retinal degeneration. *Invest. Ophthalmol. Visual Sci.* **21,** 630–657.

LaVail, M., and Hild, W. (1971). Histotypic organization of the rat retina *in vitro*. *Z. Zellforsch.* **114,** 557–579.

Letourneau, P. C. (1982). Nerve fiber growth and its regulation by extrinsic factors. *In* "Neuronal Development" (N. C. Spitzer, ed.), pp. 213–254. Plenum, New York.

Levi-Montalcini, R., and Angeletti, P. U. (1968). Nerve growth factor. *Physiol. Rev.* **48,** 534–569.

Li, H. P., and Sheffield, J. B. (1984). Isolation and preliminary characterization of flat cells, a subpopulation of the embryonic chick retina. *Tissue and Cell* **16,** 843–857.

Lindsay, R. M., Barber, P. C., Sherwood, M. R. C., Zimmer, J., and Raisman, G. (1982). Astrocyte cultures from adult rat brain. Derivation, characterization and neurotrophic properties of pure astroglial cells from corpus callosum. *Brain Res.* **243,** 329–343.

McCaffery, C. A., Bennett, M. R., and Dreher, B. (1982). The survival of neonatal rat retinal ganglion cells *in vitro* is enhanced in the presence of appropriate parts of the brain. *Exp. Brain Res.* **48,** 377–386.

Manthorpe, M., and Varon (1984). Regulation of neuronal survival and neuritic growth in the avian ciliary ganglion. *In* "Growth and Maturation Factors" (G. Guroff, ed.), Vol. 3, pp. 77–117. Wiley, New York.

Manthorpe, M., Skaper, S., Adler, R., Landa, K., and Varon, S. (1980). Cholinergic neuronotrophic factors. IV. Fractionation properties of an extract from chick embryonic eye tissues. *J. Neurochem.* **34,** 69–75.

Manthorpe, M., Adler, R., and Varon, S. (1981a). Cholinergic neuronotrophic factors. VI. Age-dependent requirements by chick embryo ciliary ganglionic neurons. *Dev. Biol.* **85,** 156–163.

Manthorpe, M., Varon, S., and Adler, R. (1981b). Neurite-promoting factor (NPF) in conditioned medium from RN22 Schwannoma cultures: Bioassay, fractionation and properties. *J. Neurochem.* **37,** 759–767.

Manthorpe, M., Engvall, E., Ruoslahti, E., Longo, F. M., Davis, G. E., and Varon, S. (1983).

Laminin promotes neuritic regeneration from cultured peripheral and central neurons. *J. Cell Biol.* **97,** 1882–1890.

Marchisio, P. C. (1969). Choline acetyltransferase (ChAc) activity in developing chick optic centres and the effects of monolateral removal of retina at an early embryonic stage and at hatching. *J. Neurochem.* **16,** 665–671.

Mathers, L. H., Jr., and Ostrach, L. H. (1979). A critical period in the development of tectal neurons in the chick, as revealed by early enucleation. *Brain Res.* **170,** 219–230.

Merin, S., and Auerbach, E. (1976). Retinitis pigmentosa. *Surv. Ophthalmol.* **20,** 303–346.

Mirsky, R., Wendon, L., Black, P., Stolkin, C., and Bray, D. (1978). Tetanus toxin: A cell surface marker for neurons in culture. *Brain Res.* **148,** 251–259.

Mizrachi, Y., and Schwartz, M. (1982). Goldfish tectal explants have a growth-promoting effect on neurites emerging from co-cultured regenerating retinal explants. *Dev. Brain Res.* **3,** 502–505.

Moscona, A. A. (1965). Recombination of dissociated cells and the development of cell aggregates. *In* "Cells and Tissues in Culture" (E. Wilmer, ed.), pp. 489–529. Academic Press, New York.

Moscona, A. A., and Linser, P. (1983). Developmental and experimental changes in retinal glia cells: Cell interactions and control of phenotype expression and stability. *Dev. Biol.* **18,** 155–188.

Narayanan, C. H., and Narayanan, Y. (1978). Neuronal adjustments in developing nuclear centers of the chick embryo following transplantation of an additional optic primordium. *J. Embryol. Exp. Morphol.* **44,** 53–70.

Nishi, R., and Berg, D. K. (1981). Two components from eye tissue that differentially stimulate the growth and development of ciliary ganglion neurons in cell culture. *J. Neurosci.* **1,** 505–513.

Nurcombe, V., and Bennett, M. R. (1981). Embryonic chick retinal ganglion cells identified "*in vitro*". *Exp. Brain Res.* **44,** 249–258.

Okada, T. S. (1980). Cellular metaplasia or transdifferentiation as a model for retinal cell differentiation. *Curr. Top. Dev. Biol.* **16,** 349–380.

Oppenheim, R. W. (1981). Neuronal cell death and some related regressive phenomena during neurogensis: A selective historical review and progress report. *In* "Studies in Developmental Neurobiology: Essays in Honor of Viktor Hamburger" (W. M. Cowan, ed.), pp. 74–133. Oxford Univ. Press, London and New York.

Pessin, M., and Adler, R. (1985). Coexistence of high affinity uptake mechanisms for putative neurotransmitter molecules in chick embryo retinal neurons in purified culture. *J. Neurosci. Res.* **14,** 317–328.

Potts, R. A., Dreher, B., and Bennett, M. R. (1982). The loss of ganglion cells in the developing retina of the rat. *Dev. Brain Res.* **3,** 481–486.

Price, D. L., Whitehouse, P. J., Struble, R. G., Clark, A. W., Coyle, J. T., DeLong, M. R., and Hedreen, J. C. (1982). Basal forebrain cholinergic systems in Alzheimer's disease and related dementias. *Neurosci. Comment.* **1,** 84–92.

Pritchard, D. J., Clayton, R. M., and DePomerai, D. I. (1978). "Transdifferentiation" of chicken neural retina into lens and pigment epithelium in culture: Controlling influences. *J. Embryol. Exp. Morphol.* **48,** 1–21.

Puro, D. G., DeMello, F. G., and Nirenberg, M. (1977). Synapse turnover: The formation and termination of transient synapses. *Proc. Natl. Acad. Sci. U.S.A.* **74,** 4977–4981.

Rager, G., and Rager, U. (1978). Systems-matching by degeneration. *Exp. Brain Res.* **33,** 65–78.

Ramirez, G., and Seeds, N. W. (1977). Temporal changes in embryonic nerve cell recognition: Correlate with cholinergic development in aggregate cultures. *Dev. Biol.* **60,** 153–162.

Rogers, S. L., Letourneau, P. C., Palm, S. L., McCarthy, J., and Furcht, L. T. (1983). Neurite extension by peripheral and central nervous system neurons in response to substratum-bound fibronectin and laminin. *Dev. Biol.* **98,** 212–220.

Sarthy, P. V., Curtis, B. M., and Catterall, W. A. (1983). Retrograde labeling, enrichment, and characterization of retinal ganglion cells from the neonatal rat. *J. Neurosci.* **3,** 2532–2544.

Schwartz, M., and Agranoff, B. (1981). Outgrowth and maintenance of neurites from cultured goldfish retinal ganglion cells. *Brain Res.* **206,** 331–343.

Schwartz, M., Mizrachi, Y., and Eshhar, N. (1982a). Factor(s) from goldfish brain induce neuritic outgrowth from explanted regenerating retinas. *Dev. Brain Res.* **3,** 29–35.

Schwartz, M., Mizrachi, Y., and Kimhi, Y. (1982b). Regenerating goldfish retinal explants induction and maintenance of neurites by conditioned medium from cells originated in the nervous system. *Dev. Brain Res.* **3,** 21–28.

Seeds, N. W. (1983). Neuronal differentiation in reaggregate cell cultures. *Adv. Cell. Neurobiol.* **4,** 57–79.

Sengelaub, D. R., and Finlay, B. I. (1981). Early removal of one eye reduces normally occurring cell death in the remaining eye. *Science* **213,** 573–574.

Sengelaub, D. R., and Finlay, B. L. (1982). Cell death in the mammalian visual system during normal development. I. Retinal ganglion cells. *J. Comp. Neurol.* **204,** 311–317.

Silver, J. (1978). Cell death during development of the nervous system. *In* "Handbook of Sensory Physiology" (M. Jacobson, ed.), Vol. 9, pp. 419–436. Springer-Verlag, Berlin and New York.

Silver, J., and Hughes, A. F. W. (1973). The role of cell death during morphogenesis of the mammalian eye. *J. Morphol.* **140,** 159–170.

Smalheiser, N. R., Crain, S. M., and Bornstein, M. B. (1981a). Development of ganglion cells and their axons in organized cultures of fetal mouse retinal explants. *Brain Res.* **204,** 159–178.

Smalheiser, N. R., Peterson, E. R., and Crain, S. M. (1981b). Neurites from mouse retina and dorsal root ganglion explants show specific behavior within co-cultured tectum or spinal cord. *Brain Res.* **208,** 499–505.

Stefanelli, A. A., Zacchei, A. M., Carsvita, S., Cataldi, A., and Teradi, L. A. (1967). New forming retinal synapses *in vitro. Experientia* **23,** 199–200.

Steinberg, M. S. (1963). Reconstitution of tissues by dissociated cells. *Science* **141,** 401–408.

Suburo, A. M., and Adler, R. (1977). Neuronal and glial differentiation in reaggregation cultures. *Cell Tissue Res.* **176,** 407–416.

Tamai, M., Takahashi, J., Noji, T., and Mizuno, K. (1978). Development of photoreceptor cells *in vitro:* Influence and phagocytic activity of homo- and heterogeneic pigment epithelium. *Exp. Eye Res.* **26,** 581–590.

Thompson, J. M., and Pelto, D. J. (1982). Attachment, survival and neurite extension of chick embryo retinal neurons on various culture substrates. *Dev. Neurosci.* **5,** 447–457.

Turner, J. E., Delaney, R. K., and Johnson, J. E. (1981). Retinal ganglion cell response to axotomy and nerve growth factor antiserum treatment in the regenerating visual system of the goldfish (*Carassius auratus*); An *in vivo* and *in vitro* analysis. *Brain Res.* **204,** 283–294.

Turner, J. E., Schwab, M. E., and Thoenen, H. (1982). Nerve growth factor stimulates neurite outgrowth from goldfish retinal explants: The influence of a prior lesion. *Dev. Brain Res.* **4,** 59–66.

Turner, J. E., Barde, Y. A., Schwab, M. E., and Thoenen, H. (1983). Extract from brain stimulates neurite outgrowth from fetal rat retinal explants. *Dev. Brain Res.* **6,** 77–83.

Varon, S. (1977). Neural growth and regeneration: A cellular perspective. *Exp. Neurol.* **54,** 1–6.

Varon, S., and Adler, R. (1980). Nerve growth factors and control of nerve growth. *Curr. Top. Dev. Biol.* **16,** 207–252.

Varon, S., and Adler, R. (1981). Trophic and specifying factors directed to neuronal cells. *Adv. Cell. Neurobiol.* **2,** 115–163.

Varon, S., Skaper, S. D., and Manthorpe, M. (1981). Trophic activities for dorsal root and sympathetic ganglionic neurons in media conditioned by Schwann and other peripheral cells. *Dev. Brain Res.* **1,** 73–87.

Varon, S., Adler, R., Manthorpe, M., and Skaper, S. (1982). Culture strategies for trophic and other factors directed to nerve cells. *In* ''Current Frontiers in Neurobiology Approached through Cell Culture'' (S. E. Pfeiffer, ed.), pp. 53–77. C. R. C. Press, Boca Raton, Florida.

Young, R. W. (1984). Cell death during differentiation of the retina in the mouse *J. Comp. Neurol.* **229,** 362–373.

CELL MOTILITY IN THE RETINA

BETH BURNSIDE AND ALLEN DEARRY

Department of Physiology–Anatomy
University of California, Berkeley
Berkeley, California

I. Introduction

Cell motility plays a crucial role in the lives of all cells. From their origin in the mitosis of preexisting cells to the intracellular transport necessary for their daily housekeeping activities, cells are actively engaged in motile processes. These fundamental motile processes are of two basic sorts: cell shape change and intracellular transport.

The basic objective of research in cell motility is to characterize the mechanism by which chemical energy is converted into mechanical energy (cf. Pollard, 1981; Raff, 1979; Stracher, 1983a,b; Sheterline, 1983). Thus for a particular motile process we want to find answers to several specific questions: What is the chemical composition of the structural machinery? What is the mechanism of force production? What is the fuel? What turns the machinery on and off? How is the machinery assembled at the appropriate time? Fortunately (for scientists), the huge varieties of motile processes observed in nature appear to depend on a relatively few basic but universal classes of force-producing mechanisms.

In this review we first summarize our recent perceptions about the machinery and force-producing mechanisms responsible for cell motility, and then examine the roles of these cytoskeletal elements and regulatory processes in retinal motility. Since retinomotor movements provide the most dramatic and thoroughly studied examples of retinal motility, we describe them in some detail. We also consider possible roles of motility in RPE cell activities, synaptic modulation, and photoreceptor alignment.

II. The Structural Machinery: The Cytoskeleton

The application of electron microscopy and of more recently developed immunolocalization techniques has permitted great advances in our understanding of the composition and distribution of structural protein fibers in the cytoplasm (cf. Schliwa *et al.*, 1982). These protein fibers, now collectively called the cytoskeleton, can be assigned to at least four major classes according to morphological criteria: microtubules (25 nm in diameter), thick filaments (10–40 nm in diameter), intermediate filaments (7–11 nm in diameter), and thin filaments (6 nm in diameter) (cf. Pollard, 1981; Schliwa *et al.*, 1982; Birchmeier, 1984). Even more delicate fibers have been described in striated flagellar rootlets (Salisbury *et al.*, 1984a,b) and in the cytoplasm of cultured cells prepared by critical point drying (microtrabeculae) (cf. Schliwa *et al.*, 1982). By *in vitro* reassembly of fibers from purified proteins, it has been possible to show that microtubules, thin filaments, and thick filaments are primarily composed of the proteins tubulin, actin, and myosin, respectively (cf. Kirshner, 1978; Raff, 1979; Pollard, 1981; Korn, 1982; Huxley, 1983). Actin filaments and microtubules also contain numerous accessory proteins, called actin-associated proteins or microtubule-associated proteins (MAPs), which influence the stability, assembly, and function of the fibers.

A. *Actin Filaments and Actin-Associated Proteins*

It has been accepted for some time that in nonmuscle cells, as in muscle, actin–myosin interactions are capable of producing contractile force (cf. Pollard, 1981; Huxley, 1983). Studies of actin function have called upon a variety of drugs which influence the state of actin assembly. The cytochalasins, particularly the more specific cytochalasin D, have been used to prevent actin filament assembly and also to disrupt intact actin filaments (cf. Tanenbaum, 1978). The mushroom toxin phalloidin has the opposite effect; it stabilizes actin filaments and favors assembly (Estes *et al.*, 1981). Results of studies using these agents suggest that disturbing the state of assembly of actin filaments interferes not only with contraction but with other cell functions as well (cf. Weeds, 1982; Pollard, 1981; Korn, 1982; Taylor and Condeelis, 1979). Thus, actin appears to have cytoskeletal (structural) functions in addition to its role in contractility (cf. Taylor and Condeelis, 1979; Taylor and Fechheimer, 1982; Stossel, 1982; Korn, 1982).

The literature now contains a growing list of cytoplasmic proteins which have been shown to interact specifically with actin monomers and actin filaments (Table I; and see reviews by Smillie, 1979; Korn, 1982; Weeds, 1982; Craig and Pollard, 1982; Schliwa, 1981; Mooseker, 1983; Birchmeier, 1984; Bennett, 1982; Stossel, 1982). Table I is only a sampling, chosen to illustrate the functional variety of the proteins which have so far been characterized; the reviews by Korn (1982) and Weeds (1982) should be consulted for a more detailed consideration. These actin-associated proteins ''modulate'' actin behavior in the cell, influencing its assembly, filament length, cross-linking, and filament stability. All these variables play central roles in the cytoskeletal and motile functions of actin. Unlike skeletal muscle where the assembly of the motile apparatus is an embryonic event and subsequent physiological regulation entails ''merely'' turning on and off the preassembled contractile machinery, nonmuscle cells usually assemble the motile machinery for a contractile event, carry it out, and then disassemble the machinery for reuse elsewhere. In this programming there are numerous intricate roles for actin-associated proteins: some stabilize actin in the monomer form and thus prevent filament assembly, some regulate filament length, some cross-link actin filaments into three-dimensional gels or stiff bundles, and others serve to attach the actin cytoskeleton to the plasma membrane or

TABLE I

ACTIN-ASSOCIATED PROTEINS

Protein	Characteristics (and references)
Proteins that stabilize G-actin and prevent filament assembly	
Profilin, DNase I	Form 1 : 1 complex with G-actin (cf. Korn, 1982; Schliwa, 1981; Weeds, 1982)
Proteins that affect actin filament length	
Acumentin, capping protein	Bind to specific end of actin filament and prevent subunit loss or addition; Ca^{2+} independent (cf. Southwick and Hartwig, 1982)
Gelsolin, villin, fragmin	Ca^{2+} dependent; cross-link actin filaments in absence of calcium; in presence of Ca^{2+} cause fragmentation of actin filaments (cf. Korn, 1982; Stossel, 1982; Hesterberg and Weber, 1983)
Proteins that cross-link actin filaments into three-dimensional gels	
Actin-binding protein, filamin	Flexible proteins that form perpendicular cross-links between actin filaments producing gel; Ca^{2+} independent (cf. Stossel, 1982; Wang and Singer, 1977)
MAP2	Microtubule-associated protein which cross-links actin filaments into gels (Griffith and Pollard, 1978, 1982)
α-Actinin	Ca^{2+} inhibits cross-linking; localized near plasma membrane (Korn, 1982)
Proteins that laterally cross-link actin filaments into bundles	
Fimbrin	Ca^{2+} inhibited; bundles actin in $<10^{-6}$ *M* Ca^{2+} (cf. Craig and Pollard, 1982)
Fascin	Ca^{2+} insensitive; cross-links actin filaments (cf. Kane, 1982; Weeds, 1982)
Vinculin	Ca^{2+} insensitive; found at focal contacts in fibroblasts (cf. Weeds, 1982)
Spectrin, fodrin, TW 240/260	Calmodulin-binding proteins that cross-link actin filaments into long flexible bundles, link cytoskeleton to plasma membrane (cf. Bennett *et al.*, 1982; Repasky *et al.*, 1982; Glenney *et al.*, 1982a,b)
Proteins that produce movement	
Myosin	An ATPase activated by interaction with actin; Ca^{2+} dependent (Pollard, 1981; Korn, 1982; Adelstein, 1980, 1982)
Proteins that bind to the sides of actin filaments	
Tropomyosin	Long molecule that lies alongside actin monomers in filament; stabilizes filament; protects against fragmentation (Smillie, 1979; Korn, 1982)

to microtubules (Table I). All these functions of actin-associated proteins necessarily have profound effects on assembly of the motile machinery, viscosity of the cell cytoplasm, and the ability of the cell to do mechanical work.

Though some actin-associated proteins may be synthesized for specific events, it seems likely that most motile processes are regulated by altering the functional

TABLE II

EFFECTS OF CALCIUM ON ACTIN CYTOSKELETON

Protein	Effects of low calcium ($<10^{-6}$ *M*)	Effects of increased calcium ($>10^{-6}$ *M*)
	Produces relaxation	*Produces contraction*
Myosin	Inhibts interaction with actin, thus blocks Mg-ATPase activity	Turns on actin-activated myosin ATPase activity
	Produces increased viscosity of cytoplasm (gel)	*Produces decreased viscosity of cytoplasm (sol)*
Gelsolin, villin, fragmin	Enhances cross-linking by these proteins of actin filaments into three-dimensional gel	Inhibits cross-linking of actin filaments by these proteins and activates severing of actin filaments into shorter lengths
	Produces assembly or stabilization of microvilli and microspikes	*Inhibits microvilli and microspike formation*
Fimbrin	Enhances lateral cross-linking of actin filaments into stiff bundles by this protein	Prevents cross-linking of actin filaments by this protein

state of some or many of these proteins in response to an appropriate signal (cf. Korn, 1982; Weeds, 1982). Thus as one might expect the activities of these actin-associated proteins are influenced by intracellular messengers, such as changes in intracellular calcium concentration (Table II; cf. Taylor and Fechheimer, 1982; Weeds, 1982; Korn, 1982), cAMP concentration (Table III), or pH (Begg and Rebhun, 1979; Tilney *et al.*, 1973; Taylor and Fechheimer, 1982). In Table II it can be seen that there is an interesting pattern of calcium effects on the actin cytoskeleton mediated by actin-associated proteins. Low calcium concentrations favor stability: cytoskeletal gels and rigid actin bundles are assembled and contraction is inhibited. Higher calcium concentrations favor motility: myosin filaments are assembled, myosin ATPase is activated, and the structural gels are made less viscous, thus facilitating movement.

TABLE III

EFFECTS OF cAMP ON ACTIN FILAMENT SYSTEMS

cAMP-dependent phosphorylation of MAP2 decreases cross-linking of actin filaments to microtubules and to each other (Selden and Pollard, 1983)
cAMP-dependent phosphorylation of myosin light chain kinase reduces its affinity for calcium / calmodulin and thus inhibits actin–myosin interaction (cf. Adelstein *et al.*, 1982)
cAMP-dependent kinase phosphorylates filamin (Wallach *et al.*, 1978)

Some effects of cAMP on actin-associated proteins and actin function are summarized in Table III. Much less is known about cAMP effects than about calcium effects on cytoskeletal actin functions; however, the molecular basis for inhibition of contraction by cAMP has been well characterized for smooth muscle (cf. Adelstein *et al.*, 1982; de Lanerolle *et al.*, 1984). Increasing evidence suggests that cAMP acts in a similar fashion in nonmuscle cells (cf. Adelstein *et al.*, 1982; Feinstein *et al.*, 1983; Porrello and Burnside, 1984).

B. Microtubules and Microtubule-Associated Proteins (MAPs)

Microtubules are found in all nucleated metazoan cells (cf. reviews by Roberts and Hyams, 1979; Sakai *et al.*, 1982; Raff, 1979; Snyder and McIntosh, 1976). Several chemical probes have been used to evaluate the role of microtubules in various types of cell motility. Colchicine, colcemid, and nocodazole bind to microtubule subunits and thus inhibit assemby (cf. Wilson and Bryan, 1974). Since microtubules appear to be in dynamic equilibrium with a pool of free subunits, these drugs will eventually completely deplete the cell of intact microtubules (cf. Raff, 1979; Kirshner, 1978). An opposite effect is produced by the drug taxol, which stabilizes the polymer state and thus tends to drive almost all the cell's tubulin into stable microtubules (cf. Kumar, 1981). Results of studies with these drugs suggest that microtubules are required for most types of extensive cell shape changes (cell elongation or elongation of cellular projections) and for several types of intracellular transport (Snyder and McIntosh, 1976; Goldman *et al.*, 1979; Schliwa, 1984; Roberts and Hyams, 1979; Sakai *et al.*, 1982). Microtubules are less likely to be required for contractile cell shape changes.

Like actin filaments, a number of accessory proteins influence the state of assembly and function of microtubules (Table IV: cf. Raff, 1979; Kirshner, 1978; Hill and Kirschner, 1983). Since MAPs play regulatory roles in microtubule assembly and function, it is not surprising that many of them are influenced by the intracellular messengers calcium and cAMP. Table V summarizes reported effects of calcium on MAPs and microtubule-associated processes. Generally, increasing calcium tends to favor microtubule disassembly and inhibition of microtubule-based motility. Increasing cAMP, on the other hand, activates ciliary and flagellar motility but also reduces the enhancement of microtubule assembly by MAPs (Table VI).

C. Intermediate Filaments

Though the 8- to 10-nm filaments found in almost all cells appear to be morphologically identical, they are, in fact, biochemically heterogeneous. Five

TABLE IV

MICROTUBULE-ASSOCIATED PROTEINS

Protein	Characteristics (and references)
High molecular weight MAPs from brain	
MAP1	Enhances microtubule assembly; not phosphorylated (Sloboda *et al.*, 1975)
MAP2	Decorates the surface of microtubule as sidearms (cf. Kirschner, 1978; Kim *et al.*, 1979)
	Enhances microtubule assembly (cf. Kirschner, 1978; Nishida *et al.*, 1981)
	Cross-links microtubules to actin filaments (Griffith and Pollard, 1978)
	Cross-links actin filaments to each other (Griffith and Pollard, 1982)
	Cross-links microtubules to secretory granules (Suprenant and Dentler, 1982)
	Is phosphorylated by intrinsic cAMP-dependent protein kinase (Sloboda *et al.*, 1975; Jameson *et al.*, 1980)
	Phosphorylation decreases ability to enhance assembly and actin binding (Selden and Pollard, 1983; Jameson *et al.*, 1980)
Low molecular weight MAPs from brain (tau proteins)	
	Enhance microtubule assembly (cf. Kirschner, 1978)
	Phosphorylated by cAMP-dependent kinase that copurifies with microtubules (Selden and Pollard, 1983)
	Phosphorylated by cAMP-independent kinase that copurifies with microtubules (Lindwall and Cole, 1984)
	Phosphorylation reduces ability to enhance assembly (Jameson *et al.*, 1980; Sloboda *et al.*, 1975)
	Binds selectively and with high affinity to calmodulin (Sobue *et al.*, 1981)
	Interacts with actin filaments (Griffith and Pollard, 1982)
Clatharin	
	Coated vesicles copurify with brain microtubules; tubulins are components of clatharin coats of coated vesicles (Imhof *et al.*, 1983)
Dynein	
	Mg-ATPase which forms mechanochemical bridges on doublet microtubules of cilia and flagella (cf. Warner and Mitchell, 1980; Bell and Gibbons, 1983)

classes of intermediate filaments have been identified according to their subunit composition: vimentin filaments, found mostly in mesenchymal tissue; glial filaments, found in neuroglia; desmin filaments, found mostly in muscle; keratin filaments, found mostly in epithelial cells; and neurofilaments, found only in neurons (cf. Birchmeier, 1984; Fuchs and Hanukoglu, 1983; Lazarides, 1982; Osborn and Weber, 1982). Though elaborate arrays of intermediate filaments have been described in many cell types (cf. Lazarides, 1982), their specific physiological functions remain unknown. Injecting a cultured cell with antibodies to intermediate filament proteins disrupts the intermediate filaments in the cell without producing discernible effects on cell shape or motility (Klymkowsky *et al.*, 1983).

TABLE V

EFFECTS OF CALCIUM ON MICROTUBULE SYSTEMS

Ca^{2+} favors microtubule disassembly
- In brain tubulin preparations *in vitro* (Weisenberg, 1972; cf. Margolis, 1983)
- In lysed and permeabilized cells (Schilwa *et al.*, 1981; Means and Dedman, 1980)
- When injected into living cells (Kiehart, 1981)
- In isolated mitotic spindles (Salmon and Segall, 1980; Pratt *et al.*, 1980)

Ca^{2+} induces changes in ciliary beat, including
- Arrest of ciliary motility (Murakami and Takahasi, 1975)
- Reversal of ciliary beat (Naitoh and Kaneko, 1973)
- Change in orientation of the cilia (Hyams and Borisy, 1978)
- Altered beat frequency (Naitoh and Kaneko, 1973)
- Altered beat pattern (Satir, 1982)

Ca^{2+} effects on microtubule-associated proteins
- MAP2 and tau proteins bind selectively and with high affinity to calmodulin (cf. Lee and Wolf, 1982)
- MAP2 is localized to dendrites, while most brain calmodulin is localized to dendrites and postsynaptic densities (Huber and Matus, 1984; Wood *et al.*, 1980a,b)
- Tubulin is phosphorylated by Ca^{2+}/calmodulin-dependent protein kinase in synaptosomes (Burke and DeLorenzo, 1981)
- Calmodulin travels with tubulin in slow axon transport (Erickson *et al.*, 1980)
- Calmodulin is bound to microtubules in ciliary axonemes (Jones *et al.*, 1978, 1980; Chafouleas *et al.*, 1979; Nagao *et al.*, 1981)
- Dynein (ciliary ATPase) binds to calmodulin affinity columns (Blum *et al.*, 1981)
- Ca^{2+}/calmodulin stimulates the ATPase of purified dynein (Blum *et al.*, 1980)
- Ca^{2+}/calmodulin stimulates the ATPase of cytoplasmic dynein (Hisanaga and Pratt, 1982)

TABLE VI

EFFECTS OF cAMP ON MICROTUBULE SYSTEMS

cAMP-dependent kinase copurifies with brain microtubules (Sloboda *et al.*, 1975)

This endogenous kinase phosphorylates MAP2 and tau proteins (Sloboda *et al.*, 1975; Jameson *et al.*, 1980)

Phosphorylated MAP2 and tau have reduced ability to enhance microtubule assembly or cross-link actin filaments to microtubules or each other (Nishida *et al.*, 1981; Seldon and Pollard, 1983)

cAMP activates sliding motility in demembranated flagellar axonemes, thus activating flagellar beating (Murofushi *et al.*, 1982; Lindemann, 1978; cf. Tash and Means, 1983)

cAMP-dependent protein kinases are localized in cilia and flagella (Schultz and Jantzen, 1980)

III. Force-Producing Mechanisms

Relatively few basic mechanisms have been described for the cellular conversion of chemical to mechanical energy. The two best studied mechanisms depend upon the sliding of fibers past one another by means of mechanochemical bridges containing ATPases. One of these systems depends on actin and myosin filaments and is most highly developed in striated muscle (cf. Huxley, 1983), while the other system depends on microtubules and is most highly developed in cilia and flagella (cf. Satir, 1979). Very recently, a third and entirely different microtubule-based mechanism has been identified which appears to mediate vesicle transport along the microtubules of axons (Vale *et al.*, 1985). This movement is ATP-dependent and mediated by a protein complex named kinesin. Others have proposed that force production can result from the assembly and disassembly of linear polymers from soluble actin or tubulin subunits (cf. Inoue and Ritter, 1975; Inoue, 1981; Hill and Kirshner, 1983; Tilney *et al.*, 1973). And finally, an actin- and microtubule-independent contractile mechanism has been described in striated rootlets and spasmonemes, where contraction results from the direct interaction of filament subunits with calcium to produce a change in subunit conformation and consequent filament shortening (cf. Salisbury *et al.*, 1984a; Routledge *et al.*, 1976).

A. Force Production by Actin–Myosin Interaction

The mechanism of actin–myosin interaction has been intensely studied in both muscle and nonmuscle cells (cf. reviews by Pollard, 1981; Huxley, 1983). ATP is utilized to produce sliding interdigitation of actin and myosin filaments by means of myosin crossbridges. The crossbridges contain Mg-ATPases which are activated on binding to actin filaments. Contraction is produced by increased overlap of actin and myosin filaments without change in filament length.

In all known cases of actomyosin-based contraction, contraction is regulated by calcium (cf. Ebashi, 1983). In vertebrate striated muscles, calcium regulation is mediated by the troponin–tropomyosin system located on the thin filament. In smooth muscle and in nonmuscle cells so far studied, calcium regulation appears to be mediated by calmodulin-dependent myosin phosphorylation (cf. Adelstein, 1980).

B. Force Production by Microtubule–Dynein Interaction

The force-producing mechanism for ciliary and flagellar movement depends upon the sliding interaction of adjacent doublet microtubules in the ciliary axoneme (cf. Satir, 1979). Sliding is produced by ATP-dependent mechanochemical conformational changes in the crossbridges which extend between the microtubules of adjacent doublets. The crossbridges contain the ATPase dynein, a very large multisubunit protein responsible for force production (cf. Bell and Gibbons, 1983). Several recent observations suggest that the microtubule–dynein system may not be confined to cilia and flagella, but also may play a role in movements dependent on cytoplasmic microtubules. Dynein-like ATPase activity has been detected in sea urchin cytoplasm and mitotic spindles (Weisenberg and Taylor, 1968; Mabuchi, 1973; Pratt, 1980; Pratt *et al.*, 1980). Monoclonal antibodies raised against sea urchin flagellar dynein subunits cross-react with proteins of similar molecular weight from sea urchin egg cytoplasm and isolated mitotic spindles (Piperno, 1984). Cytoplasmic microtubules appear to bear attachment sites for dynein, since flagellar dynein binds at regular intervals along microtubules prepared from purified brain tubulin (Haimo *et al.*, 1979). Also, physiological experiments suggest a role for dynein in microtubule-dependent movements other than those of cilia and flagella. Chromosomal movement in the isolated mitotic apparatus is inhibited by antibodies to fragments of flagellar dynein (Sakai *et al.*, 1976); and spindle elongation is inhibited by vanadate, a potent inhibitor of dyneins as well as other ATPases (Simons, 1979; Cande and Wolniak, 1978). Thus, there is growing evidence that cytoplasmic microtubule-dependent movements might utilize a sliding mechanism similar to that of cilia and flagella.

C. Force Production by Microtubule–Kinesin Interaction

The recent characterization of a translocator molecule called kinesin isolated from squid axoplasm and bovine brain has provided evidence for a novel force production mechanism dependent on cytoplasmic microtubules which does not apparently entail participation of dynein (Vale *et al.*, 1985). Kinesin is a soluble factor which attaches to microtubules and to cytoplasmic (or even exogenous) particles. When attached it can mediate movement of microtubules on glass and of axoplasmic organelles (or latex beads) on microtubules. It forms a high affinity complex with microtubules in the presence of AMP-PNP (a nonhydrolyzable analog of ATP) and is dissociated from microtubules by ATP (in contrast AMP-PNP decreases the affinity of dynein for microtubules). Kinesin appears to be a complex of two peptides (110–120 kDa and 60–70 kDa); it does not contain the high molecular weight peptides characteristic of dynein (>300 kDa). Column-purified kinesin can be added back to isolated microtubules to produce ATP-dependent movement on a glass substrate. This movement is vanadate-sensitive (at higher concentrations than dynein) but it is not sensitive to *N*-ethylmaleimide

(NEM). In contrast, NEM blocks ciliary motility, dynein ATPase activity, and the ability of dynein to bind to microtubules (cf. Bell and Gibbons, 1983). Though the ATP-induced dissociation of kinesin from microtubules and the vanadate sensitivity of movement suggest that kinesin has a nucleotide binding site, no ATPase activity has been detected with kinesin in solution (Vale *et al.*, 1985). Perhaps its ATPase must be activated by binding to microtubules, an approach not yet examined thoroughly. This form of motile force production has so far been characterized only in connection with particle transport and microtubule translocation along a surface. It is not clear whether kinesin participates in other forms of microtubule-based motility such as cell shape change.

D. Force Production by Assembly and Cross-Linking of Microtubules or Actin Filaments

Some years ago Inoue proposed that microtubule assembly (elongation) and disassembly (shortening) could provide motive force for movement (cf. Inoue and Ritter, 1975; Inoue, 1981, for review). He suggested that assembling microtubules could push organelles apart and that slowly disassembling attached microtubules could pull organelles together. Since microtubule assembly entails the hydrolysis of one GTP molecule for each dimer of tubulin added onto the polymer (cf. Kirshner, 1978; Hill and Kirshner, 1982, 1983), assembly could provide energy for mechanical work.

Force production by actin assembly and cross-linking has been suggested for a variety of phenomena in which delicate spike or leaflike projections extend from cells (cf. Korn, 1982). Like microtubule assembly, actin assembly also entails the hydrolysis of a nucleoside triphosphate (in this case, ATP) for each monomer added to the growing filament (cf. Korn, 1982). Filament assembly and bundle formation by actin-binding proteins appear to mediate the formation of structures such as microvilli, acrosomal processes, and filopodia of neuronal growth cones (cf. Tilney *et al.*, 1973; Korn, 1982; Landis, 1983). Thus, effects of pH, calcium, and cAMP on actin assembly and actin-binding proteins could play important roles in regulating formation of these structures (cf. Korn, 1982; Craig and Pollard, 1982).

E. Force Production by Calcium-Sensitive Contractile Organelles Distinct from Actin and Microtubule Systems

Several calcium-dependent motile organelles have been recognized which function in a manner completely distinct from actin and microtubule systems

(Cachon and Cachon, 1981; Amos, 1971, 1975). In the ciliate *Vorticella* extremely rapid contraction is mediated by fibrous structures called "spasmonemes," which are composed of a low molecular weight (20K) protein called "spasmin" (Amos, 1971, 1975; Routledge *et al.*, 1976). Spasmin is a calcium-binding protein which in the absence of calcium exhibits a fibrillar structure in the relaxed spasmoneme. When calcium levels approach $10^{-6}\ M$, spasmin molecules undergo dramatic conformational changes such that individual fibers are no longer visible and the spasmoneme shortens at rates up to 200 lengths/sec (as compared to 22 lengths/sec in very fast vertebrate skeletal muscle).

Recently it has been shown that striated flagellar rootlets also exhibit calcium-induced contraction and likewise contain a spasmin-like protein (cf. Salisbury and Floyd, 1978; Melkonian, 1983; Salisbury *et al.*, 1984). Cross-striated fibers are found in association with the basal apparatus of many flagellated or ciliated eukaryotic cells (cf. Pitelka, 1974; Salisbury, 1982), attached to the connecting cilium of photoreceptors (cf. Cohen, 1960; Stevens *et al.*, 1984; Spira and Milman, 1979), and in association with centrioles and primary cilia in cultured fibroblasts (Tucker *et al.*, 1979). These observations support suggestions that striated rootlets of photoreceptors might exhibit contractile activity (Spira and Milman, 1979; Stevens *et al.*, 1984).

IV. Vertebrate Retinomotor Movements

The most extensively studied examples of retinal motility are the retinomotor movements found in lower vertebrates. In fish, amphibians, and birds, the eye adjusts to changes in light intensity not by changes in pupil diameter but by means of morphological rearrangements of the photoreceptors and pigment granules of the retinal pigment epithelium (RPE) (cf. Ali, 1975; Burnside and Nagle, 1983). These rearrangements, called retinomotor movements, entail elongation and contraction of photoreceptors and migration of melanin pigment granules within the RPE (Fig. 1). In the light, cones contract, moving their outer segments toward the incoming light, and rods elongate to bury their light-sensitive outer segments in the dispersing pigment of the RPE cells. In the dark these movements are reversed.

Retinomotor movements provide particularly useful models for studying the mechanisms and control of photoreceptor and RPE motile processes. They provide examples of both major types of cell motility (cell shape change and intracellular transport), and their excursions are so dramatic that they are easily quantified. Since vertebrate retinomotor movements will occur in organ or cell culture, they are amenable to experimental manipulation (cf. Burnside and Nagle, 1983). Because retinomotor movements are regulated not only by

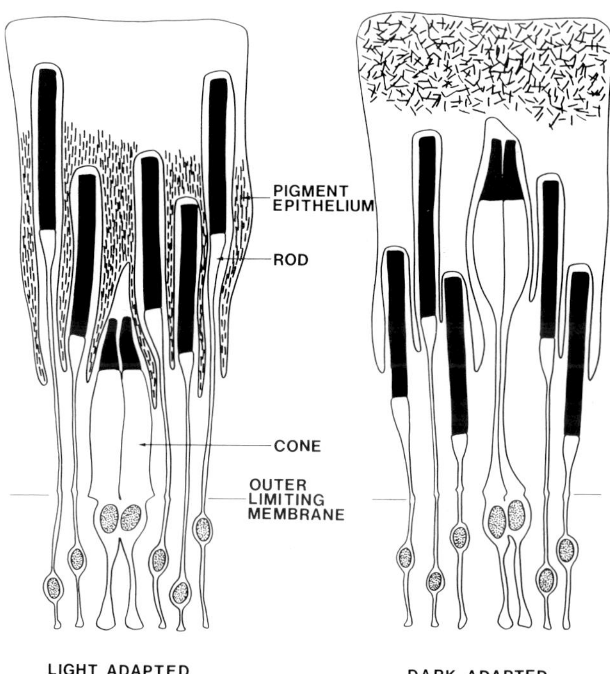

FIG. 1. Retinomotor movements in the teleost retina. In the light cones contract and rods elongate to bury their outer segments in the dispersed screening pigment of the RPE. In the dark movements are reversed: cones elongate, rods contract, and pigment granules aggregate to the base of the RPE cell.

changes in light condition but also by an endogenous circadian rhythm (cf. Burnside and Nagle, 1983; Besharse, 1982), studying their regulation can provide insight into the regulation of other diurnal and circadian aspects of retinal physiology. Though retinomotor movements are minimal or nonexistent in mammals, their study in lower vertebrates may nonetheless allow us to detect diurnal and circadian patterns of regulatory activities which influence not only retinomotor movements but also more universal retinal physiological and metabolic processes.

The large descriptive literature on vertebrate retinomotor movements has recently been reviewed in detail (Ali, 1975; Burnside and Nagle, 1983), and several reviews also describe work on invertebrate retinomotor movement (Hoglund, 1966; Goldsmith and Bernard, 1974; Walcott, 1975; Langer, 1975). In this review we will focus primarily on the mechanisms and regulation of the motile processes responsible for vertebrate cone, rod, and RPE retinomotor movements. Actin- and microtubule-dependent processes have been identified,

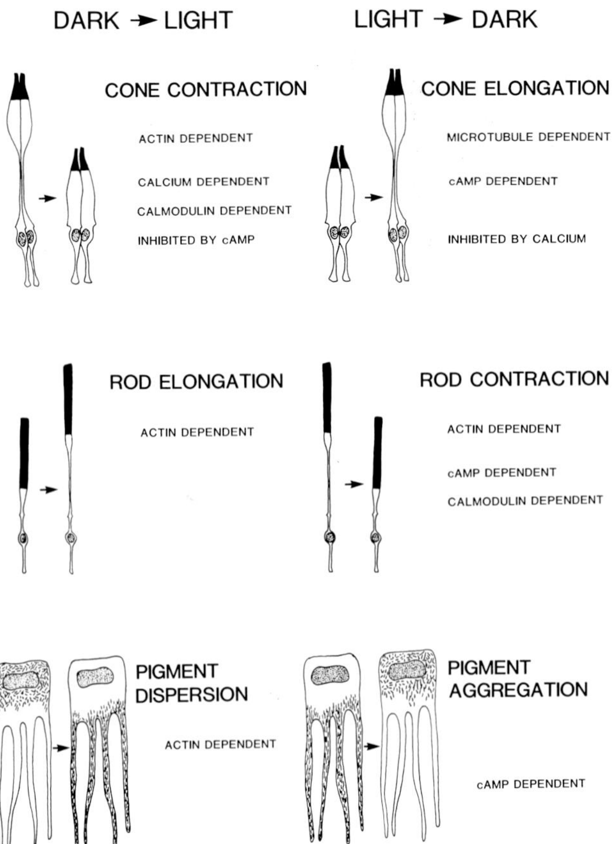

FIG. 2. Force-producing and regulatory mechanisms for vertebrate retinomotor movements. Dark-adaptive movements of cones, rods, and RPE are all induced by elevating cAMP. Calcium is required for contraction in both rods and cones. In all three cell types, light-adaptive retinomotor movements are actin dependent.

and both cAMP and calcium have been shown to play central roles in regulating these movements (Fig. 2).

A. Cone Retinomotor Movements

Cones contract in the light and elongate in the dark. Length change is mediated by the necklike myoid region of the inner segment which maintains relatively constant volume as it elongates and contracts (Fig. 3). Some fish species possess particularly athletic cone myoids which may change length by as much as 100

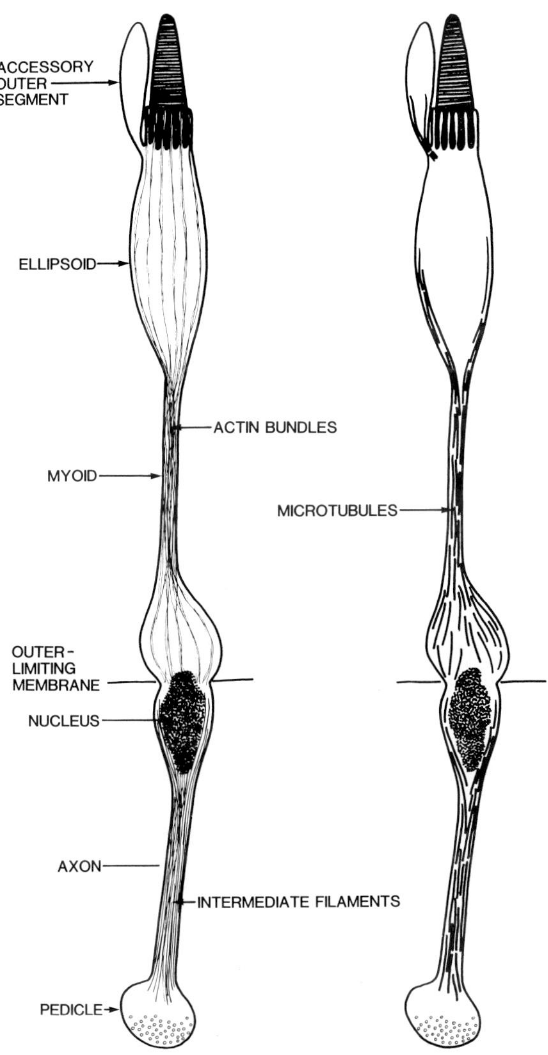

FIG. 3. The cytoskeleton of the teleost cone. The motile myoid region contains paraxially aligned microtubules (shown on the right) and actin filaments (shown on the left). Actin filament bundles originate in the calyceal process, course down the ellipsoid, and then in the myoid splay into a sleeve that lies just under the myoid plasmalemma. The noncontractile axon region contains microtubules and intermediate filaments.

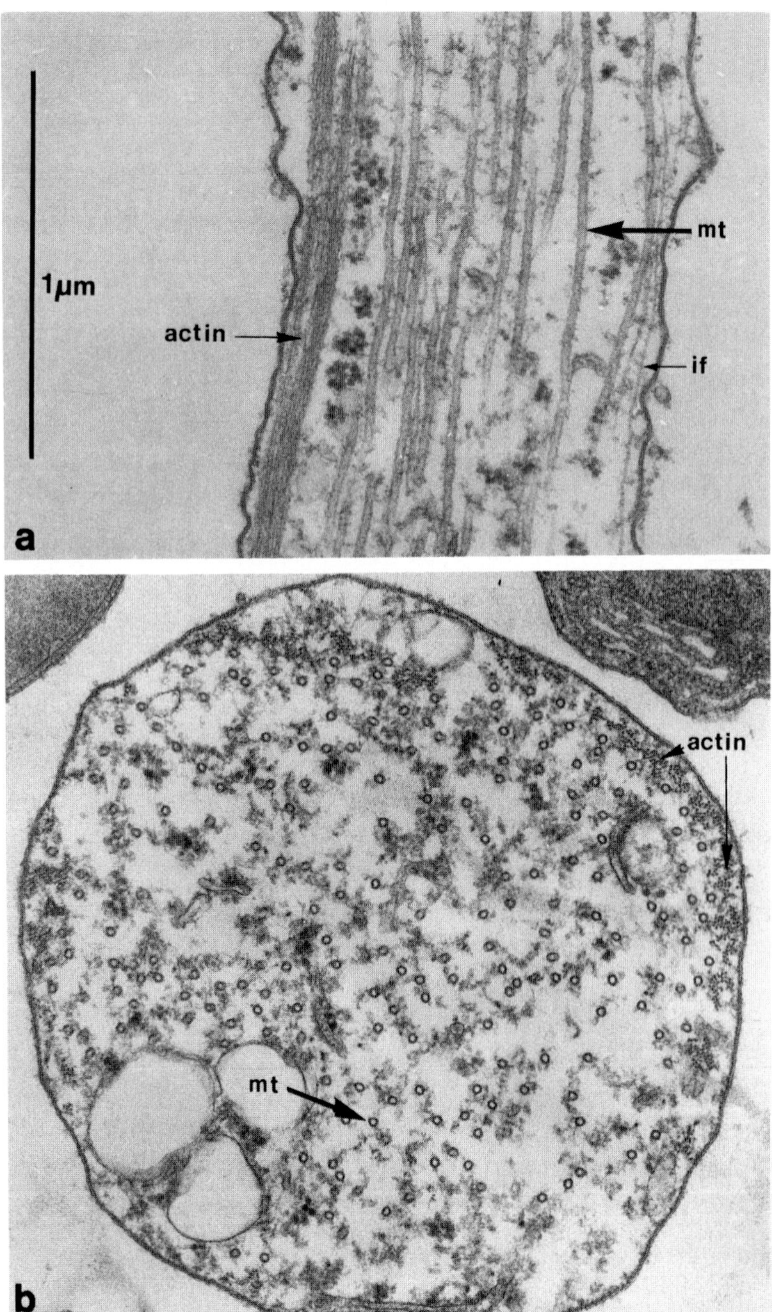
1μm
actin
mt
if
a
actin
mt
b

μm. Force production for elongation and contraction depends on the numerous actin filaments and microtubules located in the cone myoid and aligned parallel to the cone's long axis (Fig. 4). In our laboratory we have used numerous technical approaches to characterize the motile machinery, the mechanism of force production, and the regulatory mechanisms for cone contraction and elongation. Taken together the results of these studies provide a highly integrated pattern: light-induced cone contraction is ATP and actin dependent, activated by calcium, and inhibited by cAMP. In contrast, dark-induced cone elongation is ATP and microtubule dependent, activated by cAMP, and inhibited by calcium (Fig. 2). Thus, calcium and cAMP exert a rigorous, mutually antagonistic control over cone length.

1. Cone Contraction

Cone contraction is an active motile process dependent on actin filaments. Cytochalasin D disruption of myoid actin filaments prevents subsequent light-induced contraction; however, microtubule disruption with colchicine has no effect (Burnside, 1976a; Burnside *et al.*, 1983). Since disrupting microtubules in long cones did not induce shortening in the absence of a light signal, it seems clear that light-induced contraction is an active motile process rather than elastic recoil following collapse of structural support.

The mechanism of force production and the regulation of cone contraction have been studied by means of lysed cell motile models (Burnside *et al.*, 1982b; Porrello *et al.*, 1983; Porrello and Burnside, 1984). Like the use of the glycerinated myofibril in studies of muscle, this powerful approach allows us to permeabilize the cells to experimental media, while nonetheless maintaining the contractile machinery functionally intact. Detergent treatment inactivates the cell's control over its own internal milieu and makes it possible for us to analyze the effects of ions, nucleotides, and various drugs on the motile process.

After lysis with the detergent Brig-58, cones placed in media containing ATP and Ca^{2+} were reactivated to contract with rates comparable to those observed *in vivo*, i.e., approximately 1.5 μm/min (Figs. 5 and 6). Cones did not move if Ca^{2+}, ATP, or the lysis step was omitted (Fig. 6) (Burnside *et al.*, 1982b). Ultrastructural studies showed that after lysis, the cytoplasm of the cone model was extremely extracted, but actin filaments and microtubules were well preserved (Fig. 4). Though plasma membranes of cone models appeared relatively

---

FIG. 4. Electron micrographs of the teleost cone myoid in longitudinal section (a) and in cross section (b). Paraxial actin filaments are found in bundles just beneath the plasma membrane while paraxial microtubules (mt) and intermediate filaments (if) appear randomly distributed in the myoid cytoplasm. This preparation is a lysed cell model fixed during reactivated contraction (see legend of Fig. 5).

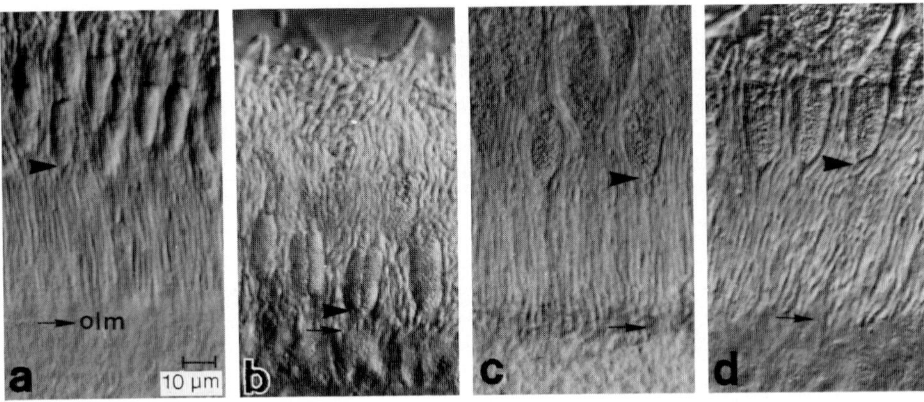

FIG. 5. Lysed cone models from green sunfish (*Lepomis cyanellus*). Nomarski interference contrast light micrographs of retinal slices of four half-retinas from the same fish fixed (a) immediately after dissection (t_0), and (b–d) after 3-min lysis in detergent (1% Brig-58) plus 15 min in one of the following reactivation media: (b) contraction medium (4 m*M* ATP, 10^{-5} *M* calcium), (c) relaxation medium (4 m*M* ATP, $<10^{-8}$ *M* calcium), and (d) rigor medium (no ATP, 10^{-5} *M* calcium). Cone myoid length was measured as the distance from the outer limiting membrane (olm, small arrows) to the base of the cone ellipsoid (large arrows). The vesiculated appearance of the cone ellipsoids in (b–d) results from detergent lysis. Lysed cones contracted only when calcium and ATP were both present. (Micrographs supplied by Kathryn Porrello; for detailed procedures see Porrello and Burnside, 1984.)

intact, they were nonetheless sufficiently permeabilized to allow entry of myosin subfragment-1 (MW 100,000) (Porrello *et al.*, 1983). In addition, the increased membrane permeability due to lysis was retained: lysed models incubated 90 min in the absence of Ca^{2+} contracted normally when Ca^{2+} was subsequently added (Porrello and Burnside, 1984).

Myosin subfragment-1 (S-1) has been used in lysed cell models to test the role of actomyosin in cone contraction (Porrello *et al.*, 1983). If myosin S-1 is modified by treatment with *N*-ethylmaleimide it retains the ability to bind to actin but loses its ATPase activity. This new species, NEM-S-1, then binds irreversibly to actin filaments and can be used as a "jammer" to inhibit access of native myosin to actin filaments and thereby interfere with contraction (Meeusen and Cande, 1979). Incubating lysed cone models with NEM-S-1 completely blocked subsequent reactivated contraction (Porrello *et al.*, 1983), thus suggesting that cone contraction not only requires actin filaments (as indicated by cytochalasin results), but also depends on actin–myosin interaction.

Reactivated cone contraction requires ATP and is regulated by calcium. The ATP requirement is highly specific; other nucleotides (ITP, CTP, GTP, AMP-PNP, ADP) fail to support movement when substituted for ATP in the procedure (Porrello and Burnside, 1984). Maximal rates of contraction are produced by ≥ 1 m*M* ATP (Porrello and Burnside, 1984).

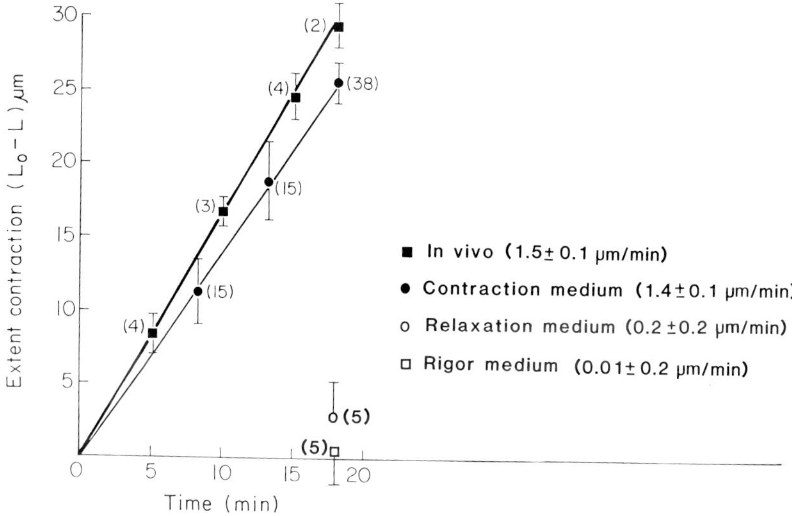

FIG. 6. Kinetics of cone contraction *in vivo* and in reactivated lysed cell models of green sunfish. Cones were induced to contract *in vivo* by exposing previously dark-adapted fish to 1.4 lumens fluorescent light/m^2. Cone models were prepared as described in Fig. 5. In lysed cone models exposed to contraction medium (4 m*M* ATP, 10^{-5} *M* calcium), the rate of reactivated contraction was not significantly different from that observed *in vivo*. In the absence of ATP (rigor medium) or calcium (relaxation medium) no contraction was observed. Points illustrate mean ± SEM; parentheses indicate numbers of retinas examined. (Modified from Burnside *et al.*, 1982b, with permission from *Journal of Cell Biology*.)

Detailed analysis of the requirements for contraction in lysed cone models strongly suggests that calcium regulation is mediated by myosin phosphorylation (Porrello and Burnside, 1984) as is the case in smooth muscle and other nonmuscle cells (cf. Kendrick-Jones and Scholey, 1981; Adelstein, 1980, 1982). In lysed cone models, dose–response studies using Ca^{2+}/EGTA buffers indicate that the rate of contraction is Ca^{2+} dependent, with maximal rates occurring at $\geq 10^{-6}$ *M* free Ca^{2+}, and no contraction occurring at $\leq 10^{-8}$ *M* free Ca^{2+}. Thus, the Ca^{2+} dose–response curve for cone contraction resembles that for smooth muscle (Ebashi *et al.*, 1978; Kerrick *et al.*, 1980). Furthermore, calcium regulation of cone contraction appears to be mediated by calmodulin; the calmodulin inhibitors trifluoperazine (TFP) and R-24571 strongly inhibit reactivated contraction in cone models (Porrello and Burnside, 1984).

More definitive support for a role of myosin phosphorylation in cone contraction is provided by the use of unregulated myosin light chain kinase (MLCK) (Burnside *et al.*, 1984). If smooth muscle MLCK is briefly treated with trypsin, the Ca^{2+}/calmodulin-binding portion of the molecule is cleaved off and the remaining catalytic portion remains permanently active even in the absence of calcium (Kendrick-Jones and Scholey, 1981; Adelstein, 1982). This fragment, called unregulated MLCK, will induce reactivated contraction in cone models in the *absence* of calcium (Burnside *et al.*, 1984). This result provides the strongest evidence so far that Ca^{2+} regulation of cone contraction is mediated via myosin phosphorylation.

Reactivated cone contraction is extremely sensitive to cAMP. Even in the presence of 10^{-6} *M* free calcium, reactivated contraction is completely inhibited by cAMP at concentrations as low as 10^{-6} *M* (Burnside *et al.*, 1984). This result is consistent with earlier findings: (1) culturing intact retinas with dibutyryl cAMP inhibits light-induced contraction in intact cones (Dearry and Burnside, 1984a), and even induces cone elongation in constant light (Burnside *et al.*, 1982a; Burnside and Ackland, 1984); (2) retinal cAMP levels are higher in the dark and fall with light onset (cf. Farber *et al.*, 1981; Burnside *et al.*, 1982a). These observations suggest that darkness is accompanied by an increase in intracellular cAMP content, and that this higher level of cAMP favors dark-adaptive cone elongation.

Similar inhibition of contraction by cAMP has been observed in smooth muscle, where it has been suggested that cAMP might act directly by modifying the contractile machinery and/or indirectly by lowering cytoplasmic calcium levels (Adelstein *et al.*, 1982; Sherry *et al.*, 1978; Anderson and Nilsson, 1977). In the case of lysed cone motile models, it is highly unlikely that cAMP could alter calcium levels since cells are lysed and free calcium levels are regulated with 10 m*M* EGTA buffer; a direct effect of cAMP on the contractile machinery seems more likely. Such an effect has been well characterized for smooth muscle and nonmuscle cells, where MLCK can be phosphorylated by cAMP-dependent protein kinase. When phosphorylated, the affinity of MLCK for calcium/calmodulin is greatly reduced (cf. Adelstein, 1980; Adelstein *et al.*, 1982). Thus, cAMP can inhibit actomyosin interaction by making MLCK less sensitive to calcium activation. These findings are consistent with the idea that cAMP inhibits reactivated cone contraction by catalyzing the phosphorylation of MLCK to reduce its affinity for calcium/calmodulin.

Our findings with lysed cell models do not rule out the possibility that cAMP also influences free calcium levels in intact cones *in vivo*. For example, in smooth muscle, cAMP accumulation increases Na^+,K^+-ATPase activity (Scheid *et al.*, 1979), which subsequently increases Na^+/Ca^{2+} exchange (van Breemen *et al.*, 1979) resulting in a decline in intracellular free calcium content (Anderson and Nilsson, 1977). In addition, dibutyryl cAMP has been found to inhibit calcium release and promote calcium sequestration in isolated aortic microsomes (Baudouin-Legros and Meyer, 1973). Thus in cones, as in intact smooth muscle, an increase in cAMP may bring about a decrease in cytoplasmic free calcium levels, indirectly affecting contraction.

Our studies with lysed cell models indicate that the physiological constraints for activation of cone contraction are that calcium levels must rise above 10^{-7} M and cAMP levels must be lower than 10^{-5} M. These findings strongly imply that *in vivo* light onset is accompanied by a rise in cytoplasmic free calcium levels and a fall in cAMP levels in the cone myoid (Fig. 2).

Several observations from our laboratory suggest that the calcium responsible for activating cone contraction *in vivo* is derived from both internal and extracellular sources. Intact cones undergo normal light-induced contraction in calcium-free culture medium containing 5 m*M* EGTA for at least 10 min after light onset. Thus, it seems clear that light can induce the release of sufficient calcium from internal stores to initiate contraction (Porrello and Burnside, 1984). However, in longer incubations (30 min) cones fail to contract to the fully light-adapted positions when whole isolated retinas are cultured in the light in calcium-free culture medium containing 1 m*M* EGTA (Dearry and Burnside, 1984a). These results suggest that cone contraction can be initiated by release of calcium from internal stores, but that an external source of calcium is required for completion of contraction over a longer time course. A similar situation has been reported for some smooth muscles, for which agonist-induced contraction can be initiated in Ca^{2+}-free medium but sustained contraction requires extracellular calcium (van Breeman *et al.*, 1982; Watkins and Davidson, 1980). The endoplasmic reticulum of frog photoreceptor inner segments has recently been shown to accumulate calcium by an ATP-dependent uptake mechanism (Ungar *et al.*, 1984). Our results suggest that light might induce calcium release from such intracellular storage sites in cones.

Since light can initiate cone contraction in the absence of extracellular calcium, the light signal must be transduced in some way to trigger calcium release from internal stores. The nature of this signal is not yet clear; however, recent observations suggest that the membrane hyperpolarization which accompanies light onset might play a role. Light-induced cone contraction is completely blocked if retinas are cultured in a medium containing elevated potassium (55 m*M*, Dearry and Burnside, 1984a), a treatment shown to depolarize the photoreceptor and prevent light-induced hyperpolarization (Cervetto, 1973; Capovilla *et al.*, 1980).

Light is not the only stimulus capable of inducing cone contraction *in vivo*. In fish maintained in constant darkness, cones contract at the time of expected dawn (Levinson and Burnside, 1981; Douglas, 1982; Burnside and Ackland, 1984). Thus, some circadian signal also appears to be capable of inducing cone contraction. In fact, in green sunfish, the species used for all the motile model studies in our laboratory, cones contract in anticipation of expected dawn on the first morning of continuous darkness (Burnside and Ackland, 1984). Thus, in the wild the circadian signal which induces cone contraction may be more important than light onset in controlling the diurnal cycles of cone retinomotor movements. Since circadian signals can induce cone movements, and we know that cAMP and calcium regulate the motile machinery, we have begun to investigate whether any of the hormones or transmitters known to influence cytoplasmic calcium and/or cAMP levels might influence retinomotor movements in cultured retinas. Dearry and Burnside (1984b, 1985a,b) have recently found that dopamine induces cone contraction when injected intraocularly into dark-adapted fish or when added to isolated dark-adapted retinas cultured in the dark. Dopamine-induced cone contraction in isolated retinas is inhibited by antagonists with the potency order sulpiride $>$ haloperidol $\gg$ domperidone $=$ metoclopramide $>$ fluphenazine $>$ SCH23390, a ranking indicative of D-2 receptor mediation. At $10^{-6}\,M$, LY171555, a D-2 dopamine agonist, induces full cone contraction, while SKF38393, a D-1 dopamine agonist, is ineffective. Sulpiride, a D-2 dopamine antagonist, partially inhibits light-induced cone contraction in previously dark-adapted retinas and induces cone elongation in light-adapted retinas. These results indicate that dopamine acts specifically via D-2 receptors to promote cone contraction, and that dopamine release may play a role in both circadian and light-induced cone contraction. Since dopamine inhibits forskolin- and IBMX-induced, but not dibutyryl cAMP-induced, cone elongation in isolated light-adapted retinas, it seems likely that dopamine inhibits adenylate cyclase activity in cones via D-2 receptors (Dearry and Burnside, 1985a). In addition, other findings suggest that GABA and serotonin influence cone retinomotor movements by modulating dopamine relase, GABA by inhibiting and serotonin by stimulating dopamine release (Dearry and Burnside, 1985c). Thus, our results suggest that dopamine serves as the final extracellular messenger directly inducing light-adaptive cone contraction and that GABA and serotonin modulate dopamine release.

2. Cone Elongation

Cone elongation is dependent on microtubules. Colchicine disruption of myoid microtubules prevented subsequent dark-induced elongation, while cytochalasin D had no effect (Burnside, 1976a; Warren and Burnside, 1978; Burnside *et al.*, 1983). Thus in contrast to contraction, which is actin dependent and microtubule independent, cone elongation requires microtubules but not actin filaments (Fig. 2).

Since cone elongation is induced by darkness, and retinal cAMP levels have been reported to be higher in darkness in several species, we tested whether we could induce cone elongation by artificially elevating cytoplasmic cAMP levels in light-adapted retinas. Using both intraocular injections and cultured retinas, we found that cAMP analogs, when administered along with phosphodiesterase inhibitors, induced cone elongation in a dose-dependent fashion (Burnside *et al.*, 1982a; Burnside and Basinger, 1983; Burnside and Ackland, 1984). Analogs of cGMP had no effect on retinomotor movements. In addition, we have recently shown that forskolin, an activator of adenylate cyclase activity, elevates cAMP levels and promotes cone elongation in light-adapted retinas cultured in constant light (Dearry and Burnside, 1985a). These results suggest that an increase in intracellular cAMP concentration is associated with dark-adaptive cone elongation. Similar activation of microtubule-dependent motility by cAMP has been observed in sperm, where elevated cAMP activates flagellar motility (cf. Murofushi *et al.*, 1982; Lindeman, 1978; Tash and Means, 1983). In the case of sperm, cAMP can activate motility in demembranated flagellar axonemes, suggesting that cAMP is affecting the motile machinery rather than ion fluxes across the membrane.

By adding cAMP to reactivation media, we have also succeeded in obtaining reactivated elongation in lysed cone models (Gilson *et al.*, 1985a). Thus it has been possible to contrast the physiological constraints of reactivated elongation with those of contraction just described. Cone models elongated approximately 20 μm when reactivated in medium containing ATP, cAMP, and very low calcium levels (Fig. 7). In reactivation medium containing 1 m*M* ATP and 1 m*M* cAMP, cone motile models elongated at a constant rate of 2 μm/min for 10 min (Fig. 7). This rate is slightly faster than the rates observed *in vivo* with dark-induced elongation (1.6 μm/min) (Gilson *et al.*, 1985a). Thus, reactivated rates of cone elongation were at least as fast as those observed *in vivo*, indicating that the microtubular elongation machinery retained its functional organization after lysis.

Dose–response studies with cAMP showed that cones elongated at a rate proportional to the cAMP concentration between 5 and 500 μ*M* (Gilson *et al.*, 1985a). It seems likely that cAMP stimulates the phosphorylation of proteins necessary for activation of elongation or for force production. Further analysis has suggested that ATP is required both for cAMP-dependent and cAMP-independent processes in cone elongation (Gilson *et al.*, 1985a). Thus, ATP could be used both for activation, via cAMP-dependent phosphorylation, and for force production, for example by dynein.

Calcium inhibits cone elongation in lysed cell models in a dose-dependent fashion between 10^{-7} and 10^{-5} *M* (Gilson *et al.*, 1985a). Maximal inhibition of reactivated cone elongation occurred at the same free calcium concentration which was shown to produce maximal contraction in lysed cone models. This observation suggested that both actin and microtubule machinery might be activated in high Ca^{2+} and high cAMP conditions with the result that competing forces prevent net movement. However, by using cytochalasin D treatment, we were able to demonstrate that calcium inhibition of reactivated elongation did not result from activating a competing contraction in the cone myoid. Even when lysed cone models were treated with a concentration of cytochalasin D which disrupted actin filaments and thus inhibited the cone's ability to contract, calcium continued to inhibit cone elongation (Gilson *et al.*, 1985a). This direct effect of

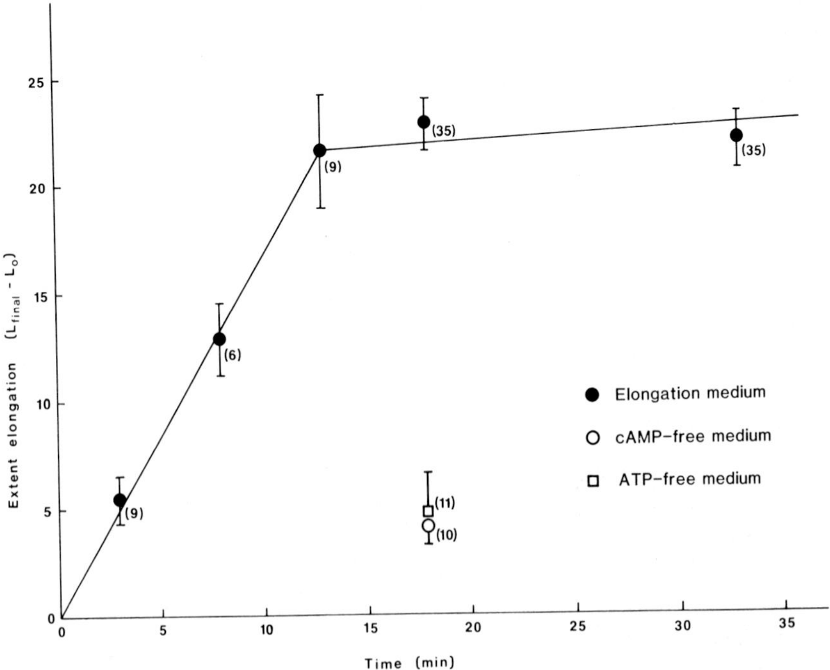

FIG. 7. Kinetics of cone elongation in reactivated lysed cell models of green sunfish. Cone models were prepared as described in Fig. 5 from fish exposed to darkness for 10–20 min to allow detachment of retina from RPE. In elongation medium (1 m*M* ATP, 1 m*M* cAMP, $<10^{-8}$ *M* calcium), lysed cones elongated more than 20 μm at rates comparable to those observed during dark-induced elongation *in vivo*. Less than 5 μm of elongation was observed if cAMP or ATP was deleted from elongation medium. (Data provided by Carol Gilson.)

calcium on the microtubule machinery appears to be mediated by calmodulin, since the calmodulin inhibitor trifluoperazine abolished the calcium block (Gilson *et al.*, 1985a). Calcium inhibition of microtubular motility has also been observed in demembranated flagellar axonemes of sperm from several animal species (Murakami and Takahashi, 1975; Satir, 1979).

Several observations suggest that force production for cone elongation results from microtubule sliding mediated by intermicrotubular mechanochemical bridges analogous to the dynein bridges of cilia and flagella (cf. Satir, 1979). Ultrastructural examination of lysed cone models showed that following detergent lysis the extent of cytoplasmic extraction was severe, yet the microtubules of the cone myoid remained intact (Gilson *et al.*, 1985a). In fact the microtubules were so stable that they were not disrupted by either calcium (10^{-5} *M*) or colchicine (2 m*M*), suggesting that they were not in equilibrium with free

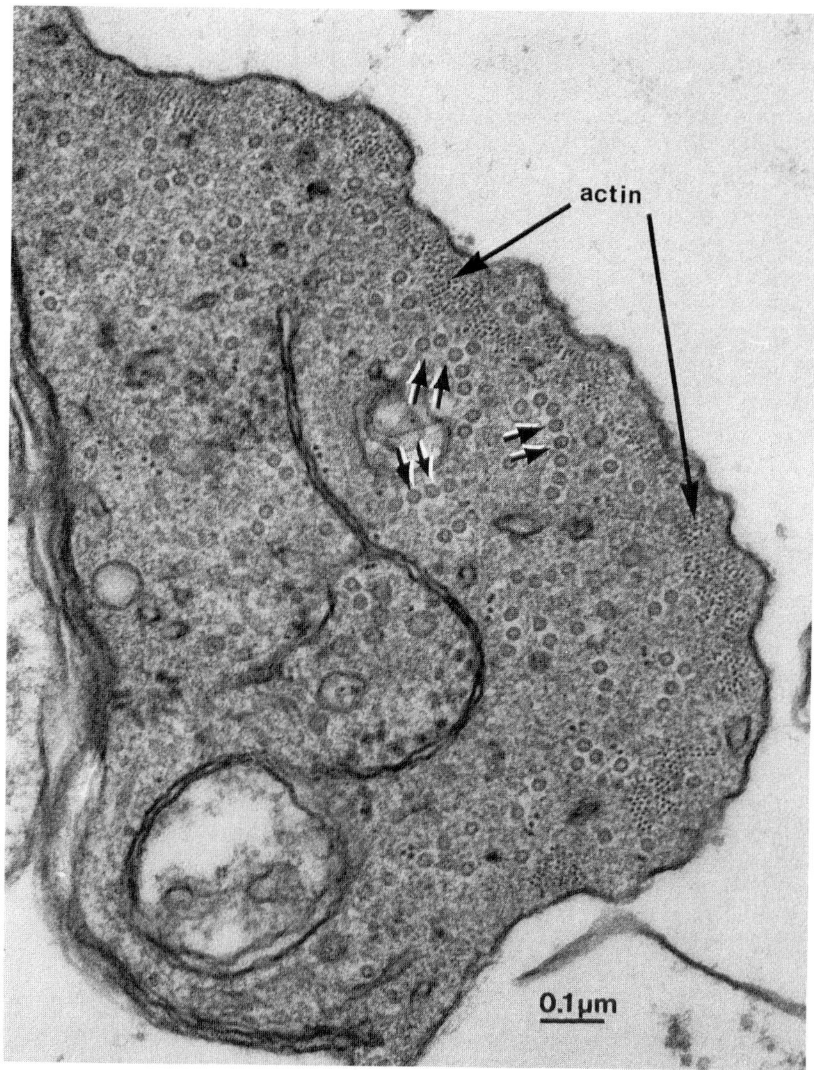

FIG. 8. Electron micrograph of a transverse section through the myoid of a lysed cone model fixed during reactivated elongation. In contrast to the random organization of microtubules in contracting cone myoids (see Fig. 4), microtubules in elongating myoids are laterally associated to form pairs, rows, and pallisades (double arrows). This appearance is consistent with the suggestion that force is produced by dynein-like crossbridges between microtubules. Lysed cone models were prepared as described in Fig. 5.

subunits (cf. Kirshner, 1978). In elongating models, but not in contracting or nonmoving models, microtubules could be found in pallisades of up to eight microtubules, with regular intertubule distances between them, as might be expected if they were connected by bridges (Fig. 8). Alternative mechanisms for microtubule-dependent elongation based on assembly have also been suggested (cf. Warren and Burnside, 1978, for discussion). However, since the lysed cell models are permeable to myosin subfragment-1 (100,000 MW), it seems likely that free tubulin subunits (dimer MW 110,000) would diffuse out of the cone models after lysis and thus be unavailable for assembly. Also an earlier morphometric analysis of microtubule number and distribution during cone elongation found no net increase in total microtubule length during elongation (Warren and Burnside, 1978). Another observation more strongly supports a role for a dynein-like ATPase in cone elongation: the sulfhydryl reagent *N*-ethylmaleimide (NEM), vanadate, and erythro-8-[3-2(hydroxynonyl)]adenine (EHNA) strongly inhibit reactivated cone elongation (Gilson *et al.*, 1985c). Though these agents affect a variety of other ATPases, they are particularly effective in inhibiting both flagellar and cytoplasmic dynein ATPases (cf. Bell and Gibbons, 1983). The sensitivity of elongation to very low dynein concentrations and to NEM argues that dynein rather than kinesin is involved (cf. Vale *et al.*, 1985). Cone elongation but not cone contraction is inhibited by vanadate (10 μM) in lysed cone models (Gilson *et al.*, 1985b). Thus, all observations to date are consistent with the idea that microtubules produce the motive force for cone elongation by sliding over one another by means of dynein-like mechanochemical cross-bridges.

Recent observations in our laboratory suggest that dark induction of cone elongation may be mediated by local release of extracellular messengers. Cone elongation can be induced in light-adapted retinas cultured in the light by prostaglandins (B. Cavallaro and B. Burnside, unpublished observations). PGE_1 induces cone movements which are dose-dependent with maximal movement occurring at 250–500 nM. PGE_2, PGE_3, and PGD_2 also induce dark-adaptive cone movements, while $PGF_{2\alpha}$ does not. Further evidence for a role of prostaglandins in regulating cone movement comes from observations using agents which inhibit prostaglandin synthesis. In retinas treated with agents which inhibit the cyclooxygenase component of the multienzyme complex prostaglandin synthetase, dark-induced cone elongation is inhibited. Indomethicin (10^{-4} M) completely blocks and acetylsalicylic acid (10^{-4} M) partially inhibits dark-induced cone movement.

Cone elongation also appears to be subject to circadian regulation. Like contraction at subjective dawn, cone elongation continues to occur at subjective dusk in fish maintained in constant darkness (see Douglas, 1982; Douglas and Wagner, 1982; Burnside and Nagle, 1983; Levinson and Burnside, 1981; Burnside and Ackland, 1984). Since intermediate cone lengths are reproducibly observed at specified times in the circadian cycle in animals kept in continuous darkness, it seems likely that cones must possess some mechanism for dictating graded lengths. In cultured retinas graded extents of cone retinomotor movement can be elicited by varying external concentrations of cAMP analog, forskolin, or calcium (Burnside *et al.*, 1982b; Dearry and Burnside, 1984a, 1985a). These results are consistent with the idea that the contributions of both light and circadian rhythm to cone retinomotor position are mediated by changes in intracellular cAMP and calcium concentrations (Fig. 2).

B. Rod Retinomotor Movements

Rods respond to changes in light conditions with retinomotor movements opposite to those of cones: in the dark rods contract and in the light they elongate (Figs. 1 and 2; cf. Ali, 1975; O'Connor and Burnside, 1981, 1982; Burnside and Nagle, 1983). As in cones, rod length changes are mediated by the myoid region of the inner segment. In most teleosts, the rod myoid is exceptionally delicate; for example in the cichlid, *Serotherodon,* the myoid diameter ranges from 0.3 μm in the long rod to 2.9 μm in the short rod (O'Connor and Burnside, 1981). In amphibians and catfish, motile rods have larger myoids (cf. Ali, 1975; Burnside and Nagle, 1983). Rod myoids contain both actin filaments (as shown by myosin subfragment-1 decoration) and microtubules, arranged parallel to the long axis of the rod (Klyne and Ali, 1980; O'Connor and Burnside, 1981, 1982).

The distributions of these elements have been described in detail for a cichlid fish by O'Connor and Burnside (1981, 1982). Actin filaments originate in the microvillus-like calyceal processes which cup the base of the outer segment and course down the ellipsoid in bundles just beneath the plasma membrane (Fig. 9). In long light-adapted rods, these actin filaments then form a circumferential ring surrounding the core of the long myoid, which also contains 3–10 paraxial microtubules. In short dark-adapted rods the ellipsoid filament bundles are present, but no filaments are visible in the short myoid. Morphometric analysis indicates that total myoid microtubule length is much greater in long than in short rod myoids (O'Connor and Burnside, 1982). Thus on the grounds of morphological observations alone, it seems likely that both actin filaments and microtubules are assembled during rod elongation and disassembled during contraction. Possible roles of microtubules and actin filaments in force production for rod elongation and contraction have been analyzed by O'Connor and Burnside (1981, 1982). These studies show that both elongation and contraction are actin-dependent rather than microtubule-dependent processes (Fig. 2).

1. ROD ELONGATION

Studies with microtubule and actin filament inhibitors indicate that actin filaments, not microtubules, are required for rod elongation (O'Connor and Burn-

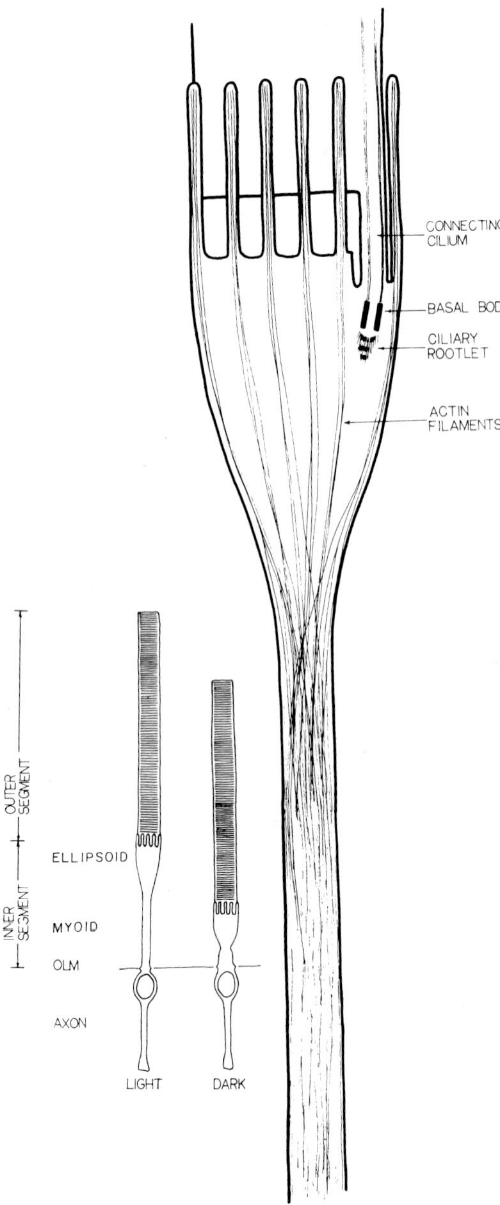

FIG. 9. The cytoskeleton of the teleost rod inner segment. Bundles of actin originate in calyceal processes, course down the ellipsoid, and then splay out in the myoid to form a sleeve of paraxially oriented filaments just beneath the myoid plasmalemma. Though two to five microtubules are present in the center of the myoid (not shown), they are not required for either elongation or contraction; both are actin dependent.

side, 1981). Rods can undergo normal light-induced elongation even if microtubule assembly is prevented by treatment with nocadazole. Thus, despite the fact that microtubules are found in the myoid and assemble there during rod elongation, they do not appear to be required for force production for elongation. Perhaps they provide structural reinforcement to help stabilize the delicate myoid once elongation is achieved.

On the other hand, disruption of actin filaments and prevention of actin assembly with cytochalasins B and D completely blocks light-induced rod elongation (O'Connor and Burnside, 1981). Since myoid actin filaments are assembled during elongation, and since cytochalasin blocks both actin assembly and rod elongation, we have suggested that the force-producing mechanism for rod elongation depends on actin filament assembly and also perhaps cross-linking of the assembled actin filaments to form stiff bundles. Similar actin filament assembly and bundling are associated with the formation of acrosomal filaments in echinoderm sperm (Tilney *et al.*, 1973) and with the microspikes found on growth cones (cf. Korn, 1982; Landis, 1983) and echinoderm eggs (Begg and Rebhun, 1979). These cellular projections are of similar dimensions to teleost rod myoids.

In the delicate rod myoid the mechanism for elongation, which is actin filament dependent, differs from that in the more robust cone myoid, which is microtubule dependent. This difference perhaps reflects the sizes of the forces necessary to produce movement. For example, in neurite outgrowth, filopodia on the growth cone can extend by actin-dependent means in the absence of microtubules (Yamada *et al.*, 1970). However, the larger parent axon itself cannot elongate in the absence of microtubules and eventually retracts, uprooting the still active growth cone (Yamada *et al.*, 1970). It would be interesting in this regard to compare the mechanism of elongation in the larger rods of catfish and amphibians to that of the delicate rods of teleost fishes described here. The actin-dependent mechanism may be size specific rather than rod specific.

Rods may possess dopaminergic D-2 receptors as do cones. Dopamine induces not only cone contraction but also rod elongation in isolated dark-adapted retinas cultured in constant darkness (Dearry and Burnside, 1985b). Light-adaptive movements in both cones and rods are blocked by the D-2 dopamine antagonist sulpiride. In addition, dopamine induces both cone contraction and rod elongation in isolated dark-adapted photoreceptor inner/outer segment fragments cultured in constant darkness. These findings suggest that dopamine induces rod elongation through a D-2 receptor-mediated mechanism and that D-2 receptors are present on rod and cone inner/outer segments. Since dibutyryl cAMP induces dark-adaptive rod contraction in isolated light-adapted retinas (O'Connor and Burnside, 1982), light-adaptive rod elongation may be associated with a relatively low level of intracellular cAMP. Our findings are consistent with the suggestion that dopamine produces light-adaptive rod elongation by binding to

D-2 receptors on rods, thereby inhibiting rod adenylate cyclase activity and lowering rod cAMP concentration.

2. Rod Contraction

Rod contraction is actin dependent. Disruption of actin filaments with cytochalasin D completely blocked dark-induced rod contraction (O'Connor and Burnside, 1982). Thus, the contractile process in both rods and cones is mediated by actin. Indeed, in virtually all cells so far studied, contractile processes are actin rather than microtubule dependent (cf. Korn, 1982; Taylor and Condelis, 1979). Nonetheless, a possible role for microtubules in rod contraction had to be considered because intraocular injections of colchicine blocked dark-induced rod contraction (Klyne and Ali, 1980; O'Connor and Burnside, 1982). Since microtubules were not completely disrupted in Klyne and Ali's study, it seemed possible that the colchicine block might result from microtubule-independent effects. Therefore we tested effects of other microtubule disruption treatments on rod contraction. We found that complete disruption of myoid microtubules with cold and nocodazole had no effect on rod contraction (O'Connor and Burnside, 1982). Thus, it seems clear that the colchicine block of rod contraction was mediated by some microtubule-independent mechanism. In fact, we have found that colchicine actually mimics light by *inducing* rod elongation in constant darkness (O'Connor and Burnside, 1982). Microtubule-independent effects of colchicine on cell shape have recently been emphasized (Beebe *et al.*, 1979), and colchicine has been found to promote light-evoked disk shedding in *Xenopus* eye cup preparations (Besharse and Dunis, 1982). Thus, experiments concerning colchicine effects on photoreceptors should be interpreted with caution.

Rod contraction appears to be regulated by calcium and cAMP, but in a different manner from that observed in cones (cf. Fig. 2). In both photoreceptor types, elevated cAMP levels induce dark-adaptive movements. However, cAMP and darkness induce contraction in rods and elongation in cones (O'Connor and Burnside, 1982; Burnside *et al.*, 1982b). Since rod contraction is an actin-dependent, contractile process, it seems likely that it is dependent on calcium-mediated actin–myosin interaction, as has been observed for all contractile processes so far characterized in metazoan cells (cf. Taylor and Condelis, 1979; Pollard, 1981). Support for this suggestion comes from the observation that rod contraction is blocked by the calmodulin inhibitor R24571 (O'Connor, 1982). Thus, in terms of mechanism of force production, regulation by calcium, and rate of movement, rod contraction appears to resemble cone contraction (O'Connor and Burnside, 1982; Burnside *et al.*, 1982b).

Rods differ from cones, however, in that cAMP and calcium levels appear to be elevated simultaneously. Since rod contraction is promoted by cAMP analogs and inhibited by calmodulin inhibitors, it seems likely that both cAMP and calcium levels are at least transiently increased. In cones cAMP and calcium are antagonistic: elevated calcium is associated with actin-dependent contraction while elevated cAMP is associated with microtubule-dependent elongation (Porrello and Burnside, 1984). In fact when cAMP and calcium are simultaneously elevated in lysed cone models, all movement is blocked. However in rods, we must hypothesize that cAMP either induces an increase in free calcium levels or acts cooperatively with calcium to induce contraction. Contrasting effects of cAMP have also been observed in other cell types, including muscle, where agonists that elevate cAMP induce relaxation in some smooth muscles (cf. Adelstein, 1982; Sherry *et al.*, 1978; Anderson and Nilsson, 1977) and enhance contraction in cardiac muscle (cf. Katz, 1979). It seems likely that different cascades of phosphorylation have different effects on the motile machinery in different cell types. For example, cAMP might activate rod contraction by phosphorylation of some component of the contractile machinery.

Several cytoskeletal proteins have been found to be phosphorylated by cAMP-dependent kinases (Wallach *et al.*, 1978; Selden and Pollard, 1983). One such protein is filamin, an actin-binding protein which cross-links actin filaments and also interferes with actin activation of myosin ATPase (Davies *et al.*, 1977; Wallach *et al.*, 1978). If in rods cAMP-dependent phosphorylation decreased the affinity of some filamin-like protein for actin, activation of contraction might result. Alternatively, cAMP might regulate rod contraction indirectly by altering calcium fluxes across external and internal membranes (Anderson and Nilsson, 1977). Lysed cell model studies, like those with cones, should be able to clarify the roles of these messengers in regulating rod motility.

As a step toward understanding the role of calcium in regulating rod motility and other physiological processes, Nagle and Burnside (1984) have begun to characterize the calmodulin-binding proteins found in detached rod fragments containing inner and outer segments (RIS/ROS) isolated from teleost retinas. The detached rod fragment retained the motile myoid portion of the rod, which could be induced by light to elongate after detachment (Fig. 10).

Using gel overlay techniques with ^{125}I-labeled calmodulin, Nagle and Burnside (1984) have compared calmodulin-binding proteins from whole teleost retina, isolated RIS/ROS, and Triton-extracted RIS/ROS cytoskeletons. Isolated whole retinas have six prominent calmodulin-binding proteins, migrating at 240K, 190K, 150K, 61K, and a doublet at 18/19K. In contrast, detached RIS/ROS have three different prominent calmodulin-binding proteins migrating at 330K, 33K, and 31K. The RIS/ROS cytoskeletons, which are composed of actin filaments, microtubules, and the connecting cilium, contain calmodulin-binding proteins migrating at 240K and 18/19K. Several observations suggest that the 240K calmodulin-binding protein may be the brain spectrin-like protein called fodrin: the 240K retinal calmodulin-binding protein comigrates with the α subunit of brain fodrin and cross-reacts with antibodies to rat brain fodrin (Nagle and Burnside, 1984).

Several roles for spectrin-like proteins in rod motility might be considered.

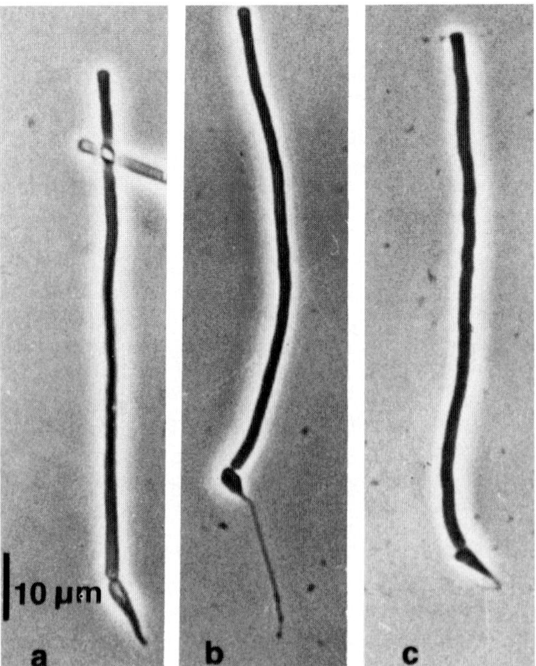

FIG. 10. Phase-contrast light micrographs of detached rod fragments composed of inner and outer segment (RIS/ROS) obtained from dark-adapted teleost retinas by mechanical agitation. Rod inner segments were short immediately after detachment (a). After 30 min culture in the light, myoids elongated more than 15 μm (b). When 1 m*M* dibutyryl cAMP and 1 m*M* IBMX were included in the incubation medium, light-induced myoid elongation was blocked (c). (Reprinted from Nagle and Burnside, 1984, with permission from *European Journal of Cell Biology*.)

Spectrin-like proteins have been shown by immunofluorescence to be localized to the inner surface of the neuronal plasma membrane (Glenney *et al.*, 1982a; Repasky *et al.*, 1982), where they are thought to cooperate with actin to provide a membrane cytoskeleton similar to that in erythrocytes. In rods, such a membrane cytoskeleton might play a role in the assembly of rod disks or provide scaffolding for the great membrane surface of the long delicate teleost rods. Results of recent immunolocalization studies which show actin concentrated at the site of forming disks in frog rods (Chaitin *et al.*, 1984) are consistent with the first suggestion. Spectrin and its nonerythrocyte analogs have also been shown to cross-link actin filaments *in vitro* (Brenner and Korn, 1979; Glenney *et al.*, 1982a). Thus spectrin-like molecules might contribute to cytoplasmic as well as membrane support. O'Connor and Burnside (1981) have proposed that an actin filament cross-linking mechanism provides structural support for the long delicate myoid. A spectrin-like protein could provide calcium-sensitive cross-links.

In addition, brain fodrin can stimulate actomyosin ATPase (Shimo-Oka and Watanabe, 1981). This result suggests that a rod spectrin-like protein could function not only as a structural element but also as an active regulator of the contractile process.

C. RPE Retinomotor Movements

In the RPE of most lower vertebrates, melanin pigment granules migrate in and out of the apical projections in response to changes in light conditions (Figs. 1 and 2; cf. Ali, 1975; Burnside and Nagle, 1983). They migrate out into the apical projections in the light and aggregate back toward the basal (scleral) end of the RPE cell in the dark. As has been shown for vertebrate melanophores (cf. Schliwa, 1984), pigment migration in RPE cells occurs within the confines of relatively fixed apical projections (cf. Burnside and Nagle, 1983). Thus, RPE retinomotor movements represent a form of intracellular transport rather than an extension of amoeboid apical projections.

The mechanism of force production for pigment granule transport has not yet been definitively elucidated for either RPE cells or melanophores (cf. Burnside and Nagle, 1983; Schliwa, 1984). However, several observations suggest that both microtubules and actin filaments may play active roles. Both actin filaments and microtubules have been reported to be present in apical RPE projections as well as RPE cell bodies (Murray and Dubin, 1975; Burnside, 1976b; Klyne and Ali, 1980; Burnside *et al.*, 1983). Disrupting actin filaments with cytochalasin D completely blocks light-induced pigment dispersion in fish RPE cells (Burnside *et al.*, 1983), and in some (but not all) dermal melanophores (cf. Schliwa, 1984). In fact cytochalasin treatment also induces pigment aggregation in fully light-adapted fish (Burnside *et al.*, 1983). These results suggest that actin filaments are required not only for pigment dispersion but also for maintaining the dispersed state. This conclusion is supported by morphological observations that actin filaments are prominent in the apical projections, often in close association with the membranes surrounding pigment granules and with the plasma membrane (Murray and Dubin, 1975; Burnside *et al.*, 1983; Burnside and Nagle, 1983). On the other hand, actin filaments do not appear to be necessary for pigment aggregation since their disruption with cytochalasin D fails to block dark-induced aggregation (Burnside *et al.*, 1983).

Microtubule disruption with colchicine blocks both dispersion and aggregation of granules within the RPE cell body. Since microtubules of the apical projections are not disrupted by colchicine, it is not possible to ascertain whether they are needed for pigment granule migration within these projections (cf. Klyne and Ali, 1980; Burnside *et al.*, 1983). Since microtubules do appear to be required for pigment movement in the RPE cell body and also in several types of fish chromatophores (cf. Schliwa, 1984), microtubule participation in pigment movements in the apical projections is a likely possibility.

The locus of control for regulating pigment granule migration in RPE cells has been the subject of numerous investigations (cf. Burnside and Nagle, 1983, for review). Two observations indicate that the control is local: (1) pigment dispersion can be induced by light in cultured, enucleated eyes (Burnside and Basinger, 1983); and (2) a tiny spot of light directed onto the dark-adapted retina induces pigment dispersion (and also cone movement) only in the spot illuminated, the rest of the retina remaining dark adapted (Easter and Macy, 1978). However, three observations suggest that the light signal does not act directly on the RPE cells themselves but rather appears to be mediated by rods: (1) light does not induce pigment movement in RPE if retina is not present (Burnside and Basinger, 1983; Snyder and Zadunaisky, 1976); (2) the action spectrum for RPE pigment dispersion in frogs and fishes resembles that of rods (Liebman *et al.*, 1969; Ali and Crouzy, 1968); and (3) RPE pigment movements do not appear during development until after rods develop, although well-developed cones and RPE are present earlier (cf. Blaxter, 1970). The nature of the signal which passes from rods to RPE is not known.

Recently we have found that dopamine can induce pigment dispersion in dark-adapted fish after intraocular injection in the absence of light (Dearry and Burnside, 1984b, 1985b) and in pure suspensions of isolated RPE cells (Bruenner and Burnside, 1985). This latter finding indicates that there is no mechanical requirement for the retina to be present to permit pigment dispersion, and also strongly suggests that RPE cells have a dopamine receptor on their surfaces. Since dopamine is released from retina by light stimulation both *in vivo* and from isolated retinas (Kramer, 1971; Reading, 1983), a role for dopamine in retina-to-RPE communication is an interesting possibility. This hypothesis is supported by the work of Dawis and Niemeyer (1984), who suggested that dopamine release from the neural retina may contribute to generating the RPE light peak potential. RPE pigment migration not only is regulated by light but in some species also exhibits a circadian rhythm (cf. Burnside and Nagle, 1983; Douglas, 1982). Thus, there is some precedent for looking for a diffusible factor which can influence pigment dispersion in the absence of light.

Both cAMP and calcium may play roles in regulating RPE retinomotor movements (Fig. 2). Treatments which elevate cytoplasmic cAMP induce dark-adaptive pigment granule aggregation in light-adapted fish *in vivo* (Burnside *et al.*, 1982a), in cultured light-adapted RPE-retinas (Burnside and Basinger, 1983; Dearry and Burnside, 1984a, 1985a; Burnside and Ackland, 1984), and in dissociated light-adapted RPE cells in culture (Bruenner and Burnside, 1985).

Several observations suggest that RPE cells possess their own adenylate cyclase system capable of elevating cytoplasmic cAMP levels. Koh and Chader 1984) have recently demonstrated that cultured chick RPE cells are capable of generating cAMP in response to the β-adrenergic agonist isoproterenol. Forskolin, a stimulator of adenylate cyclase, induces dark-adaptive pigment aggregation in isolated RPE cells (U. Bruenner and B. Burnside, unpublished observation) and in light-adapted RPE-retinas cultured in the light (Dearry and Burnside, 1985a). The proximal signal which stimulates the activation of adenylate cyclase at darkness onset *in vivo* is not known. Since elevating extracellular calcium levels inhibits dark-induced RPE aggregation in cultured retinas (Dearry and Burnside, 1984a), and since forskolin can overcome this inhibitory effect of calcium (Dearry and Burnside, 1985a), it seems possible that calcium and cAMP act antagonistically in RPE in a manner analogous to that in cones, where cAMP favors dark-adaptive movement and inhibits light-adaptive movement while calcium favors light-adaptive movement and inhibits dark-adaptive movement.

V. Invertebrate Retinomotor Movements

Invertebrate eyes exhibit two major types of retinomotor phenomena: screening pigment migration and changes in rhabdom shape (cf. reviews by Hoglund, 1966; Goldsmith and Bernard, 1974; Walcott, 1975; Langer, 1975). Most experimental investigations of these movements have dealt with screening pigment migration since it is more extensive and easily quantifiable than rhabdom movement.

Several aspects of invertebrate movements provide interesting parallels with those of vertebrates just described. In most species, screening pigment movement reduces the amount of light reaching the rhabdom in light-adapted eyes (Hoglund, 1966; Daw and Pearlman, 1974; Walcott, 1974; Wilson, 1975). As in the vertebrate RPE (Liebman *et al.*, 1969), invertebrate pigment movement appears to be activated by visual pigment bleaching: in several species the action spectrum for pigment migration matches the visual pigment absorption spectrum (Franceschini, 1972; Menzel and Knaut, 1973; Daw and Pearlman, 1974; Kolb and Autrum, 1974; Stavenga *et al.*, 1975). However, Hamdorf and Hoglund (1981) have proposed that a separate ultraviolet- or blue-sensitive system located within the pigment cells may partially mediate pigment migration in the moth.

In invertebrates light-induced membrane potential change may play a role in signaling pigment movement, as has been suggested for vertebrate cone movement (Dearry and Burnside, 1984a). In crayfish, replacement of extracellular Na^+ by K^+ mimics the normal light-induced depolarization, promotes light-adaptive pigment dispersion in constant darkness, and prevents dark-induced pigment aggregation (Frixione and Arechiga, 1981). Additionally, several workers have suggested that an increase in free intracellular calcium triggers light-adaptive pigment dispersion in invertebrate eyes (Olivo and Larsen, 1978;

Kirschfeld and Vogt, 1980; Frixione and Arechiga, 1981; Lo and Pak, 1981; Howard, 1984).

Invertebrate retinomotor movements are regulated not only by light/dark stimuli, but also by an endogenous circadian rhythm (Kleinholz, 1937; Welsh, 1930; Smith, 1948; Bennitt, 1932; Parker, 1932; Page and Larimer, 1975). In *Limulus,* morphological pigment and rhabdom changes continue in complete darkness and follow the circadian rhythm of optic nerve activity (Barlow and Chamberlain, 1980; Barlow *et al.,* 1980). Efferent input from the optic nerve induces dark-adapted pigment distribution and rhabdom shape and position. This efferent input may be mediated by release of octopamine or a substance P-like peptide from efferent terminals. Octopamine may stimulate photoreceptor adenylate cyclase activity, and the resultant increase in intracellular cAMP level may produce dark-adaptive changes in photoreceptor morphology and physiology (Battelle *et al.,* 1982; Kass and Barlow, 1981, 1982, 1984; Pelletier *et al.,* 1984; O'Day and Lisman, 1984). Substance P has also been reported to induce structural and physiological changes associated with dark adaptation (Mancillas and Brown, 1984; Mancillas and Selverston, 1984). Thus, the induction of dark-adaptive movements in invertebrate eyes by agents known to increase intracellular cAMP content resembles the ability of forskolin and cAMP analogs to promote dark-adaptive movements in vertebrate photoreceptors and RPE (O'Connor and Burnside, 1982; Burnside *et al.,* 1982a; Burnside and Basinger, 1983; Dearry and Burnside, 1984a, 1985a). Despite their gross anatomical differences, it appears that both invertebrate and vertebrate retinomotor movements may be regulated through light- and circadian-induced changes in cytoplasmic calcium and cAMP levels.

The mechanism of force production for invertebrate pigment granule transport appears to be microtubule dependent. In *Limulus,* colchicine induces light-adaptive pigment dispersion, suggesting a microtubular mechanism for maintaining the aggregated state (Miller and Cawthorn, 1974). In crayfish, microtubules appear to be necessary for both aggregative movement and maintenance of the aggregated state (Frixione *et al.,* 1979). Cytochalasin B has no effect on either dark- or light-induced movement in crayfish (Frixione *et al.,* 1979), suggesting that actin filaments are not directly involved in pigment migration.

VI. Cytoskeleton and Motility in RPE Cells

A. Maintenance of RPE Cell Shape and Epithelial Integrity

Regardless of species, the cells of the vertebrate retinal pigment epithelium have characteristic shapes and associations. The epithelium is composed of a single layer of hexagonally packed cells, joined at their apices by junctional complexes. Basal infoldings cover the surface adjacent to Bruch's membrane, and the apical membrane is thrown into projections of various shapes according to species and location in the eye. The intimate association of these apical projections with the photoreceptor outer segments almost certainly contributes to such critical functions as retinal attachment, photoreceptor turnover, and photoreceptor alignment.

Since the cytoskeleton of the RPE cell can be expected to contribute to the development and maintenance of its characteristic shape as well as to mediate various cellular motile processes, numerous workers have examined the cytoskeleton in RPE cells of both normal and dystrophic retinas. Prominent components of the RPE cytoskeleton have been shown to include actin (Burnside, 1976b; Murray and Dubin, 1975; Crawford *et al.*, 1972; Chaitin and Hall, 1983a,b; Burnside *et al.*, 1983; Owaribe *et al.*, 1981; Drenckhahn and Groschel-Stewart, 1977; Haley *et al.*, 1983a,b; Philp and Nachmias, 1985; Owaribe and Eguchi, 1985), microtubules (Burnside, 1976b; Burnside and Laties, 1979; Klyne and Ali, 1981; Irons and Kalnins, 1984), and intermediate filaments (Takeuchi and Takeuchi, 1979; Burnside and Laties, 1979; Klyne and Ali, 1981).

Apical projections of RPE cells range in size and shape from short microvillus-like and leaf-like processes less than 5 μm in length to extremely long delicate projections more than 100 μm in length (Burnside, 1976b; Burnside and Laties, 1979; Steinberg and Wood, 1979; Burnside *et al.*, 1983; Philp and Nachmias, 1985; Owaribe and Eguchi, 1985). Mammalian RPE cells possess two distinct populations of processes: short processes often associated with rods, and long leaf-like projections which reach down to ensheath cone outer segments (Steinberg and Wood, 1974). The cytoskeletal mechanisms responsible for development and maintenance of these apical projections are not known. However, several morphological studies suggest that actin filaments may play an important role, especially in mammalian RPE. Leaf-like projections of mammalian RPE contain numerous microfilaments aligned in rows and closely associated with the plasma membrane (Burnside and Laties, 1976, 1979; Burnside, 1976b). These filaments have been shown to be actin by myosin subfragment-1 decoration (Burnside and Laties, 1976). Since these actin filaments are laterally associated with the plasma membranes of the apical projections, it seems likely that they lend mechanical strength to the projections. In addition, they may contribute to motile activities.

Associated with the apical junctional complexes in each RPE cell is a dense circumferential bundle of actin filaments (Burnside, 1976b; Crawford, 1979; Owaribe *et al.*, 1981). Crawford (1979) observed that these bundles are very pronounced during the conversion of cultured RPE cells from a flat monolayer to a more typical epithelial appearance and suggested that these actin bundles act like a purse string to contribute to the columnar or cuboidal shape of the RPE cells in the epithelium. Owaribe and co-workers (1981, 1982) have carried out an elegant series of studies which show that the RPE apical circumferential actin bundles are in fact contractile and thus could contribute to RPE shape as Crawford suggested. They have shown that the apices of RPE cells can be reactivated to contract after glycerination by addition of Mg-ATP (Owaribe *et al.*, 1981) and that the actin filament bundles themselves can be isolated from glycerinated RPE cells without losing their ability to contract. Contraction of isolated bundles was inhibited by NEM-modified subfragment-1, suggesting that actin–myosin interaction mediates contraction (Owaribe and Masuda, 1982).

Recently three research groups have begun to examine the cytoskeletons of chick retinal epithelial cells using gel electrophoresis and immunological techniques. Opas and colleagues (1985) have examined the locations of two actin-building proteins, spectrin and vinculin, in the cytoskeletons of cultured chick RPE cells during maturation of the culture. The cells occupying the center of the colony resemble RPE cells *in vivo* and are cuboidal, pigmented, and relatively nonadherent, whereas those toward the periphery are flatter, nonpigmented, motile, and strongly adherent to the substratum. Immunofluorescence localization studies using antiserum against chicken erythrocyte α-spectrin revealed that this protein was present in the cortex of the RPE cells in all parts of the colony. Vinculin, however, was localized differently in cuboidal and flat RPE cells: in the flat cells it was associated with cell–substrate attachment sites, while in the cuboidal cells it was found at the circumferential band of microfilaments associated with the zonula adherens of the cells. Owaribe and Eguchi (1985) have examined developmental changes in the RPE cytoskeleton by examining RPE taken directly from chick embryos at different stages. They found that actin content in the cells markedly increased between the 15-day embryo and hatching stages. Using immunofluorescence, they have shown that the increase in actin is associated with the formation of apical projections and the development of the core bundle of actin filaments found in these projections. The actin filaments of this bundle are regularly packed into a paracrystalline array. Philp and Nachmias (1985) have examined the components of the chick RPE cytoskeleton using electrophoresis and antibodies to several cytoskeletal proteins. By relative mobility and immunoblotting they have identified actin, vimentin, myosin, spectrin, and α-actinin as major components. By immunofluorescence they showed that the apical circumferential bundle of filaments appeared to be similar to that seen in intestinal epithelial cells, i.e., it contained myosin, actin, spectrin, and α-actinin. In the apical projections they observed staining for myosin, actin, spectrin, α-actinin, and vinculin. Thus the localization of vinculin appears to differ in freshly isolated RPE cells as compared to cultured RPE cells as described by Opas and colleagues (1985). Vinculin has been found in many cell types to be localized at sites of actin-membrane attachment.

B. Roles of Actin and Microtubules in RPE Phagocytosis

One of the major RPE functions is to phagocytose and degrade the shed tips of photoreceptor outer segments (Young and Bok, 1969; Bok and Young, 1979). Since the detailed biology of photoreceptor turnover is discussed elsewhere in this volume, we consider here only a brief analysis of the motile aspects of RPE phagocytosis. In RPE as in other phagocytes, microtubules and actin filaments appear to play important roles (Stossel, 1974). Disrupting actin filaments with cytochalasin D blocks RPE phagocytosis by preventing ingestion of phagosomes (Besharse and Dunis, 1982). Colchicine disruption of microtubules does not block ingestion but does block transport of newly ingested phagosomes to the base of the RPE cell where they fuse with lysosomes for subsequent digestion (Beauchemin and Leuenberger, 1977; Herman and Steinberg, 1982). Thus, actin appears to be required for phagosome ingestion and microtubules for intracellular transport of phagosomes. Clearly the failure of either of these cytoskeletal systems could block phagocytosis.

In RCS rats with inherited retinal degeneration, RPE phagocytosis is defective (Bok and Hall, 1971; Edwards and Szamier, 1977; Goldman and O'Brien, 1978). Therefore, several workers have closely examined actin and microtubules in the RPE cells of these animals. Using immunofluorescence studies, Chaitin and Hall (1983a,b) have shown that although ingestion is greatly reduced in cultured RPE cells from dystrophic RCS rats, some phagocytosis does occur. The few successfully ingested disk packages exhibit recruitment of an actin feltwork at the site of ingestion in a manner indistinguishable from that observed in control cultured RPE cells. This finding suggests that the defect in phagocytosis in dystrophic RCS rats might occur at the recognition and attachment stage rather than in the motile machinery. An immunofluorescence study of cultured normal and dystrophic RPE cells revealed no detectable differences in microtubule assembly and distribution (Irons and Kalnins, 1984).

VII. Morphological Aspects of Retinal Synaptic Modulation

Several workers have suggested that morphological alterations of synaptic contacts between retinal neurons might have a significant influence on the efficiency of synaptic communication and thereby contribute to synaptic modulation (Wagner, 1973; Raynauld *et al.*, 1979; Cooper and McLaughlin, 1982; Weiler and Wagner, 1984). For example, considerable interest has been focused on shape changes of dendritic spines (cf. Crick, 1982). The shapes of these spines, which serve as the primary sites of synaptic contact on many types of neurons, are sensitive to presynaptic input and can therefore be modified by experience or experimental manipulation (Coss and Globus, 1978). Since spines appear to be dynamic structures capable of rapid morphological changes, their shape change could influence synaptic transmission by affecting the length constant and resistance of the spine (Fifkova and Delay, 1982).

The cytoskeleton has been attributed a role in the shape changes of dendritic spines. Immunofluorescence and myosin S-1 subfragment decoration have demonstrated that actin is organized in filaments in dendritic spines and that these filaments appear to be associated with the plasma membrane and postsynaptic density (Fifkova and Delay, 1982; Caceres *et al.*, 1983). Microtubule-associated protein 2 (MAP2) has also been localized by immunofluorescence to dendritic spines (Caceres *et al.*, 1983). MAP2 has been shown to cross-link actin filaments into bundles or gels in a calcium- and cAMP-dependent manner (Griffith and Pollard, 1982; Selden and Pollard, 1983). Thus, postsynaptic changes in intracellular calcium and/or cAMP could lead to changes in actin filament organization and consequent changes in dendritic spine structure.

In teleost retinas, dendritic spines on horizontal cells change shape in response to light and dark signals. During light adaptation, H1 and H2 horizontal cell terminals that invaginate cone synaptic pedicles give rise to 0.3-μm-long spines; during dark adaptation these spines disappear (Wagner, 1980). A 1- to 2-min light exposure is as effective as continual illumination in eliciting light-adaptive spine formation (Wagner and Douglas, 1983). Spines form or disappear over a 60-min period following light or dark onset. These changes thus follow a time course similar to that of photoreceptor and RPE retinomotor movements. Microfilaments observed below the plasma membrane of retinal horizontal cell spines (Wagner, 1980) could play a role in spine shape changes. Weiler and Wagner (1984) have recently recorded intracellular potentials from carp horizontal cells and found that the depolarization of H2 horizontal cells to a red light stimulus is correlated with the number of spines arising from H1 horizontal cell terminals during light adaptation. Djamgoz *et al.* (1985) have suggested that the appearance of these spines during light adaption may provide the synaptic sites responsible for inhibitory feedback from horizontal cells to cones.

In addition to dendritic spine shape changes, light and dark onset may induce changes in the degree of horizontal cell invagination into photoreceptor synaptic endings. These morphological alterations occur in the synaptic contacts of both rods and cones (Cragg, 1969; Raynauld *et al.*, 1979; Cooper and McLaughlin, 1982). Schaeffer and Raviola (1975, 1978) found that horizontal cell processes are extruded from turtle cone synaptic endings in the light. This change in synaptic configuration was attributed to changes in addition/retrieval of synaptic vesicle membrane to the plasma membrane. It was hypothesized that in the dark, when the cone is relatively depolarized, the rates of transmitter release and synaptic vesicle fusion are high and more membrane is therefore incorporated into the plasma membrane. This results in expansion of the presynaptic membrane, and the postsynaptic processes are thus more deeply invaginated.

Stevens and co-workers (1984) have recently proposed that such diurnal variations in photoreceptor synaptic structure may represent an active "retinomotor" movement rather than simple membrane addition/subtraction. Using serial electron micrographs and computer reconstruction, they demonstrated that cat cone pedicles contain a 4- to 6-μm-diameter ring of cross-striated fibrils that surrounds the invaginating bipolar and horizontal cell synaptic contacts. Such cross-striated structures have been observed in photoreceptor inner segments from a number of species (Cohen, 1960; Cragg, 1969; Spira and Milman, 1979, 1982). They resemble the cross-striated ciliary and flagellar rootlets described above and thus might also exhibit calcium-mediated contraction (cf. Salisbury *et al.*, 1984). The properties of cross-striated rootlets are consistent with the suggestion (Stevens *et al.*, 1984) that the ring of cross-striated fibrils in cat cone pedicles serves a contractile function and thus participates in determining cone pedicle shape and the degree of postsynaptic invagination.

Several observations suggest that the cytoskeleton may play some direct role in synaptic vesicle release. The idea that actin may be an integral part of synaptic vesicle exocytosis and recycling (Berl *et al.*, 1973; Puszkin and Schook, 1979) is supported by the preferential concentration of actin at photoreceptor synapses (Drenckhahn and Kaiser, 1983) and of fodrin in synaptic membrane preparations (Sobue *et al.*, 1979, 1982). In addition, a close structural relationship is observed between microtubules, synaptic vesicles, and membrane densities during development of cortical synapses, suggesting that microtubules may have a role in forming presynaptic dense projections and in directing synaptic vesicles to the site of transmitter release (Westrum *et al.*, 1983). Gray (1983) has suggested that synaptic vesicles move along the surface of presynaptic microtubules and that these microtubules are arranged in a precise geometrical array in order to channel vesicles to the presynaptic active zone.

Synaptic microfilaments and microtubules may also have a role in mediating shape changes of synaptic structures. Several workers have observed diurnal variations in photoreceptor synaptic ribbons and vesicles. For example, in turtle rods, synaptic ribbons appear as single stick-shaped profiles during the night and grow into large multilayered ribbon complexes during the day (Abe and Yamamoto, 1984). The number of synaptic ribbons in fish photoreceptors also declines in the dark (Wagner, 1973). In newt photoreceptors, the number of dense-cored synaptic vesicles decreases during the night and increases during the day (Ball and Dickson, 1983). These observations are consistent with the notion that transmitter release is high during the night, while increasing quantities of transmitter are stored in the synapse during the day.

VIII. Alignment of Photoreceptors

Observations obtained with histological, psychophysical, and X-ray diffraction techniques indicate that vertebrate retinal photoreceptors are preferentially aligned toward the pupil of the eye (Laties *et al.*, 1968; Laties, 1969; Laties and Enoch, 1971; Webb, 1977; Enoch *et al.*, 1979; see reviews by Enoch, 1978, 1979). Photoreceptor inner and outer segments vary in orientation with respect to the outer limiting membrane in a graded manner across the retina. Inner and outer segments of photoreceptors near the retinal periphery subtend greater angles than those in the central retina, so that outer segments are directed toward the center of the exit pupil of the eye (Laties and Enoch, 1971; Enoch and Hope, 1972a,b). Directional sensitivity significantly contributes to the waveguide properties of the photoreceptor (Enoch, 1975). Most of the light striking a rod outer segment obliquely will be excluded from reaching the rhodopsin-bearing disk membranes and consequently have a low probability of being detected. Effects of spherical aberration and scattered light are thereby reduced.

Normal receptor alignment may be partially dependent upon interaction between photoreceptors and RPE. Receptor orientation is lost following detachment of the retina but can recover following reattachment (Frankhauser *et al.*, 1961; Enoch *et al.*, 1973; Fitzgerald *et al.*, 1980). Different areas of the retina may regain alignment at different times (Campos *et al.*, 1978), a finding which implies that local changes can occur in receptor alignment.

The ability to reestablish retinal directional sensitivity following trauma suggests that the capacity for reorienting photoreceptor alignment is continuously present. This idea is supported by the work of Applegate and Bonds (1981). They created a "new" pupil by placing a soft contact lens containing an artificial pupil over a subject's eye. Over 5 days, retinal directional sensitivity was shifted in the direction of the new pupil. This alignment persisted as long as the new pupil was worn (1 month), and alignment returned to its original configuration upon removal of the new pupil. These results suggest that an active mechanism is responsible for receptor alignment toward the pupillary aperture.

The nature of such a mechanism is unknown. It has been proposed that microfilaments, microtubules, or the striated ciliary rootlet in photoreceptor inner segments may be involved in mediating receptor alignment (Laties and Burnside, 1979; Enoch, 1979; Applegate and Bonds, 1981). Although it seems likely that one or more or these cytoskeletal elements has a role in establishing receptor orientation, no direct demonstration of their involvement has yet been obtained.

IX. Concluding Remarks

In this review we have described a variety of motile events which play integral roles in retinal function. We have described numerous microtubule- and actin-dependent processes and have emphasized the reported ways in which the cytoskeleton is influenced by the intracellular messengers calcium and cAMP. Since these messengers play central roles in regulating motility in almost all cell types so far considered, it is not unrealistic to expect that they play similar regulatory roles in the retina. Many aspects of retinal physiology exhibit dramatic diurnal and circadian cycles of activity (cf. LaVail, 1976, 1980; Besharse, 1982). In this review, we have described in detail how patterned changes in cytoplasmic cAMP and calcium levels regulate diurnal and circadian cycles of teleost retinomotor movements (Fig. 2). We suggest that similar changes in cAMP and calcium levels are likely to be associated with light and dark onset and with circadian signals in all retinas and that these changes may coordinate not only motile processes but other physiological and metabolic processes as well.

Acknowledgments

The authors extend their appreciation to Carol Gilson for drawing figures, to Julieta Gonzalez and Marisa D'Souza for patiently typing and retyping the manuscript, and to Barbara Nagle and Carol Gilson for critical readings of the manuscript. The authors' own work is supported by NIH Grants EY03575 and GM32566.

References

Abe, H., and Yamamoto, Y. (1984). Diurnal changes in synaptic ribbons of rod cells of the turtle. *J. Ultrastruct. Res.* **86,** 246–251.

Adelstein, R. S. (1980). Phosphorylation of muscle proteins. *Fed. Proc. Fed. Am. Soc. Exp. Biol.* **39,** 1544–1549.

Adelstein, R. S. (1982). Calmodulin and the regulation of actin–myosin interaction in smooth muscle and non-muscle cells. *Cell* **30,** 349–350.

Adelstein, R. S., Sellers, J. R., Conti, M. A., Pato, M. D., and deLanerolle, P. (1982). Regulation of smooth muscle contractile proteins by calmodulin and cyclic AMP. *Fed. Proc. Fed. Am. Soc. Exp. Biol.* **41,** 2873–2878.

Ali, M. A. (1975). Retinomotor responses. *In* "Vision in Fishes" (M. A. Ali, ed.), Series A, pp. 313–355. Plenum, New York.

Ali, M. A., and Crouzy, R. (1968). Action spectrum and quantal thresholds of retinomotor responses in the brook trout *Salvelinus fontinalis* (Mitchill). *Z. Vergl. Physiol.* **59,** 86–89.

Amos, W. B. (1971). A reversible mechanochemical cycle in the contraction of *Vorticella. Nature (London)* **229,** 127–129.

Amos, W. B. (1975). Contraction and calcium binding in vorticellid ciliates. *In* "Molecules and Cell Movement" (R. E. Stephens, ed.), pp. 411–436. Raven, New York.

Anderson, R. G. G., and Nilsson, K. B. (1977). Role of cyclic nucleotide metabolism and mechanical activity in smooth muscle. *In* "The Biochemistry of Smooth Muscle" (L. Stephens, ed.), pp. 236–291. Univ. Park Press, Baltimore, Maryland.

Applegate, R., and Bonds, A. (1981). Induced movement of receptor alignment toward a new pupillary aperture. *Invest. Ophthalmol. Visual Sci.* **21,** 869–873.

Ball, A., and Dickson, H. (1983). Diurnal variations in photoreceptor synaptic terminals of the newt retina. *Am. J. Anat.* **168,** 305–320.

Barlow, R., and Chamberlain, S. (1980). Light and a circadian clock modulate structure and function in *Limulus* photoreceptors. *In* "The Effects of Constant Light on Visual Processes" (T. Williams and B. Baker, eds.), pp. 247–269. Plenum, New York.

Barlow, R., Chamberlain, S., and Levinson, J. (1980). *Limulus* brain modulates the structure and function of the lateral eyes. *Science* **210,** 1037–1039.

Battelle, B., Evans, J., and Chamberlain, S. (1982). Efferent fibers to *Limulus* eyes synthesize and release octopamine. *Science* **216,** 1250–1252.

Baudouin-Legros, M., and Meyer, P. (1973). Effects of angiotensin, catecholamines, and cyclic AMP on calcium storage in aortic microsomes. *Br. J. Pharmacol.* **47,** 377–385.

Beauchemin, M. L., and Leuenberger, P. M. (1977). Effects of colchicine on phagosome–lysosome interaction in retinal pigment epithelium. I. *In vivo* observations in albino rats. *Abstr. von Graefes Arch. Klin. Exp. Ophthalmol.* **203,** 237–241.

Beebe, D. C., Feagans, D. E., Blanchette-Mackie, J., and Nan, M. (1979). Lens epithelial cell elongation in the absence of microtubules: Evidence for a new effect of colchicine. *Science* 206, 836–839.

Begg, D. A., and Rebhun, L. I. (1979). pH regulates polymerization of actin in the sea urchin egg. *J. Cell Biol.* **83,** 241–248.

Bell, W., and Gibbons, I. R. (1983). Preparation and properties of dynein ATPase. *In* "Muscle and Nonmuscle Motility" (A. Stracher, ed.), pp. 1–36. Academic Press, New York.

Bennett, V. (1982). The molecular basis for membrane–cytoskeletal association in human erythrocytes. *J. Cell. Biochem.* **18,** 49–65.

Bennett, V., Davis, J., and Fowler, W. E. (1982). Immunoreactive forms of erythrocyte spectrin and ankyrin in brain. *Philos. Trans. R. Soc. London Ser. B* **299,** 301–312.

Bennitt, R. (1932). Diurnal rhythm in the proximal pigment cells of the crayfish retina. *Physiol. Zool.* **5,** 65–69.

Berl, S., Puszkin, S., and Nicklas, W. (1973). Actomyosin-like protein in brain. Actomyosin-like protein may function in the release of transmitter material at synaptic endings. *Science* **179,** 441–446.

Besharse, J. C. (1982). The daily light–dark cycle and rhythmic metabolism in the photoreceptor–pigment epithelial complex. *In* "Progress in Retinal Research" (N. Osborne and G. Chader, eds.), Vol. 1, pp. 81–124. Pergamon, Oxford.

Besharse, J. C., and Dunis, D. A. (1982). Rod photoreceptor disk shedding *in vitro:* Inhibition by cytochalasins and activation by colchicine. *In* "The Structure of the Eye" (J. G. Hollyfield, ed.), pp. 85–96. Elsevier, New York.

Birchmeier, W. (1984). Cytoskeleton structure and function. *Trends Biochem. Sci.* **9,** 192–195.

Blaxter, J. H. S. (1970). Light: Fishes. *In* "Marine Ecology" (O. Kinne, ed.), pp. 213–320. Wiley (Interscience), New York.

Blum, J. J., Hayes, A., Jamieson, G. A., and Vanaman, T. C. (1980). Calmodulin confers calcium sensitivity on ciliary dynein ATPase. *J. Cell Biol.* **87,** 386–397.

Blum, J. J., Hayes, A., Jamieson, G. A., and Vanaman, T. C. (1981). Interrelationships between thermal, *N*-ethylmaleimide and Ca^{++}-calmodulin mediated activation–inactivation of dynein ATPase activities. *Arch. Biochem. Biophys.* **210,** 363–371.

Bok, D., and Hall, M. O. (1971). The role of the pigment epithelium in the etiology of inherited retinal dystrophy in the rat. *J. Cell Biol.* **49,** 664–682.

Bok, D., and Young, R. W. (1979). Phagocytic properties of the retinal pigment epithelium. *In* "The Retinal Pigment Epithelium" (K. M. Zinn and M. F. Marmor, eds.), pp. 148–174. Harvard Univ. Press, Cambridge, Massachusetts.

Brenner, S. L., and Korn, E. D. (1979). Spectrin–actin interaction. Phosphorylated and dephosphorylated spectrin tetramers cross-link F-actin. *J. Biol. Chem.* **254,** 8620–8627.

Bruenner, U., and Burnside, B. (1985). Pigment granule migration in isolated cells of the teleost retinal pigment epithelium. *Invest. Ophthalmol. Visual Sci.,* in press.

Burke, B. E., and DeLorenzo, R. J. (1981). Ca^{++} stimulated and calmodulin stimulated endogenous phosphorylation of neurotubulin. *Proc. Natl. Acad. Sci. U.S.A.* **78,** 991–995.

Burnside, B. (1976a). Microtubules and actin filaments in teleost visual cone elongation and contraction. *J. Supramol. Struct.* **5,** 257–275.

Burnside, B. (1976b). Possible roles of microtubules and actin filaments in retinal pigmented epithelium. *Exp. Eye Res.* **23,** 257–275.

Burnside, B., and Ackland, N. (1984). Effects of circadian rhythm and cAMP on retinomotor movements in the green sunfish, *Lepomis cyanellus. Invest. Ophthalmol. Visual Sci.* **25,** 539–545.

Burnside, B., and Basinger, S. (1983). Retinomotor pigment migration in the teleost retinal pigment epithelium. II. Cyclic-3′,5′-adenosine monophosphate induction of dark-adaptive movement *in vitro. Invest. Ophthalmol. Visual Sci.* **24,** 16–23.

Burnside, B., and Laties, A. M. (1976). Actin filaments in apical projections of the primate pigmented epithelial cell. *Invest. Ophthalmol. Visual Sci.* **15,** 570–575.

Burnside, B., and Laties, A. M. (1979). Pigment movements and cellular contractility in the retinal pigment epithelium. *In* "The Retinal Pigment Epithelium" (K. M. Zinn and M. F. Marmor, eds.), pp. 175–191. Harvard Univ. Press, Cambridge, Massachusetts.

Burnside, B., and Nagle, B. (1983). Retinomotor movements of photoreceptors and retinal pigment epithelium: Mechanisms and regulation. *In* "Progress in Retinal Research" (N. Osborn and G. Chader, eds.), Vol. 2, pp. 67–109. Pergamon, Oxford.

Burnside, B., Evans, M., Fletcher, R. T., and Chader, G. J. (1982a). Induction of dark-adaptive retinomotor movement (cell elongation) in teleost retinal cones by cyclic adenosine 3′,5′-monophosphate. *J. Gen. Physiol.* **79,** 759–774.

Burnside, B., Smith, B., Nagata, M., and Porrello, K. (1982b). Reactivation of contraction in teleost retinal cones. *J. Cell Biol.* **92,** 199–206.

Burnside, B., Adler, R., and O'Connor, P. (1983). Retinomotor pigment migration in the teleost retinal pigment epithelium. I. Roles for actin and microtubules in pigment granule transport and cone movement. *Invest. Ophthalmol. Visual Sci.* **24,** 1–15.

Burnside, B., Ackland, N., and Cande, Z. (1984). Calcium-independent activation of reactivated contraction in lysed-cone models by unregulated myosin light chain kinase and high magnesium. *J. Cell Biol.* **99,** 183a.

Caceres, A., Payne, M., Binder, L., and Steward, O. (1983). Immunocytochemical localization of actin and microtubule-associated protein MAP2 in dendritic spines. *Proc. Natl. Acad. Sci. U.S.A.* **80,** 1738–1742.

Cachon, J., and Cachon, M. (1981). Movement by non-actin filament mechanisms. *Biosystems* **14,** 313–326.

Campos, E., Bedell, H., Enoch, J., and Fitzgerald, C. (1978). Retinal receptive field-like properties and Stiles–Crawford effect followed in a patient with traumatic choroidal rupture. *Doc. Ophthalmol.* **45,** 381–395.

Cande, W. Z., and Wolniak, S. M. (1978). Chromosome movement in lysed mitotic cells is inhibited by vanadate. *J. Cell Biol.* **79,** 573–580.

Capovilla, M., Cervetto, L., and Torre, V. (1980). Effects of changing external potassium and chloride concentrations on the photoresponses of *Bufo bufo* rods. *J. Physiol. (London)* **307,** 529–551.

Cervetto, L. (1973). Influence of sodium, potassium, and chloride ions on the intracellular responses of turtle receptors. *Nature (London)* **241,** 401–403.

Chafouleas, J. G., Dedman, J. R., Manjaal, R. P., and Means, A. R. (1979). Calmodulin-development and application of a sensitive radioimmunoassay. *J. Biol. Chem.* **254,** 10262–10267.

Chaitin, M. H., and Hall, M. O. (1983a). Defective ingestion of rod outer segments by cultured dystrophic rat pigment epithelial cells. *Invest. Ophthalmol. Visual Sci.* **24,** 812–820.

Chaitin, M. H., and Hall, M. O. (1983b). The distribution of actin in cultured normal and dystrophic rat pigment epithelial cells during the phagocytosis of rod outer segments. *Invest. Ophthalmol. Visual Sci.* **24,** 821–831.

Chaitin, M. H., Schneider, B. G., Hall, M. O., and Papermaster, D. S. (1984). Actin in the photoreceptor connecting cilium: Immunocytochemical localization to the site of outer segment disk formation. *J. Cell Biol.* **99,** 237–247.

Cohen, A. (1960). The ultrastructure of the rods of the mouse retina. *Am. J. Anat.* **107,** 23–38.

Cooper, N., and McLaughlin, B. (1982). Structural correlates of physiological activity in chick photoreceptor synaptic terminals: Effects of light and dark stimulation. *J. Ultrastruct. Res.* **79,** 58–73.

Coss, R., and Globus, A. (1978). Spine stems on tectal interneurons in jewel fish are shortened by social stimulation. *Science* **200,** 787–790.

Cragg, B. (1969). Structural changes in naive retinal synapses detectable within minutes of first exposure to daylight. *Brain Res.* **15,** 79–96.

Craig, S. W., and Pollard, T. D. (1982). Actin-binding proteins. *Trends Biochem. Sci.* **7,** 88–92.

Crawford, B. (1979). Cloned pigmented retinal epithelium. The role of microfilaments in the differentiation of cell shape. *J. Cell Biol.* **81,** 301–315.

Crawford, B., Cloney, R. A., and Cahn, R. D. (1972). Cloned pigmented cells: The effects of cytochalasin B on ultrastructure and behavior. *Z. Zellforsch. Mikrosk. Anat.* **130,** 135–151.

Crick, F. (1982). Do dendritic spines twitch? *Trends Neurosci.* **5,** 44–46.

Davies, P., Shizua, Y., Olden, K., Gallo, M., and Pastan, I. (1977). Phosphorylation of filamin and other proteins in cultured fibroblasts. *Biochem. Biophys. Res. Commun.* **74,** 300–307.

Daw, N., and Pearlman, A. (1974). Pigment migration and adaptation in the eye of the squid, *Loligo pealei. J. Gen. Physiol.* **63,** 22–36.

Dawis, S., and Niemeyer, G. (1984). Dopamine: A messenger from the neural retina to the retinal pigment epithelium? *Invest. Ophthalmol. Visual Sci. Suppl.* **25,** 289a.

Dearry, A., and Burnside, B. (1984a). Effects of extracellular Ca^{++}, K^{+}, and Na^{+} on cone and RPE retinomotor movements in isolated teleost retinas. *J. Gen. Physiol.* **83,** 589–611.

Dearry, A., and Burnside, B. (1984b). Influence of catecholamines on cone and RPE retinomotor movements in teleost fish. *Invest. Ophthalmol. Visual Sci. Suppl.* **25,** 65a.

Dearry, A., and Burnside, B. (1985a). Dopamine inhibits forskolin- and IBMX-induced dark-adaptive retinomotor movements in isolated teleost retinas. *J. Neurochem.* **44,** 1753–1763.

Dearry, A., and Burnside, B. (1985b). Dopaminergic regulation of cone retinomotor movement in isolated teleost retinas. I. Induction of cone contraction is mediated by D-2 receptors. *J. Neurochem.,* in press.

Dearry, A., and Burnside, B. (1985c). Dopaminergic regulation of cone retinomotor movement in isolated teleost retinas. II. Modulation by GABA and serotonin. *J. Neurochem.,* in press.

de Lanerolle, P., Nishikawa, M., Yost, D., and Adelstein, R. (1984). Increased phosphorylation of myosin light chain kinase after an increase in cyclic AMP in intact smooth muscle. *Science* **223,** 1415–1417.

Djamgoz, M., Downing, J., Prince, D., and Wagner, H. (1985). Physiological and ultrastructural evidence for light-dependent plasticity of horizontal cell-cone photoreceptor feedback interactions in the isolated retina of a cyprinid fish. *J. Physiol. (London)* **362,** 20p.

Douglas, R. (1982). An endogenous crepuscular rhythm of rainbow trout (*Salmo gairdneri*) photomechanical movements. *J. Exp. Biol.* **96,** 377–388.

Douglas, R., and Wagner, H. (1982). Endogenous patterns of photomechanical movements in teleosts and their relation to activity rhythms. *Cell Tissue Res.* **266,** 133–142.

Drenckhahn, D., and Groschel-Stewart, U. (1977). Localization of myosin and actin in ocular non-muscle cells: Immunofluorescence-microscopic, biochemical and electron-microscopic studies. *Cell Tissue Res.* **181,** 493–503.

Drenckhahn, D., and Kaiser, H. (1983). Evidence for the concentration of F-actin and myosin in synapses and in the plasmalemmal zone of axons. *Eur. J. Cell Biol.* **31,** 235–240.

Easter, S. S., and Macy, A. (1978). Local control of retinomotor activity in the fish retina. *Vision Res.* **18,** 937–942.

Ebashi, S. (1983). Regulation of contractility. *In* "Muscle and Nonmuscle Motility" (A. Stracher, ed.), pp. 217–232. Academic Press, New York.

Ebashi, S., Mikawa, T., Hirata, M., and Nonomura, Y. (1978). The regulatory role of calcium in muscle. *Ann. N.Y. Acad. Sci.* **307,** 451–461.

Edwards, R. B., and Szamier, R. B. (1977). Defective phagocytosis of isolated rod outer segments by RCS rat retinal pigment epithelium in culture. *Science* **197,** 1001–1003.

Enoch, J. M. (1975). Vertebrate rods are directionally sensitive. *In* Photoreceptor Optics" (A. Snyder and R. Menzel, eds.), pp. 17–37. Springer-Verlag, Berlin and New York.

Enoch, J. M. (1978). The relationship between retinal receptor orientation and photoreceptor optics. *Int. Ophthalmol. Clin.* **18,** 41–80.

Enoch, J. M. (1979). Vertebrate receptor optics and orientation. *Doc. Ophthalmol.* **48,** 373–388.

Enoch, J. M., and Hope, G. (1972a). An analysis of retinal receptor orientation. III. Results of initial psychophysical tests. *Invest. Ophthalmol. Visual Sci.* **11,** 765–782.

Enoch, J. M., and Hope, G. (1972b). An analysis of retinal receptor orientation. IV. Center of the entrance pupil and the center of convergence of orientation and directional sensitivity. *Invest. Ophthalmol. Visual Sci.* **11,** 1017–1021.

Enoch, J. M., Van Loo, J., and Okun, E. (1973). Realignment of photoreceptors distributed in orientation secondary to retinal detachment. *Invest. Ophthalmol. Visual Sci.* **12,** 849–853.

Enoch, J. M., Birch, D., and Birch, E. (1979). Monocular light exclusion for a period of days reduces directional sensitivity of the human retina. *Science* **206,** 705–707.

Erickson, P. F., Seamon, K. B., Moore, B. W., Lasher, R. S., and Minier, L. N. (1980). Axonal transport of the Ca^{++} dependent protein modulator of 3′-5′ cyclic AMP. *J. Neurochem.* **35,** 242–248.

Estes, J. E., Selden, L. A., and Gershman, L. C. (1981). Mechanism of action of phalloidin on the polymerization of muscle actin. *Biochemistry* **20,** 708–712.

Farber, D. (1981). Cyclic nucleotides in cone dominant retinas. Reduction of cyclic AMP levels by light and by cone degeneration. *Invest. Ophthalmol. Visual Sci.* **20,** 24–31.

Feinstein, M., Egan, J., and Opas, E. (1983). Reversal of thrombin-induced myosin phosphorylation and the assembly of cytoskeletal structures in platelets by the adenylate cyclase stimulants prostaglandin D_2 and forskolin. *J. Biol. Chem.* **258,** 1260–1267.

Fifkova, E., and Delay, R. (1982). Cytoplasmic actin in neuronal processes as a possible mediator of synaptic plasticity. *J. Cell Biol.* **95,** 345–350.

Fitzgerald, C., Enoch, J. M., Birch, D., Benedetto, M., Tenme, L., and Dawson, W. (1980). Anomalous pigment epithelium/photoreceptor relationships and receptor orientation. *Invest. Ophthalmol. Visual Sci.* **19,** 956–966.

Franceschini, N. (1972). Pupil and pseudopupil in the compound eye of Drosophila. *In* "Information Processing in the Visual System of Arthropods" (R. Rehner, ed.), pp. 75–82. Springer-Verlag, Berlin and New York.

Frankhauser, F., Enoch, J. M., and Cibis, P. (1961). Receptor orientation in retinal pathology. *Am. J. Ophthalmol.* **52,** 767–783.

Frixione, E., and Arechiga, H. (1981). Ionic dependence of screening pigment migrations in crayfish retinal photoreceptors. *J. Comp. Physiol.* **144,** 35–43.

Frixione, E., Arechiga, H., and Tsutsumi, V. (1979). Photomechanical migrations of pigment granules along the retinula cells of the crayfish. *J. Neurobiol.* **10,** 573–590.

Fuchs, E., and Hanukoglu, I. (1983). Unraveling the structure of intermediate filaments. *Cell* **34,** 332–334.

Gilson, C., Ackland, N., and Burnside, B. (1985a). Reactivated elongation in lysed-cell models of teleost retinal cones. Activation by cAMP and inhibition by calcium. *J. Cell Biol.*, in press.

Gilson, C., Ackland, N., and Burnside, B. (1985b). Vanadate and EHNA inhibit elongation in lysed cell models of teleost retinal cones (abstract). *J. Cell Biol.*, in press.

Glenney, J. R., Glenny, P., and Weber, K. (1982a). F-actin binding and cross-linking properties of porcine brain fodrin, a spectrin-related molecule. *J. Biol. Chem.* **257,** 9781–9787.

Glenney, J. R., Glenney, P., and Weber, K. (1982b). Erythroid spectrin, brain fodrin, and intestinal brush border proteins (TW 260/240) are related molecules containing a common calmodulin-binding subunit bound to a variant cell-type specific subunit. *Proc. Natl. Acad. Sci. U.S.A.* **79,** 4002–4005.

Goldman, A. I., and O'Brien, P. (1978). Phagocytosis in the retinal pigment epithelium of the RCS rat. *Science* **201,** 1023–1025.

Goldman, R. D., Milsted, A., Schloss, J. A., Starger, J., and Yerna, M. (1979). Cytoplasmic fibers in mammalian cells: Cytoskeletal and contractile elements. *Annu. Rev. Physiol.* **41,** 703–741.

Goldsmith, T., and Bernard, G. (1974). The visual system of insects. *In* "Physiology of Insects" (M. Rockstein, ed.), Vol. II, pp. 165–272. Academic Press, New York.

Gray, E. (1983). Neurotransmitter release mechanisms and microtubules. *Proc. R. Soc. London Ser. B* **218,** 253–258.

Griffith, L. M., and Pollard, T. D. (1978). Evidence for actin filament–microtubule interaction mediated by microtubule-associated proteins. *J. Cell Biol.* **78,** 958–965.

Griffith, L. M., and Pollard, T. D. (1982). The interaction of actin filaments with microtubules and microtubule-associated proteins. *J. Biol. Chem.* **257,** 9143–9151.

Haimo, L. T., Telzer, B. R., and Rosenbaum, J. L. (1979). Dynein binds to and cross-bridges cytoplasmic microtubules. *Proc. Natl. Acad. Sci. U.S.A.* **76,** 5759–5763.

Haley, J. E., Flood, M. T., Gouras, P., and Kjeldbye, H. M. (1983a). Proteins from the human retinal pigment epithelial cells: Evidence that a major protein is actin. *Invest. Ophthalmol. Visual Sci.* **24,** 803–811.

Haley, J. E., Flood, M. T., Kjeldbye, H. M., Maiello, E. M., Bilck, M. K., and Gouras, P. (1983b). Two dimensional electrophoresis of proteins in human retinal pigment epithelial cells: Identification of cytoskeletal proteins. *Electrophoresis* **4,** 133–137.

Hamdorf, K., and Hoglund, G. (1981). Light induced retinal screening pigment migration independent of visual cell activity. *J. Comp. Physiol.* **143,** 305–309.

Herman, K. G., and Steinberg, R. H. (1982). Phagosome movement and the diurnal pattern of phagocytosis in the tapetal retinal pigment epithelium of the opossum. *Invest. Ophthalmol. Visual Sci.* **23,** 277–290.

Hesterberg, L. K., and Weber, K. (1983). Ligand-induced conformational changes in villin, a calcium-controlled actin-modulating protein. *J. Biol. Chem.* **258,** 359–364.

Hill, T. L., and Kirschner, M. W. (1983). Regulation of microtubule and actin filament assembly–disassembly by associated small and large molecules. *Int. Rev. Cytol.* **84,** 185–234.

Hisanaga, S., and Pratt, M. M. (1982). Cytoplasmic dynein and flagellar dynein activation by calmodulin. *J. Cell Biol.* **94,** 322A.

Hoglund, G. (1966). Pigment migration, light screening, and receptor sensitivity in the compound eye of nocturnal lepidoptera. *Acta Physiol. Scand.* **69** (Suppl. 282), 1–56.

Howard, J. (1984). Calcium enables photoreceptor pigment migration in a mutant fly. *J. Exp. Biol.* **113,** 471–475.

Huber, G., and Matus, A. (1984). Differences in the cellular distributions of two microtubule-associated proteins, MAP1 and MAP2, in rat brain. *J. Neurosci.* **4,** 151–160.

Huxley, H. E. (1983). Molecular basis of contraction in cross-striated muscles and relevance to motile mechanisms in other cells. *In* "Muscle and Nonmuscle Motility" (A. Stracher, ed.), pp. 1–104. Academic Press, New York.

Hyams, J. S., and Borisy, G. G. (1978). Isolated flagellar apparatus of *Chlamydomonas*—characterization of forward swimming and alteration of waveform and reversal of motion by calcium ions *in vitro*. *J. Cell Sci.* **33,** 235–253.

Imhof, B., Marti, U., Boller, K., Frank, H., and Birchmeier, W. (1983). Association between coated vesicles and microtubules. *Exp. Cell Res.* **145,** 199–207.

Inoue, S. (1981). Cell division and mitotic spindle. *J. Cell Biol.* **91,** 131s–147s.

Inoue, S., and Ritter, H. (1975). Dynamics of mitotic spindle organization and function. *In* "Molecules and Cell Movement" (S. Inoue and R. E. Stephens, eds.), pp. 3–30. Raven, New York.

Irons, M. J., and Kalnins, V. I. (1984). Distribution of microtubules in cultured RPE cells from normal and dystrophic rats. *Invest. Ophthalmol. Visual Sci.* **25,** 434–439.

Jameson, L., Frey, T., Zeeberg, B., Daldorf, F., and Caplow, M. (1980). Inhibition of microtubule assembly by phosphorylation of microtubule-associated proteins. *Biochemistry* **19,** 2472–2479.

Jones, H. P., Bradford, M. M., McRorie, R. A., and Cormier, M. J. (1978). High levels of a calcium-dependent modulator protein in spermatozoa and its similarity to brain modular protein. *Biochem. Biophys. Res. Commun.* **82,** 1264–1272.

Jones, H. P., Lenz, R. W., Palevotz, B. A., and Cormier, M. J. (1980). Calmodulin localization in mammalian spermatozoa. *Proc. Natl. Acad. Sci. U.S.A.* **77,** 2772–2776.

Kane, R. E. (1982). Structural and contractile roles of actin in sea urchin egg cytoplasmic extracts. *Cell Differ.* **11,** 258–287.

Kass, L., and Barlow, R. (1981). Octopamine increases the ERG of the *Limulus* lateral eye. *Biol. Bull.* **159,** 487a.

Kass, L., and Barlow, R. (1982). Efferent neurotransmission of circadian rhythms in *Limulus* lateral eye: Single cell studies. *Biol. Bull.* **163,** 386a.

Kass, L., and Barlow, R. (1984). Efferent neurotransmission of circadian rhythms in *Limulus* lateral eye. I. Octopamine-induced increases in retinal sensitivity. *J. Neurosci.* **4,** 908–917.

Katz, A. M. (1979). Response of the heart to catecholamines. *Adv. Cyclic Nucleotide Res.* **11,** 303–343.

Kendrick-Jones, J., and Scholey, J. M. (1981). Myosin-linked regulatory systems. *J. Mus. Res. Cell Motility* **2,** 347–372.

Kerrick, W. G. L., Hoar, P. E., Cassidy, P. S., and Malencik, D. A. (1980). Ca^{++} regulation of contraction in skinned muscle fibers. *In* "Regulation of Muscle Contraction: Excitation–Contraction Coupling" (A. D. Grinnell and M. A. B. Brazier, eds.), pp. 227–239. Academic Press, New York.

Kiehart, D. P. (1981). Studies on the *in vivo* sensitivity of spindle microtubules to calcium ions and evidence for a vesicular calcium-sequestering system. *J. Cell Biol.* **88,** 604–617.

Kim, H., Binder, L., and Rosenbaum, J. (1979). The periodic association of MAP2 with brain microtubules *in vitro*. *J. Cell Biol.* **80,** 266–276.

Kirschfeld, K., and Vogt, K. (1980). Calcium ions and pigment migration in fly photoreceptors. *Naturwissenschaften* **67,** 516–517.

Kirschner, M. W. (1978). Microtubule assembly and nucleation. *Int. Rev. Cytol.* **54,** 1–71.

Kleinholz, L. (1937). Studies in the pigmentary system of crustacea. II. Diurnal movements of the retinal pigments of Bermudian decapods. *Biol. Bull.* **72,** 176–189.

Klymkowsky, M. W., Miller, R. H., and Lane, E. B. (1983). Morphology, behavior, and interaction of cultured epithelial cells after the antibody-induced disruption of keratin filament organization. *J. Cell Biol.* **96,** 494–509.

Klyne, M. A., and Ali, M. A. (1980). Microtubules and 10 nm filaments in the retinal pigment epithelium during the light–dark cycle. *Cell Tissue Res.* **214,** 397–406.

Koh, S., and Chader, G. (1984). Retinal pigment epithelium in culture demonstrates a distinct beta-adrenergic receptor. *Exp. Eye Res.* **38,** 7–14.

Kolb, G., and Autrum, H. (1974). Selektive Adaptation and Pigmentwanderung in den Sehzellen des Bienenauges. *J. Comp. Physiol.* **94,** 1–6.

Korn, E. D. (1982). Actin polymerization and its regulation by proteins from nonmuscle cells. *Physiol. Rev.* **62,** 672–737.

Kramer, S. (1971). Dopamine: A retinal neurotransmitter. I. Retinal uptake, storage, and light-stimulated release of ^{3}H-dopamine *in vivo. Invest. Ophthalmol. Visual Sci.* **10,** 438–452.

Kumar, N. (1981). Taxol induced polymerization of purified tubulin. *J. Biol. Chem.* **256,** 10435–10441.

Landis, S. C. (1983). Neuronal growth cones. *Annu. Rev. Physiol.* **45,** 567–580.

Langer, H. (1975). Properties and functions of screening pigments in insect eyes. *In* "Photoreceptor Optics" (A. Snyder and R. Menzel, eds.), pp. 429–455. Springer-Verlag. Berlin and New York.

Laties, A. (1969). Histological techniques for study of photoreceptor orientation. *Tissue Cell* **1,** 63–81.

Laties, A., and Burnside, B. (1979). The maintenance of photoreceptor orientation. *In* "Motility in Cell Function" (F. Pepe, J. Sanger, and V. Nachmias, eds.), pp. 285–298. Academic Press, New York.

Laties, A., and Enoch, J. M. (1971). An analysis of retinal receptor orientation. I. Angular relationships of neighboring photoreceptors. *Invest. Ophthalmol. Visual Sci.* **10,** 69–77.

Laties, A. M., Liebman, P. A., and Campbell, C. E. M. (1968). Photoreceptor orientation in the primate eye. *Nature (London)* **218,** 172–174.

LaVail, M. M. (1976). Rod outer segment disc shedding in rat retina: Relationship to cyclic lighting. *Science* **194,** 1071–1073.

LaVail, M. M. (1980). Circadian nature of rod outer segment disc shedding in the rat. *Invest Ophthalmol. Visual Sci.* **19,** 407–413.

Lazarides, E. (1982). Intermediate filaments: A chemically heterogeneous developmentally regulated class of proteins. *Annu. Rev. Biochem.* **51,** 219–250.

Lee, U. C., and Wolff, J. (1982). Two opposing effects of calmodulin on microtubule assembly depend on the presence of microtubule-associated proteins. *J. Biol. Chem.* **257,** 6306–6310.

Levinson, G., and Burnside, B. (1981). Circadian rhythms in teleost retinomotor movements. *Invest. Ophthalmol. Visual Sci.* **20,** 294–303.

Liebman, P. A., Carroll, S., and Laties, A. (1969). Spectral sensitivity of retinal screening pigment migration in the frog. *Vision Res.* **9,** 377–384.

Lindeman, C. B. (1978). cAMP induced increase in motility of demembranated bull sperm models. *Cell* **13,** 9–18.

Lindwall, G., and Cole, R. D. (1984). Phosphorylation affects the ability of tau protein to promote microtubule assembly. *J. Biol. Chem.* **259,** 5301–5305.

Lo, M., and Pak, W. (1981). Light-induced pigment granule migration in the retinular cells of *Drosophila melanogaster*. Comparison of wild-type with ERG-defective mutants. *J. Gen. Physiol.* **77,** 155–175.

Mabuchi, I. (1973). ATPase in the cortical layer of sea urchin egg: Its properties and interaction with cortex protein. *Biochim. Biophys. Acta* **297,** 317–322.

Mancillas, J., and Brown, M. (1984). Neuropeptide modulation of photosensitivity. I. Presence, distribution, and characterization of a substance P-like peptide in the lateral eye of *Limulus. J. Neurosci.* **4,** 832–846.

Mancillas, J., and Selverston, A. (1984). Neuropeptide modulation of photosensitivity. II. Physiological and anatomical effects of substance P on the lateral eye of *Limulus. J. Neurosci.* **4,** 847–859.

Margolis, R. (1983). Calcium and microtubules. *In* "Calcium and Cell Function" (W. Cheung, ed.), pp. 313–335. Academic Press, New York.

Means, A. R., and Dedman, J. R. (1980). Calmodulin: An intracellular calcium receptor. *Nature (London)* **285,** 73–77.

Meeusen, R. L., and Cande, W. Z. (1979). *N*-Ethylmaleimide-modified heavy meromyosin—a probe for actomyosin interactions. *J. Cell Biol.* **82,** 57–65.

Melkonian, M. (1983). Functional and phylogenetic aspects of the basal apparatus in algal cells. *J. Submicrosc. Cytol.* **15,** 121–125.

Menzel, R., and Knaut, R. (1973). Pigment movement during light and chromatic adaptation in the retinula cells of *Formica polyctena* (Hymenoptera, Formicidae). *J. Comp. Physiol.* **86,** 125–138.

Miller, W., and Cawthorn, D. (1974). Pigment granule movement in *Limulus* photoreceptors. *Invest. Ophthalmol. Visual Sci.* **13,** 401–405.

Mooseker, M. S. (1983). Actin-binding proteins of the brush border. *Cell* **35,** 11–13.

Murakami, A., and Takahashi, K. (1975). Correlation of electrical and mechanical responses in nervous control of cilia. *Nature (London)* **257,** 48–49.

Murofushi, H., Ishiguro, K., and Sakai, H. (1982). Involvement of cyclic AMP-dependent protein kinase and a protein factor in the regulation of the motility of sea urchin and starfish spermatozoa. *In* "Biological Functions of Microtubules and Related Structures" (H. Sakai, H. Mobri, and G. Borisky, eds.), pp. 163–176. Academic Press, New York.

Murray, R. L., and Dubin, M. W. (1975). The occurrence of actinlike filaments in association with migrating pigment granules in frog retinal pigment epithelium. *J. Cell Biol.* **64,** 705–710.

Nagao, S., Banno, Y., Mozawa, Y., Sobue, K., Yamazakil, R., and Kakiushi, S. (1981). Subcellular distribution of calmodulin and calmodulin binding sites in *Tetrahymena pyriformis. J. Biochem.* **90,** 897–899.

Nagle, B. W., and Burnside, B. (1984). Calmodulin-binding proteins in teleost retina, rod inner and outer segments, and rod cytoskeletons. *Eur. J. Cell Biol.* **33,** 248–257.

Naitoh, Y., and Kaneko, H. (1973). Control of ciliary activities by adenosine triphosphate and divalent cations in Triton extracted models of *Paramecium caudatum. J. Exp. Biol.* **58,** 657–662.

Nishida, E., Kuwaki, T., and Sakai, H. (1981). Phosphorylation of microtubule-associated proteins (MAPs) and pH of the medium control interaction between MAPs and actin filaments. *J. Biochem.* **90,** 575–578.

O'Connor, P. M. (1982). The role of the cytoskeleton in photomechanical movements of teleost retinal rods. Ph.D. dissertion, University of California, Berkeley.

O'Connor, P. M., and Burnside, B. (1981). Actin-dependent cell elongation in teleost retinal rods: Requirement for actin filament assembly. *J. Cell Biol.* **89,** 517–524.

O'Connor, P. M., and Burnside, B. (1982). Elevation of cyclic AMP activates an actin-dependent contraction in teleost retinal rods. *J. Cell Biol.* **95,** 445–452.

O'Day, P., and Lisman, J. (1984). Octopamine enhances dark-adaptation rates in *Limulus* ventral photoreceptors. *Invest. Ophthalmol. Visual Sci. Suppl.* **25,** 235a.

Olivo, R., and Larsen, M. (1978). Brief exposure to light initiates screening pigment migration in the retinula cells of the crayfish *Procambarus. J. Comp. Physiol.* **125,** 91–96.

Opas, M., Turksen, K., and Kalnins, V. (1985). Adhesiveness and distribution of vinculin and spectrin in retinal pigmented epithelial cells during growth and differentiation *in vitro. Dev. Biol.* **107,** 269–280.

Osborn, M., and Weber, K. (1982). Intermediate filaments: Cell-type specific markers in differentiation and pathology. *Cell* **31,** 303–306.

Owaribe, K., and Eguchi, G. (1985). Increase in actin contents and elongation of apical projections in retinal pigmented epithelial cells during development of the chicken eye. *J. Cell Biol.* **101,** 590–596.

Owaribe, K., and Masuda, H. (1982). Isolation and characterization of circumferential microfilament bundles from retinal pigmented epithelial cells. *J. Cell Biol.* **95,** 310–315.

Owaribe, K., Arabi, M., Hatano, S., and Eguchi, G. (1981). Demonstration of contractility of circumferential actin bundles and its morphological significance in pigmented epithelium *in vitro* and *in vivo*. *J. Cell Biol.* **90,** 507–514.

Page, T., and Larimer, J. (1975). Neural control of circadian rhythmicity in the crayfish. II. The ERG amplitude rhythm. *J. Comp. Physiol.* **97,** 81–96.

Parker, G. (1932). The movements of the retinal pigment. *Ergeb. Biol.* **9,** 239–291.

Pelletier, J., Kass, L., Renninger, G., and Barlow, R. (1984). cAMP and octopamine partially mimic a circadian clock's effect on *Limulus* photoreceptors. *Invest. Ophthalmol. Visual Sci. Suppl.* **25,** 288a.

Philp, N. J., and Nachmias, V. T. (1985). Components of the cytoskeleton in the retinal pigmented epithelium of the chick. *J. Cell Biol.* **101,** 358–362.

Piperno, G. (1984). Monoclonal antibodies to dynein subunits reveal the existence of cytoplasmic antigens in sea urchin egg. *J. Cell Biol.* **98,** 1842–1850.

Pitelka, D. R. (1974). Basal bodies and root structures. *In* "Cilia and Flagella" (E. M. Sleigh, ed.), pp. 437–464. Academic Press, New York.

Pollard, T. D. (1981). Cytoplasmic contractile proteins. *J. Cell Biol.* **91,** 156s–165s.

Porrello, K., and Burnside, B. (1984). Regulation of reactivated contraction in teleost retinal cones by calcium and cAMP. *J. Cell Biol.* **98,** 2230–2238.

Porrello, K., Cande, W. Z., and Burnside, B. (1983). *N*-Ethylmaleimide-modified subfragment-1 and heavy meromyosin inhibit reactivated contraction in motile models of retinal cones. *J. Cell Biol.* **96,** 449–454.

Pratt, M. M. (1980). Identification of a dynein ATPase in unfertilized sea urchin eggs. *Dev. Biol.* **74,** 364–378.

Pratt, M. M., Otter, T., and Salmon, E. D. (1980). Dynein-like Mg^{2+}-ATPase in mitotic spindles isolated from sea urchin embryos (*Strongylocentrotus droebachiensis*). *J. Cell Biol.* **86,** 738–745.

Puszkin, S., and Schook, W. (1979). The role of cytoskeleton in neuron activity. *Methods Achiev. Exp. Pathol.* **9,** 87–111.

Raff, E. C. (1979). The control of microtubule assembly *in vivo*. *Int. Rev. Cytol.* **59,** 1–95.

Raynauld, J., Laviolette, J., and Wagner, H. (1979). Goldfish retina: A correlate between cone activity and morphology of the horizontal cell in cone pedicles. *Science* **204,** 1436–1438.

Reading, W. (1983). Dopaminergic receptors in bovine retina and their interaction with thyrotropin-releasing hormone. *J. Neurochem.* **41,** 1587–1595.

Repasky, E. A., Granger, B. L., and Lazarides, E. (1982). Widespread occurrence of avian spectrin in non-erythroid cells. *Cell* **29,** 821–833.

Roberts, K., and Hyams, J. S. (1979). "Microtubules," p. 573. Academic Press, New York.

Routledge, L. M., Amos, W. B., Yew, F. F., and Weis-Fogh, T. (1976). New calcium-binding contractile proteins. *In* "Cell Motility" (R. D. Goldman, T. D. Pollard, and J. Rosenbraum, eds.), Vol. 3, Book A, pp. 93–114. Cold Spring Harbor Conferences on Cell Proliferation, Cold Spring Harbor, New York.

Sakai, H., Mabuchi, I., Shimoda, S., Kuriyama, R., Ogawa, K., and Mohri, H. (1976). Induction of chromosome motion in the glycerol-isolated mitotic apparatus: Nucleotide specificity and effects of anti-dynein and myosin sera on the motion. *Dev. Growth Differ.* **18,** 211–219.

Sakai, H., Mohri, H., and Borisy, G. (1982). "Biological Functions of Microtubules and Related Structures." Academic Press, New York.

Salisbury, J. L. (1982). Calcium-sequestering vesicles and contractile flagellar roots. *J. Cell Sci.* **58,** 433–443.

Salisbury, J. L., and Floyd, G. L. (1978). Calcium induced contraction of the rhizoplast of a quadriflagellate green alga. *Science* **202,** 975–977.

Salisbury, J. L., Surek, B., and Melkonian, M. (1984). Striated flagellar roots: Isolation and partial characterization of a calcium-modulated contractile organelle. *J. Cell Biol.* **99,** 962–970.

Salmon, E. D., and Segall, R. R. (1980). Calcium labile mitotic spindles isolated from sea urchin eggs (*Lytechinus variegatus*). *J. Cell Biol.* **86,** 355–365.

Satir, P. (1979). Basis of flagellar motility in spermatozoa: Current status. *In* "The Spermatozoon" (D. W. Fawcett and J. M. Bedford, eds.), pp. 81–90. Urban & Schwarzenberg, Munich.

Satir, P. (1982). Mechanisms and controls in cilia. *In* "Prokaryotic and Eurkaryotic Flagella." *Symp. Soc. Exp. Biol.* **35,** 179–201.

Schaeffer, S., and Raviola, E. (1975). Ultrastructural analysis of functional changes in the synaptic endings of turtle cone cells. *Cold Spring Harbor Symp. Quant. Biol.* **40,** 521–528.

Schaeffer, S., and Raviola, E. (1978). Membrane recycling in the cone cell endings of the turtle retina. *J. Cell Biol.* **79,** 802–825.

Scheid, C., Honeyman, T., and Fay, F. (1979). Mechanisms of beta-adrenergic relaxation of smooth muscle. *Nature (London)* **277,** 32–36.

Schliwa, M. (1981). Proteins associated with cytoplasmic actin. *Cell* **25,** 587–590.

Schliwa, M. (1984). Mechanisms of intracellular organelle transport. *In* "Cell and Muscle Motility" (J. Shay, ed.), Vol. 5, pp. 1–82. Plenum, New York.

Schliwa, M., Euteuer, U., Bulinski, J. C., and Izant, J. G. (1981). Calcium lability of cytoplasmic microtubules and its modulation by microtubule-associated proteins. *Proc. Natl. Acad. Sci. U.S.A.* **78,** 1037–1041.

Schliwa, M., van Blerkon, J., and Pryzwansky, K. B. (1982). Structural organization of cytoplasm. *Cold Spring Harbor Symp. Quart. Biol.* **46,** 51–67.

Schultz, J. E., and Jantzen, H. M. (1980). Cyclic nucleotide dependent protein kinases from cilia of *Paramecium tetraurelia:* Partial purification and characterization. *FEBS Lett.* **116,** 75–78.

Selden, S. C., and Pollard, T. D. (1983). Phosphorylation of microtubule-associated proteins regulates their interaction with actin filaments. *J. Biol. Chem.* **258,** 7064–7071.

Sherry, J. M. F., Gorecka, A., Askoy, O. Dabrowska, R., and Hartshorne, D. J. (1978). Roles of calcium and phosphorylation in the regulation of the activity of gizzard myosin. *Biochemistry* **17,** 4411–4418.

Sheterline, P. (1983). "Mechanisms of Cell Motility." Academic Press, New York.

Shimo-Oka, T., and Watanabe, Y. (1981). Stimulation of actomyosin Mg^{2+}-ATPase activity by a brain microtubule-associated protein fraction. High molecular weight actin-binding protein is the stimulating factor. *J. Biochem.* **90,** 1297–1307.

Simons, T. J. B. (1979). Vanadate—a new tool for biologists. *Nature (London)* **281,** 337–338.

Sloboda, R. D., Rudolf, S. A., Rosenbaum, J. L., and Greengard, P. (1975). Cyclic AMP-dependent endogenous phosphorylation of a microtubule-associated protein. *Proc. Natl. Acad. Sci. U.S.A.* **72,** 177–181.

Smillie, L. B. (1979). Structure and functions of tropomyosin from muscle and non-muscle sources. *Trends Biochem. Sci.* **4,** 151–155.

Smith, R. (1948). The role of the sinus glands in retinal pigment migration in grapsoid crabs. *Biol. Bull.* **95,** 169–185.

Snyder, J. A., and McIntosh, J. R. (1976). Biochemistry and physiology of microtubules. *Annu. Rev. Biochem.* **45,** 699–720.

Snyder, W. Z., and Zadunaisky, J. A. (1976). A role for calcium in the migration of retinal screening pigment in the frog. *Exp. Eye Res.* **22,** 377–388.

Sobue, K., Muramoto, Y., Yamazaki, R., and Kakiuchi, S. (1979). Distribution in rat tissues of modulator-binding protein of particulate nature. Studies with ^{3}H-modulator protein. *FEBS Lett.* **105,** 105–109.

Sobue, K., Fujita, M., Muramoto, Y., and Kakiuchi, S. (1981). The calmodulin binding protein in microtubules is tau factor. *FEBS Lett.* **132,** 137–140.

Sobue, K., Kanda, K., Yamagami, K., and Kakiuchi, S. (1982). Ca^{++}- and calmodulin-dependent phosphorylation of calspectrin (spectrin-like calmodulin-binding protein; fodrin) by protein kinase systems in synaptasomal cytosol and membranes. *Biomed. Res.* **3,** 561–570.

Southwick, F. S., and Hartwig, J. H. (1982). Acumentin, a protein in macrophages which caps the "pointed" end of actin filaments. *Nature (London)* **297,** 303–307.

Spira, A., and Milman, G. (1979). The structure and distribution of the cross-striated fibril and associated membranes in guinea pig photoreceptors. *Am. J. Anat.* **155,** 319–337.

Spira, A., and Milman, G. (1982). Filament arrays in the photoreceptor cell of the human, monkey, and guinea pig. *In* "The Structure of the Eye" (J. Hollyfield, ed.), pp. 1–10. Elsevier, Amsterdam.

Stavenga, D., Flokstra, J., and Kuiper, J. (1975). Photopigment conversions expressed in pupil mechanism of blowfly visual sense cells. *Nature (London)* **253,** 740–742.

Steinberg, R. H., and Wood, I. (1974). Pigment epithelial cell ensheathment of cone outer segments in the retina of the domestic cat. *Proc. R. Soc. London* **187,** 461–478.

Stevens, J., Jacobs, R., and Jackson, M. (1984). Rings of cross-striated fibrils within the cat cone pedicle: A computer-assisted serial EM analysis. *Invest. Ophthalmol. Visual Sci.* **25,** 201–208.

Stossel, T. P. (1974). Phagocytosis. *N. Engl. J. Med.* **290,** 833–839.

Stossel, T. P. (1982). The structure of cortical cytoplasm. *Philos. Trans. R. Soc. London Ser. B* **299,** 275–289.

Stracher, A. (1983a). "Muscle and Nonmuscle Motility," Vol. 1. Academic Press, New York.

Stracher, A. (1983b). "Muscle and Nonmuscle Motility," Vol. 2. Academic Press, New York.

Suprenant, K. A., and Dentler, W. L. (1982). Association between endocrine pancreatic secretory granules and *in vitro* assembled microtubules is dependent on microtubule associated proteins. *J. Cell Biol.* **93,** 164–174.

Takeuchi, I. K., and Takeuchi, Y. K. (1979). Intermediate filaments in the retinal pigment epithelial cells of the goldfish. *J. Electron Microsc.* **28,** 134–142.

Tanenbaum, S. W. (1978). "Cytochalasins, Biochemical and Cell Biological Aspects." Elsevier, Amsterdam.

Tash, J. S., and Means, A. R. (1983). Cyclic adenosine 3′,5′ monophosphate, calcium and protein phosphorylation in ciliary motility. *Biol. Reprod.* **28,** 75–104.

Taylor, D. L., and Condeelis, J. S. (1979). Cytoplasmic structure and contractility in amoeboid cells. *Int. Rev. Cytol.* **56,** 57–144.

Taylor, D. L., and Fechheimer, M. (1982). Cytoplasmic structure and contractility: The solation–contraction coupling hypothesis. *Philos. Trans. R. Soc. London Ser. B* **299,** 185–197.

Tilney, L. G., Haiano, S., Ishikawa, H., and Mooseker, M. (1973). The polymerization of actin: Its role in the generation of the acrosomal process of certain echinoderm sperm. *J. Cell Biol.* **59,** 109–126.

Tucker, R. W., Pardee, A. B., and Fujiwara, K. (1979). Centriole ciliation is related to quiescence and DNA synthesis in 3T3 cells. *Cell* **17,** 527–535.

Ungar, F., Piscopo, I., Letizia, J., and Holtzman, E. (1984). Uptake of calcium by the endoplasmic reticulum of the frog photoreceptor. *J. Cell Biol.* **98,** 1645–1655.

Vale, R. D., Reese, T. S., and Sheetz, M. P. (1985). Identification of a novel force-generating protein, kinesin, involved in microtubule-based motility. *Cell* **42,** 39–50.

van Breemen, C., Aaronsson, P., and Loutzenhiser, R. (1979). Sodium–calcium interactions in mammalian smooth muscle. *Pharmacol. Rev.* **30,** 167–208.

van Breemen, C., Aaronson, P., Loutzenhiser, R., and Meishiri, K. (1982). Calcium fluxes in isolated rabbit aorta and guinea pig tenia coli. *Fed. Proc. Fed. Am. Soc. Exp. Biol.* **41,** 2891–2897.

Wagner, H. (1973). Darkness-induced reduction of the number of synaptic ribbons in fish retina. *Nature (London) New Biol.* **246,** 53–55.

Wagner, H. (1980). Light-dependent plasticity of the morphology of horizontal cell terminals in cone pedicles of fish retinas. *J. Neurocytol.* **9,** 573–590.

Wagner, H., and Douglas, R. (1983). Morphologic changes in teleost primary and secondary retinal cells following brief exposure to light. *Invest. Ophthalmol. Visual Sci.* **24,** 24–29.

Walcott, B. (1974). Unit studies on light-adaptation in the retina of the crayfish, *Cherax destructor. J. Comp. Physiol.* **94,** 207–218.

Walcott, B. (1975). Anatomical changes during light-adaptation in insect compound eyes. *In* "The Compound Eye and Vision of Insects" (G. Horridge, ed.), pp. 20–36. Oxford Univ. Press (Clarendon), London and New York.

Wallach, D., Davies, P., Bechtel, P., Willingham, M., and Pastan, I. (1978). Cyclic AMP-dependent phosphorylation of the actin-binding protein filamin. *Adv. Cyclic Nucleotide Res.* **9,** 371–392.

Wang, K., and Singer, S. J. (1977). Interaction of filamin with F-actin in solution. *Proc. Natl. Acad. Sci. U.S.A.* **74,** 2021–2025.

Warner, F. D., and Mitchell, D. R. (1980). Dynein: The echanochemical coupling adenosine triphosphatase of microtubule-based sliding filament mechanisms. *Int. Rev. Cytol.* **66,** 1–43.

Warren, R. H., and Burnside, B. (1978). Microtubules in cone myoid elongation in the teleost retina. *J. Cell Biol.* **78,** 247–259.

Watkins, R., and Davidson, I. (1980). Contractile velocity analyses of norepinephrine and angiotensin II activation of vascular smooth muscle. *Eur. J. Pharmacol.* **62,** 177–189.

Webb, N. (1977). Orientation of retinal rod receptor membranes in the intact eye using X-ray diffraction. *Vision Res.* **17,** 625–631.

Weeds, A. (1982). Actin-binding proteins. Regulators of cell architecture and motility. *Nature (London)* **296,** 811–816.

Weiler, R., and Wagner, H. (1984). Light-dependent change of cone horizontal cell interactions in carp retina. *Brain Res.* **298,** 1–9.

Weisenberg, R. C. (1972). Microtubule formation *in vitro* in solutions containing low calcium concentrations. *Science* **177,** 1104–1105.

Weisenberg, R., and Taylor, E. W. (1968). Studies on ATPase activity of sea urchin eggs and the isolated mitotic apparatus. *Exp. Cell Res.* **53,** 372–384.

Welsh, J. (1930). Diurnal rhythm of the distal pigment cells in the eyes of certain crustaceans. *Proc. Natl. Acad. Sci. U.S.A.* **16,** 386–395.

Westrum, L., Gray, E., Burgoyne, R., and Barron, J. (1983). Synaptic development and microtubule organization. *Cell Tissue Res.* **231,** 93–102.

Wilson, L., and Bryan, J. (1974). Biochemical and pharmacological properties of microtubules. *Adv. Cell Biol.* **3,** 21–36.

Wilson, M. (1975). Angular sensitivity of light and dark-adapted locust retinula cells. *J. Comp. Physiol.* **97,** 323–328.

Wood, J. G., Wallace, R. W., and Cheung, W. Y. (1980a). Immunocytochemical studies of the localization of calmodulin and CaM-BP_{80} in brain. *In* "Calcium and Cell Function" (W. Y. Cheung, ed.), Vol. 1, pp. 291–303. Academic Press, New York.

Wood, J. G., Wallace, R. W., Whitaker, J. N., and Cheung, W. Y. (1980b). Immunocytochemical localization of calmodulin and a heat labile calmodulin binding protein in basal ganglia of mouse brain. *J. Cell Biol.* **84,** 66–76.

Yamada, K. M., Spooner, B. S., and Wessells, N. K. (1970). Axon outgrowth: Roles of microfilaments and microtubules. *Proc. Natl. Acad. Sci. U.S.A.* **66,** 1206–1212.

Yamada, K. M., Spooner, B. S., and Wessells, N. K. (1971). Ultrastructure and function of growth cones and axons in cultured nerve cells. *J. Cell Biol.* **49,** 614–635.

Young, R. W., and Bok, D. (1969). Participation of the retinal pigment epithelium in the rod outer segment renewal precess. *J. Cell Biol.* **42,** 392–403.

Zisapel, N., Levi, M., and Gozes, D. (1980). Tubulin: An integral protein of mammalian synaptic vesicle membranes. *J. Neurochem.* **34,** 26–32.

MOLECULAR DYNAMICS OF THE ROD CELL

PAUL A. HARGRAVE

Department of Ophthalmology and
Department of Biochemistry and Molecular Biology
College of Medicine
University of Florida
Gainesville, Florida

When light strikes a retinal rod photoreceptor cell, rhodopsin molecules are photolyzed and a series of biochemical events rapidly follows. Several enzymes are stimulated which hydrolyze cyclic GMP, GTP, and phosphatidylinositol bisphosphate. Some rod outer segment proteins become phosphorylated, and others are dephosphorylated. Na^+ channels are closed in the plasma membrane and Ca^{2+} is mobilized and extruded from the outer segment. Current evidence suggests that these events occur rapidly enough to be a part of the excitation sequence that leads eventually to rod cell hyperpolarization. Other events occur on a longer time scale and are important for the renewal of cellular components which carry out these and other functions. It is a goal of visual cell biology to provide a molecular level understanding of the structure and dynamics of photoreceptor cells in order to characterize their functions.

I. Structure of the Rod Cell

Vertebrate retinas generally possess two types of photoreceptor cells, rods and cones, whose functions are complementary. Rods are sensitive in dim light and are responsible for scotopic vision, while cones function in daylight and are responsible for color vision. Because rod cells are larger and more numerous in most animals, they have proven easier to isolate and study biochemically. For that reason there is far more information available concerning rod cell composition and function than for cone cells. This article will be concerned with the molecular properties of vertebrate rod cells. For more detailed coverage of some of these topics the reader should consult Rosenkranz (1977), O'Brien (1978), Olive (1980), Papermaster and Schneider (1982), and Hargrave (1982).

The rod cell is a specialized elongated cell whose outer segment is connected to the inner segment by a modified ciliary process (Fig. 1). Typical rods have been considered to have cylindrical inner and outer segments whose diameters are similar, but there is considerable variation. The size of rod cells varies according to the species. Frog rod cells, measured in suspension, are approximately 7 μm in diameter and from 43 to 64 μm in length, depending upon the species of frog (Rosenkranz, 1977). Human rod cells are smaller in both dimensions; mudpuppy rods are wider.

The inner segment contains the major metabolic machinery of the cell. The apex of the inner segment, called the "ellipsoid," is densely packed with mitochondria. These powerhouses for ATP production are thus pivotally situated in order to provide energy for both inner segment and outer segment needs. The apex of the inner segment also contains vesicles bearing opsin destined for transport to the outer segment (see article by Besharse, this volume). The proximal portion of the inner segment has been designated the "myoid" region. In lower vertebrate cones it is this cellular region which in many fish and amphibians can elongate or shorten in response to light (see article by Burnside and Dearry, this volume). Bundles of actin filaments and microtubules are arranged longitudinally along the cellular axis in the myoid of some rods. Rough and

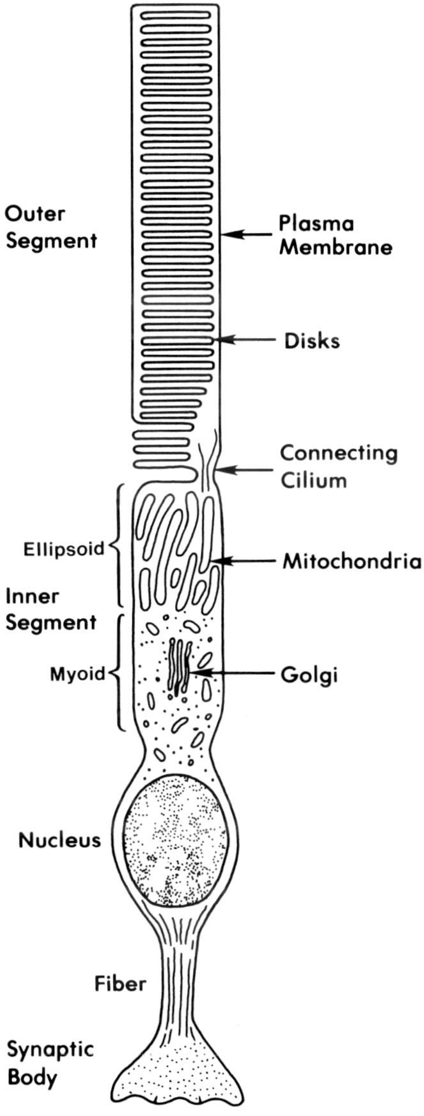

FIG. 1. Schematic diagram of vertebrate rod cell.

smooth endoplasmic reticulum and the Golgi complex are the sites of protein synthesis in the rod cell and lie adjacent to the nucleus in the myoid. How the diverse soluble and membrane proteins are delivered from here to their functional site in this highly compartmentalized cell remains a challenge for the cell biologist.

Below the level of the nucleus, the photoreceptor axon or fiber proceeds to the cell terminal. The fiber appears to contain the same array of microtubular elements found in the axons of conventional nerve cells (Kuwabara, 1965). The synaptic terminals of the rod cells form synapses with second-order neurons and must contain all of the specialized biochemical machinery needed for neural cell signaling.

At the apex of the inner segment is the site of origin of the ciliary process which serves as the sole cytoplasmic bridge between the inner and outer segment. The cilium originates from a centriole or basal body situated ~0.5 μm from the plasma membrane of the inner segment. From this basal body and an adjacent centriole extend filamentous bundles similar to those found in conventional cilia (Cohen, 1972). The cilium contains the usual nine pairs of circumferential tubules but lacks the central pair possessed by motile cilia. The length which the cilium extends into the outer segment is very species dependent. Surrounding the cilium at the apex of the inner segment is an exciting, newly described structure, the periciliary ridge complex (Peters *et al.*, 1983; Andrews and Cohen, 1980; Andrews, 1982; Besharse, this volume). This specialized cellular structure may be involved in regulating the access of newly synthesized membrane material to the outer segment. Actin has been localized in the periciliary ridge complex, the basal disks, and the distal end of the ciliary process (Chaitin *et al.*, 1984). Such localization suggests that actin plays a role in opsin transport and incorporation into the outer segment membrane, and is implicated in disk formation. F-actin filaments, identified by phallotoxin binding, have been observed surrounding portions of the inner segment (Del Priore *et al.*, 1985). Another contractile protein, myosin, can be immunochemically localized throughout the ciliary extension (Chaitin and Bok, 1984). Calmodulin, a Ca^{2+}-binding protein which is functionally linked to myosin light chain, is also found associated with the ciliary extension (Chaitin and Bok, 1984; Roof and Applebury, 1984).

The rod outer segment consists of the plasma membrane enclosing a stack of flattened sacs or disk membranes. Disks arise by evagination of the ciliary plasma membrane and disks at the base of the outer segment are still continuous with the plasma membrane and are not fully formed (see Besharse, this volume). Fish, amphibian, and primate rod disks are lobulated, some with deep incisures (Fig. 2). In other species such as monkey, man, and amphibia the incisures are superficial. Rodent, cat, dog, and bovine disks are cleaved by a single deep incisure (Cohen, 1972). The incisures are aligned along the disk stack and often enclose a bundle of microtubule-like elements (Young, 1971; Roof and Applebury, 1984). A large intrinsic membrane glycoprotein (M_r 290,000) has been immunochemically localized to the disk margin and to the incisures of frog rods and cones (Papermaster *et al.*, 1978, 1982). It is present in 1000–3000 molecules per disk, which represents a molar ratio to rhodopsin of between 1 : 300 and 1 : 900. Its function, and indeed the function of incisures, is currently unknown.

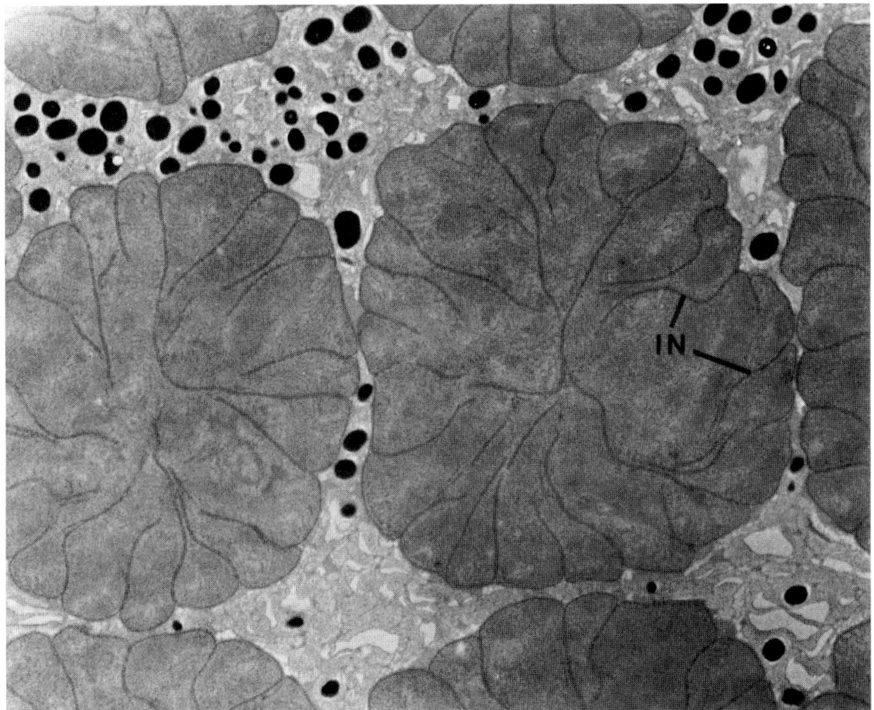

FIG. 2. Horizontal cross section of a frog rod outer segment. The incisures (IN) are deep invaginations of the disk membrane which serve to divide it into lobes. Epon-embedded section; ×7800. (Reproduced from *The Journal of Cell Biology*, 1978, **78**, 415–425 by copyright permission of The Rockefeller University Press.)

Bovine disk rims appear to contain at least two other proteins in addition to the high molecular weight glycoprotein described above (R. Molday, personal communication). Monoclonal antibodies have been prepared which identify a protein of molecular weight similar to that of the protein characterized by Papermaster. This protein does not appear to contain carbohydrate and is antigenically related to spectrin. Molday also finds that an M_r 32K protein is immunochemically localized at the disk rim (R. Molday, personal communication).

A. *The Rod Outer Segment Has a Cytoskeletal Structure*

Disk edges have previously been recognized as having distinctive properties, and there has been suggestive evidence for association between disks and the plasma membrane (Falk and Fatt, 1969). Although disks in the outer segment have been described as "free floating," disks have been shown to hold together

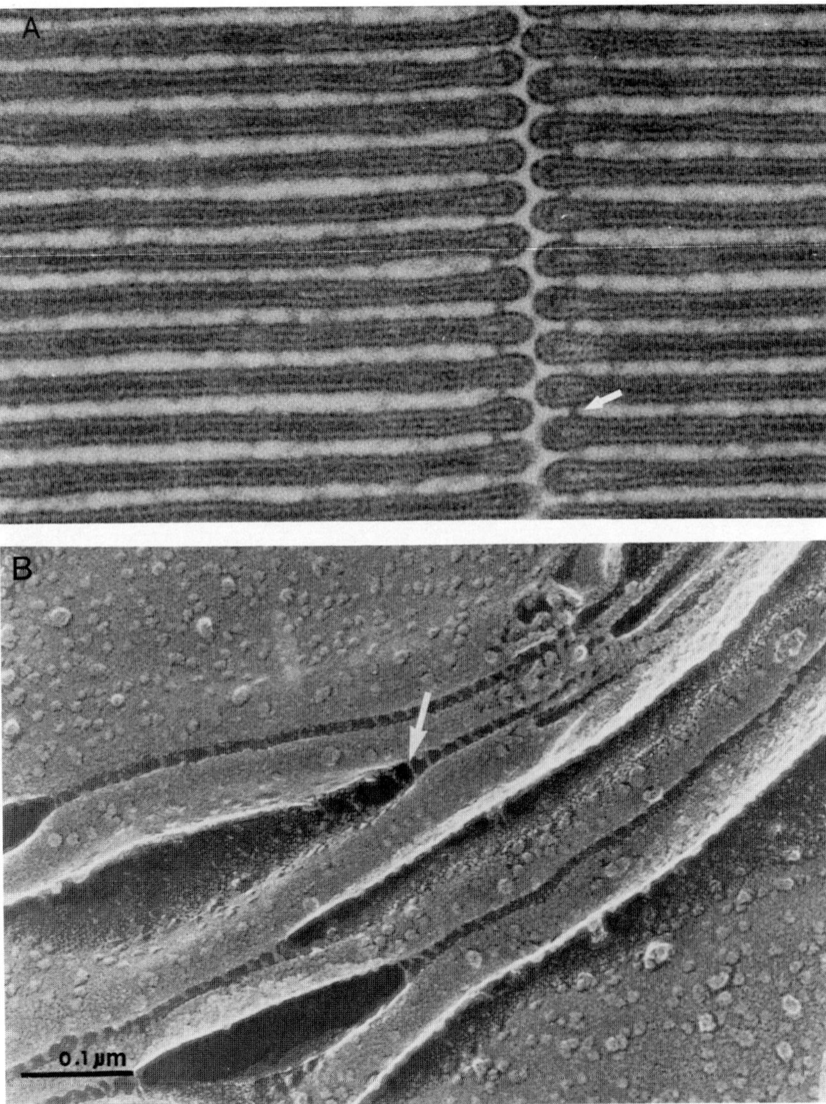

FIG. 3. (A) Thin-section electron micrograph of frog freeze-substituted rod outer segment illustrating filaments linking disks (arrow points to filaments). ×214,620. (Reproduced with permission from Usukura and Yamada, 1981.) (B) Disk rims from fragmented, swollen toad rod outer segments. ROS were rapidly frozen, fractured at −100°C, etched, and shadowed. Bar, 0.1 µm. ×181,700. (Reproduced from *The Journal of Cell Biology*, 1982, **95,** 487–500 by copyright permission of The Rockefeller University Press.)

in stacks even when the plasma membrane has been removed (Cohen, 1971). Thin filaments have occasionally been seen between disks in fixed cross sections prepared for transmission electron microscopy (Usukura and Yamada, 1981; Fig. 3A). Such filaments have been found by freeze fracture of intact retina at 14-nm intervals along the disk edge and recessed from the disk edge by 15 nm (Roof and Heuser, 1982; Fig. 3B). These filaments are also found to bridge incisures but seldom are found between nonrim areas of disks.

Disks have also been observed to persist in contact with the plasma membrane following cellular disruption (Cohen, 1971). Filaments have been identified which mediate the disk to plasma membrane contact in both amphibian and mammalian rods (Roof and Heuser, 1982). The length and irregular attachment points and appearance of these filaments suggest that they are different from the types of filaments which bridge disk rims.

B. Rod Outer Segments Are Easily Isolated

In order to study the composition and properties of complete rod cells it would be necessary to obtain them in pure form and in reasonable quantity. This has not proved feasible, although the isolation of whole rod cells has been attempted using both enzymatic and mechanical methods (Dudley *et al.*, 1979; Schaeffer, 1983). When the retina is mechanically agitated, even by gentle shaking, breakage of rod cells occurs at or near the cilium releasing the outer segments. Pinching off may also occur below the ellipsoid region of the inner segment to produce a truncated rod cell with the outer segment attached to a portion of inner segment (Biernbaum *et al.*, 1985). Rod cell outer segments are easily isolated following shaking or homogenizing of the retina, by differential centrifugation and density gradient centrifugation (Papermaster and Dreyer, 1974; reviewed by Hargrave, 1982). Early preparative procedures were most concerned with preparation of outer segments which were free of other contaminating cells or organelles. Only in recent years has attention focused on preparation of pure outer segments which retain their full complement of cytoplasmic components.

Intact resealed outer segments have been prepared from cattle retinas by sucrose–Ficoll gradient centrifugation (Schnetkamp *et al.*, 1979). The rod outer segment plasma membranes remain intact and appear to retain most soluble proteins and small molecules as evaluated by several criteria. Endogenous cofactors and enzyme systems are able to reduce photolytically produced retinal to retinol as would occur *in vivo*. Exogenous [γ^{32}P]ATP is unable to penetrate the plasma membrane and participate in opsin phosphorylation. The complement of proteins present in these preparations has been only partially described. Such apparently intact bovine outer segments have proved useful for several studies, e.g., investigation of the rates of enzymatic reactions under conditions simulating physiological conditions (Sitaramayya and Leibman, 1983a).

Intact sealed rat rod outer segments have been prepared on Percoll gradients (Shuster and Farber, 1984). These preparations are free of inner segments as demonstrated by microscopy and by the absence of Na,K-ATPase and cytochrome *c* oxidase activities. Additional enzyme assays showed the rod outer segments to be impermeant to small molecules such as ATP, glucose 6-phosphate, and NADP.

C. *An Inventory of Rod Outer Segment Proteins*

Intact frog rod outer segments have been purified by Percoll density gradient centrifugation (Hamm and Bownds, 1985; Fig. 4). Such outer segments are both morphologically and osmotically intact and exclude exogenous dyes and ATP. The small percentage of these outer segments which contain the attached mitochondria-enriched ellipsoid are electrophysiologically active. More than 70 different polypeptides can be detected by electrophoretic analysis of this type of preparation of frog rod outer segments (Fig. 5). Twenty of the detected proteins are present in at least one copy per 1000 rhodopsin molecules (Table I). When all of the outer segment protein is measured, rhodopsin comprises 70% or 3×10^9 copies per frog outer segment. In hypotonically washed membranes rhodopsin accounts for 98% of the protein. The M_r 220,000 protein found at disk rims and incisures is the next most abundant membrane protein (Papermaster *et al.*, 1978; Molday and Molday, 1979). Four other proteins, which are presumably intrinsic membrane proteins, have also been detected (Table I).

The most quantitatively significant soluble or peripheral membrane proteins are those involved in cyclic nucleotide metabolism. The G protein (also known as transducin, GTPase, Γ, GTP-binding protein) is composed of an αβγ trimer and is present in approximately 10 copies per 100 rhodopsins. G protein accounts for ~17% of total rod outer segment protein (Hamm and Bownds, 1985). A M_r 50,000 protein which accounts for ~2.5% of rod outer segment protein binds reversibly to opsin (Kühn, 1981). This appears, to be the protein recently identified as the retinal S antigen (Pfister *et al.*, 1984) and may also be the same as a protein identified by its light-dependent binding of ATP (Zuckerman *et al.*, 1984). cGMP-phosphodiesterase (PDE) is an αβγ trimer which is present in less than one copy per 100 rhodopsins. The PDE accounts for ~1.5% of rod outer segment protein. The many remaining proteins which comprise the outer segment appear to be present in smaller quantities. They are responsible for other important functions of the outer segment such as structural properties of the disks (Roof and Heuser, 1982), metabolism of retinal (Zimmerman *et al.*, 1975), and phosphorylation and dephosphorylation of rhodopsin (Kühn, 1974).

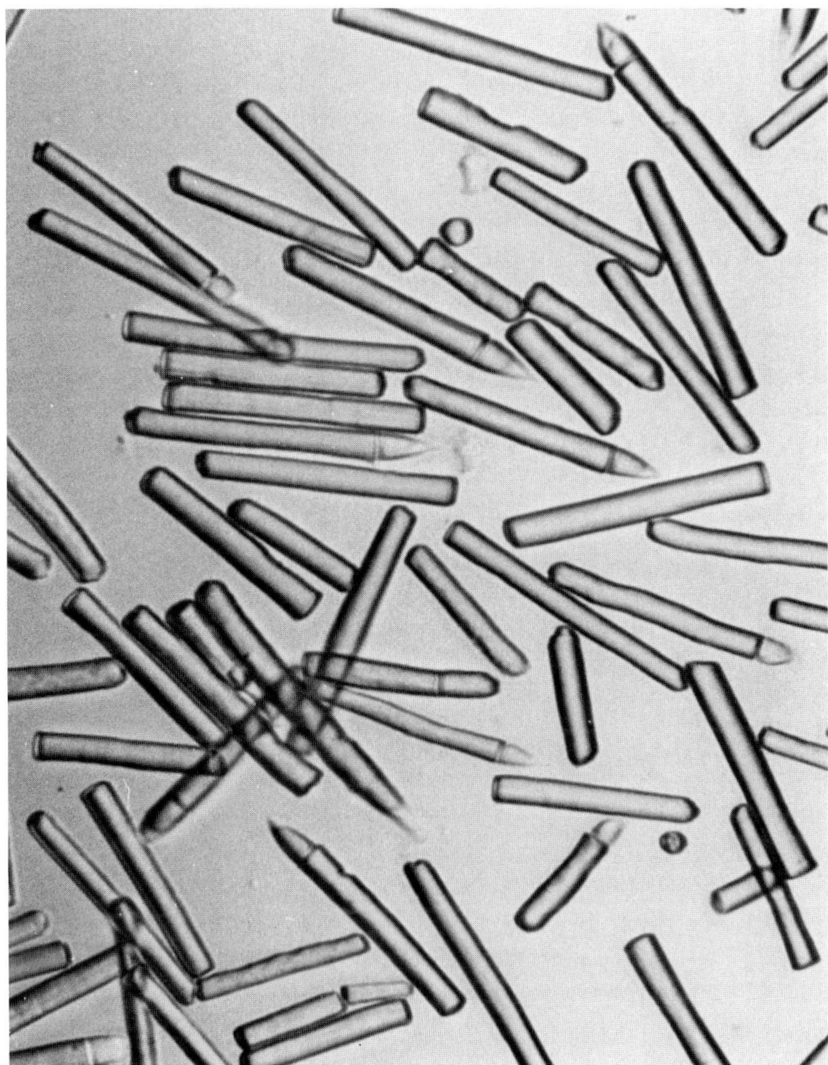

FIG. 4. Intact frog rod outer segments prepared by Percoll density gradient centrifugation. ROS containing ellipsoids are present in 5–10% of the outer segments (Hamm and Bownds, 1985).

TABLE I

QUANTITATION OF MAJOR ROS PROTEINS[a]

Protein	Identification number	Apparent MW	Percentage of stain: Coomassie	Percentage of stain: Silver	Molar ratio: Coomassie	Molar ratio: Silver	Copies ROS: Coomassie	Copies ROS: Silver
Rhodopsin	3.8	38,000–40,000	69 ± 6.7	68 ± 2.6	1000	1000	3×10^9	3×10^9
G protein							$\sim 3 \times 10^8$	$\sim 3 \times 10^8$
α	D3.71	40,000	9.0 ± 1.18	11 ± 1.6	130	166		
β	E3.72	38,000	7.3 ± 0.82	Nonstaining	100	—		
γ	C5.9	10,000	0.59 ± 0.04	1.4 ± 0.69	30	70		
50K	E3.10	50,000	2.6 ± 0.08	2.6 ± 0.48	30	30	8.4×10^7	8.4×10^7
PDE							1.5×10^7	2.2×10^7
α	E2.1	95,000	0.87 ± 0.14	1.4 ± 0.54	5	9		
β	E2.2	94,000	0.78 ± 0.14	1.4 ± 0.54	5	9		
220K	D1.1	220,000	2.6 ± 0.43	11 ± 1.7	6	28	1.8×10^7	8.4×10^7
	F5.6	14,000	0.43 ± 0.02	0.47 ± 0.09	15	19	4.5×10^7	5.7×10^7
	D,E5.1[b]	29,000	0.57 ± 0.14	0.23 ± 0.05	11	5	3.2×10^7	1.3×10^7
	D,F4.9[b]	31,000	0.71 ± 0.07	0.74 ± 0.05	12	13	3.6×10^7	3.9×10^7
	E3.4	45,000	0.37 ± 0.08	—	4	—	1.2×10^7	—
	C,D,E,F3.3[b]	48,000	0.77 ± 0.06	0.10	8	12	2.4×10^7	3.6×10^7
	E2.9	55,000	0.77 ± 0.06	0.10	7	3	2.4×10^7	7.5×10^6
	E2.7	68,000	0.14 ±	—	1	—	3×10^6	—
	E2.6	70,000	1.1 ± 0.06	0.23 0.05	9	2	2.7×10^7	5.7×10^6
	E2.5	80,000	0.47 ± 0.04	—	4	—	1.3×10^7	—
	5.3	20,000	0.06 ± 0.007	0.15	2	4	5×10^6	1.2×10^7
	4.3	36,000	0.18 ± 0.04	0.26	3	5	9×10^6	1.4×10^7
	3.6	43,000	0.69	0.03	8	0.3	2.4×10^7	9.3×10^5
	2.8	56,000	0.15 ± 0.01	0.08	1	1	3×10^6	1.7×10^6

[a]The proteins in frog rod outer segments were quantified following separation by SDS–polyacrylamide gel electrophoresis (Hamm and Bownds, 1985). Gels were stained with either silver or Coomassie blue and scanned densitometrically. The relative abundance of each protein is calculated with respect to rhodopsin.

[b]Represents one band in one-dimensional gels, multiple spots in two-dimensional gels.

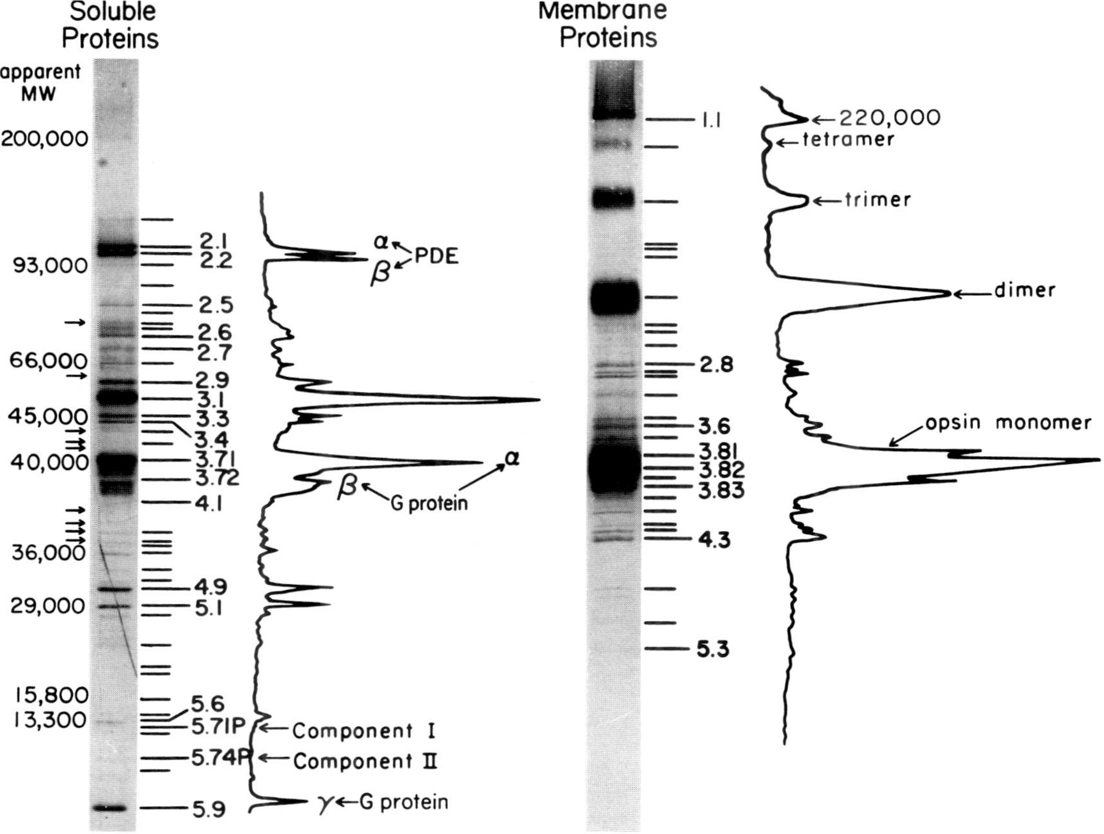

FIG. 5. Separation of frog rod outer segment membrane proteins and soluble proteins by sodium dodecyl sulfate–polyacrylamide gel electrophoresis. Gels were silver stained and scanned by densitometer (Hamm and Bownds, 1985). Oligomers of opsin are artifacts of sample preparation or of the gel system (see Papermaster and Dreyer, 1974).

II. Disk Membrane of the Rod Cell Outer Segment

A. *Structure of the Disk Membrane*

As previously mentioned, outer segments vary in length and in the number of disk membranes which they contain. Disks are densely packed, showing a repeat distance of 295 Å by X-ray diffraction (Chabre, 1981). The disk membranes contain rhodopsin molecules embedded in a matrix of lipid, the 45-Å-wide lipid bilayer. Each rhodopsin molecule has half of its mass surrounded by lipid hydrocarbon chains with the remainder exposed to lipid polar head groups and water at both membrane aqueous surfaces. There are approximately 80 lipid molecules per rhodopsin, which includes approximately 65 phospholipids and nine cholesterols (Dratz *et al.*, 1979). This means that rhodopsins are very close to one another, the most probable distance between molecules being 56 Å (Dratz and Hargrave, 1983). The rhodopsin molecules themselves appear to be cylindrical or ovoid, having a diameter of ~20–25 Å and a length of 60–65 Å (Chabre, 1981; Dratz *et al.*, 1979; Corless *et al.*, 1982).

B. *The Disk Membrane Bilayer Is Highly Fluid*

In vertebrate photoreceptor membranes, phospholipids make up from 85–90% of the total ROS lipids (Anderson and Andrews, 1982). The remainder is neutral lipids, predominantly cholesterol. Phosphatidylethanolamine and phosphatidylcholine account for the bulk of the phospholipids (up to 70%), but significant amounts of phosphatidylserine are also present (as much as 12%). Sphingomyelins and phosphatidylinositol are both present at only 1–2%, but we shall see later that this small amount of phosphatidylinositol serves an important function.

Rod cell membranes have a very high percentage of polyunsaturated fatty acids present in their phospholipids. The principal such polyunsaturated fatty acid is docosahexaenoic acid (C22 : 6) which accounts for up to 50% of the total essential fatty acid (Stone *et al.*, 1979; Anderson and Andrews, 1982). This highly unsaturated fatty acid is present primarily, although not exclusively, in the 2 position of the glycerol backbone (Miljanich *et al.*, 1979). The presence of a high content of unsaturated fatty acids in the phospholipids comprising the lipid bilayer results in less rigid lipid packing and greater relative fluidity of the bilayer.

Studies on rhodopsin in the disk membrane have been important to an understanding of properties of membranes in general. It has been little more than a decade since it became clear that proteins were able to freely rotate about their axis and to diffuse laterally within the plane of the lipid bilayer (Frye and Edidin, 1970; Singer and Nicholson, 1972). Since rhodopsin has a built-in chromophore and since it is effectively the only intrinsic membrane protein in disks, certain optical experiments were facilitated. Brown (1972) found that when dark-adapted disk membranes were cross-linked with glutaraldehyde and then partially bleached with polarized light, the membranes became highly dichroic. Membranes which were not cross-linked were only transiently dichroic because rhodopsin was then free to undergo rotational motion about its axis perpendicular to the plane of the membrane (Cone, 1972). By bleaching a portion of rhodopsin in the large mudpuppy rod and by following rhodopsin's movement, evidence was obtained for lateral diffusion of rhodopsin in the plane of the membrane (Poo and Cone, 1974; Liebman and Entine, 1974; Wey *et al.*, 1981; Drzymala *et al.*, 1984). Such measurements allowed calculation of the first diffusion coefficient to be determined for any membrane protein. The rate of diffusion of rhodopsin was found to be at least an order of magnitude higher than that obtained for proteins in other membranes and is that expected to be due only to the limits imposed by lipid bilayer viscosity (Axelrod, 1983). Fluidity of the disk membrane is equivalent to that of olive oil (Cone, 1972). It is reasonable to expect that this high rate of diffusion which is unique to vertebrate rhodopsin must be important for its function. This may be relevant for a first stage of amplification following bleaching by allowing a high rate of collision of bleached rhodopsin with the peripheral membrane protein, G protein (Liebman and Sitaramayya, 1984).

III. Rhodopsin: The Photoreceptor Protein

A. Molecular Properties of Rhodopsin

Vertebrate rhodopsins are about 41,000 in molecular weight, of which 39,000 is protein and 2000 is carbohydrate. Each rhodopsin molecule contains a covalently bound molecule of 11-*cis*-retinal which confers sensitivity to light to the protein. Rhodopsins are glycoproteins, having two sites of asparagine-linked oligosaccharide attachment on Asn-2 and Asn-15 (Hargrave, 1977). The oligosaccharides, of composition $Man_3GlcNAc_3$, are smaller than those typically found in this type of linkage, although the functional significance of this small oligosaccharide size is not known (Fukuda *et al.*, 1979; Liang *et al.*, 1979). However, the addition of carbohydrate to opsin is required in order for the protein to be incorporated into the disk membrane, inasmuch as this process is blocked in the presence of tunicamycin (Fliesler and Basinger, 1985).

The amino acid sequence has been determined for bovine rhodopsin (Ovchinnikov *et al.*, 1982; Hargrave *et al.*, 1983; Nathans and Hogness, 1983), sheep (Brett and Findlay, 1983), and human (Nathans and Hogness, 1984) rhodopsins. We can look forward to the completion of additional sequences, including that of the human cone proteins, in the coming years. All of the vertebrate rhodopsins whose sequences have been determined to date have 348 amino acids and show a high degree of sequence homology. A recent important contribution has been the determination of the sequence of an opsin gene from an invertebrate, *Drosophila* (O'Tousa *et al.*, 1985; Zuker *et al.*, 1985). The derived amino acid sequence is 373 residues long, and although the degree of amino acid sequence homology with bovine rhodopsin is only 22%, the degree of gross structural homology is considerable. Detailed comparisons will greatly aid us in understanding structure–function relationships. However, at the present time, since bovine rhodopsin has been the most extensively studied (reviewed in Hargrave, 1982), most of our comments will refer to findings on this particular rhodopsin species.

Like all integral membrane proteins, rhodopsin is synthesized on ribosomes of the rough endoplasmic reticulum. In contrast to most secreted and many membrane proteins, rhodopsin lacks an amino-terminal signal or leader sequence (Schechter *et al.*, 1979). For that reason the study of rhodopsin biosynthesis and membrane insertion is of particular interest and may lead to a better general understanding of how membrane proteins are incorporated into the lipid bilayer (Goldman and Blobel, 1981). Newly synthesized rhodopsin becomes glycosylated via a dolichol-phosphate-sugar (Plantner *et al.*, 1980) and the oligosaccharide chains are further modified by the action of glycosidases and glycosyltransferases in the Golgi complex. In amphibia, rhodopsin-containing vesicles appear to bud off from the Golgi and move to the apex of the rod inner segment, possibly under the direction of a molecular coding system which remains to be elucidated. From there the vesicles containing membrane-bound rhodopsin appear to fuse near the base of the connecting cilium in the grooves of the periciliary ridge complex (Peters *et al.*, 1983; Nir and Papermaster, 1983). Some means of transfer of rhodopsin along the ciliary plasma membrane is accompanied by a restraint of lateral diffusion of opsin into the inner segment plasma membrane (Nir *et al.*, 1984). Evagination and subsequent pinching off of the plasma membrane lead to compartmentation of recently synthesized rhodopsin in the membrane of newly formed disks at the base of the outer segment. (This topic is reviewed in detail by Besharse, this volume). Opsin appears to acquire its retinal chromophore only after insertion into the disk membrane (Defoe and Bok, 1983).

B. A Model for Rhodopsin in the Disk Membrane

Inspection of rhodopsin's amino acid sequence reveals regions of predominantly polar and charged amino acids which are separated by seven stretches of

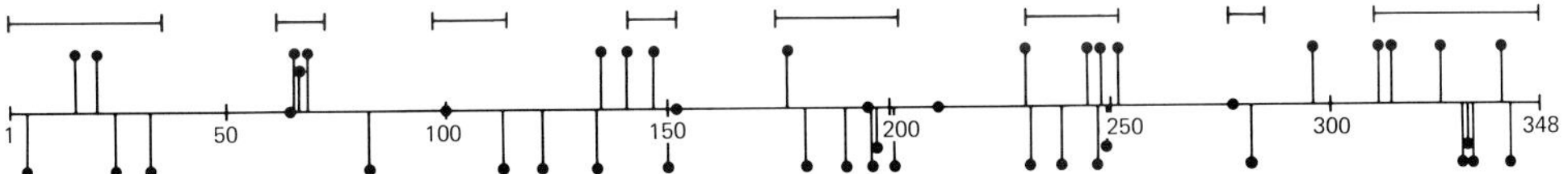

FIG. 6. Distribution of charged amino acids in the sequence of bovine rhodopsin. The line represents the polypeptide chain of rhodopsin from the amino-terminus (1) at the left to the carboxyl-terminus (348) at the right. Charged amino acids are represented by balls at the appropriate positions in the sequence. Positively charged amino acids (lysine, arginine) are represented by a ball above the line, negatively charged amino acids (aspartic acid, glutamic acid) by a ball below the line, and histidines by a ball on the line. Closely spaced balls were shifted vertically so that they would not overlap (there is no significance to the distance of balls from the line). Bars above the line show those regions which are predicted to fall outside of the transmembrane helical segments of the polypeptide chain [Reprinted with permission from (*Vision Research,* **24,** Hargrave *et al.*, Rhodopsin's protein and carbohydrate structure: Selected aspects), Copyright (1984), Pergamon Press.]

predominantly hydrophobic amino acids, 21–28 amino acids in length (Fig. 6). Studies on the topography of rhodopsin have shown that it is a transmembrane protein (Fung and Hubbell, 1978) whose carboxyl-terminus is located at the external or cytoplasmic surface of the disk membrane (Hargrave and Fong, 1977; Dratz *et al.*, 1979). Rhodopsin's amino-terminus is located at the intraluminal surface of the disk (Adams *et al.*, 1978; Clark and Molday, 1979). Results from a variety of topographic studies (see Hargrave, 1982; Ovchinnikov, 1982; Findlay, 1984) allow us to construct the model shown in Fig. 7. Here the rhodopsin polypeptide chain is shown traversing the lipid bilayer seven times. Half of rhodopsin's mass is embedded in the hydrophobic portion of the lipid bilayer with the remaining half equally divided among the membrane aqueous surfaces. Lysine-296, to which the retinal chromophore is attached, is located deep within the hydrophobic portion of the lipid bilayer (Thomas and Stryer, 1982). A few other charged amino acids appear to be buried in what is otherwise a very nonpolar region of the molecule.

We may anticipate that vertebrate rhodopsins will contain their seven helices arranged in a bundle which serves as a container for the retinal (Fig. 8). This model serves as a working hypothesis and suggests that amino acid side chains from internal surfaces of the transmembrane helices will form the environment of the binding pocket for retinal. It is likely that the particular location of the buried charged residues with respect to the retinal will be largely responsible for the unique spectrum of light sensitivity of each visual pigment (Honig *et al.*, 1979). Therefore it is of interest to determine what amino acids in the rhodopsin sequence form the retinal binding pocket. Much valuable information has already been obtained by many laboratory groups from the spectral analysis of rhodopsin containing different retinal isomers and their analogs (reviewed by Mathies *et al.*, 1985). Based upon these results, a model for the retinal binding region has been proposed (Liu *et al.*, 1984). A detailed understanding of the binding pocket

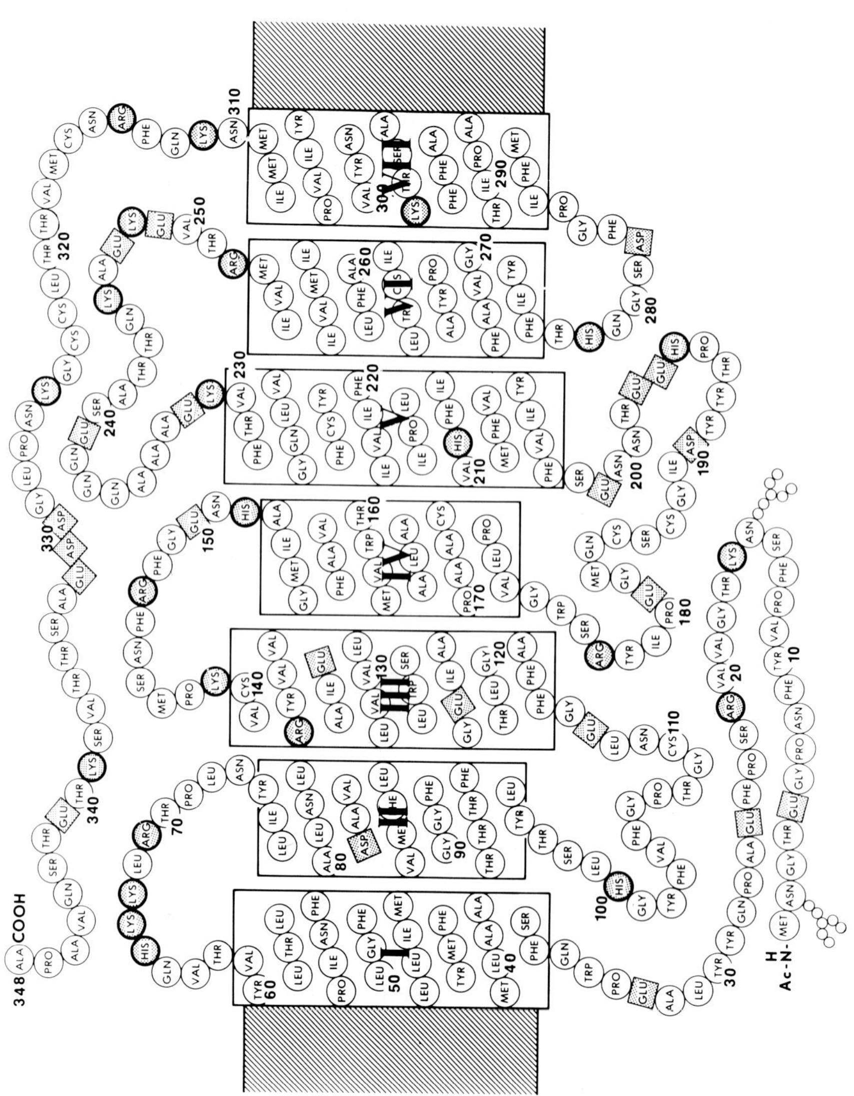

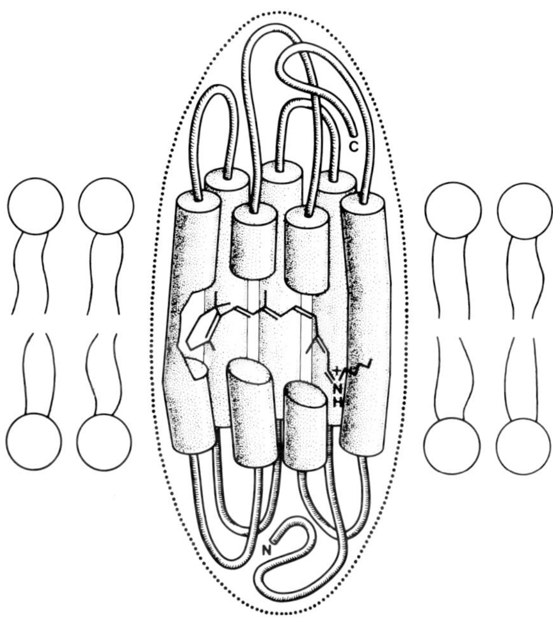

FIG. 8. Helical bundle model for rhodopsin. Rhodopsin is hypothesized to consist of an elongated bundle of seven helices whose connecting segments and chain termini form the aqueous-exposed surface regions. Helices known to contain prolines are shown with bends. Parts of helices are cut away to show the retinal (not to scale) which is attached to the side chain of Lys-296, midway in helix VII. The binding pocket for retinal is shown as consisting of portions of the inside surface of several helices [Reprinted with permission from (*Vision Research*, **24,** Hargave *et al.*, Rhodopsin's protein and carbohydrate structure: Selected aspects), Copyright (1984), Pergamon Press.]

FIG. 7. A model for the organization of rhodopsin's polypeptide chain in the disk membrane lipid bilayer. Ac-Met, the amino-terminus, is exposed at the internal aqueous surface to the disk membrane. Rhodopsin's polypeptide chain is shown to traverse the lipid bilayer seven times. The carboxyl-terminus, Ala-348, is exposed to the disk membrane external aqueous surface. Positively charged amino acids (Lys, Arg, His) are shown as shaded circles. Negatively charged amino acids (Glu, Asp) are shaded squares. Striped areas at each side of the figure represent the low-polarity portion of the lipid bilayer (the phospholipid fatty acid groups and the lower half of the phospholipid head groups) [Reprinted with permission from (*Vision Research*, **24,** Hargrave *et al.*, Rhodopsin's protein and carbohydrate structure: Selected aspects), Copyright (1984), Pergamon Press.]

should be greatly helped by the use of photoactivated retinals (Singh *et al.*, 1984) and by ^{13}C solid state NMR of [^{13}C]retinal reconstituted rhodopsins (Mathies *et al.*, 1985).

C. *Photolysis of Rhodopsin Causes Biochemical Changes*

How does the absorption of light energy by rhodopsin's chromophore lead to the many light-dependent biochemical changes which occur in the rod cell? We are just beginning to be able to provide answers to this question.

Absorption of light by rhodopsin's 11-*cis*-retinal chromophore leads to its isomerization to the all-*trans* form and its eventual dissociation from opsin. Most of the changes in the structure of retinal and in the conformation of opsin are quite rapid at physiological temperature (Fig. 9). Metarhodopsin II is formed in

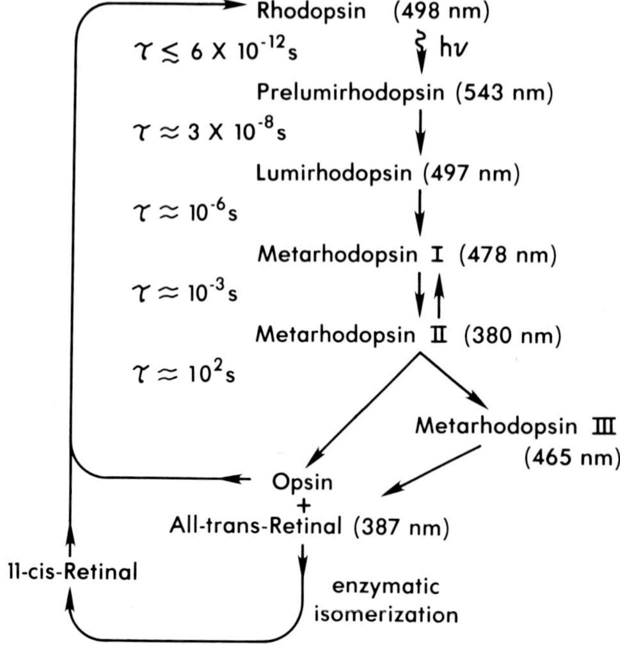

FIG. 9. Rhodopsin photolysis intermediates and the visual cycle. Following absorption of a photon by 11-*cis*-retinal, rhodopsin is converted through a series of spectrally identifiable intermediates to all-*trans*-retinal and opsin. The wavelength of maximum absorption of each intermediate and its halftime at 20°C are shown (adapted from Dratz, 1977). The enzymatic isomerization to regenerate 11-*cis*-retinal is complex. It appears to involve reduction to retinol and its esterification to palmitate, followed by hydrolysis and oxidation to yield retinal, with the isomerization occurring at a step yet to be identified (Bridges *et al.*, 1983).

milliseconds. At this point the previously inaccessible Schiff-base linkage by which the all-*trans*-retinal is bound to opsin becomes exposed to small water-soluble molecules (Bownds, 1967). The cytoplasmic surface of rhodopsin, in metarhodopsin II, is altered in conformation (Kühn *et al.*, 1982). This change in a protein structure occurs, presumably, as a result of movement of rhodopsin's transmembrane helices in response to a change in the geometry of the bound retinal.

D. Rhodopsin's Surface Mediates Effects of Light

When rhodopsin in disk membranes is bleached, several proteins in the rod outer segment bind to it strongly (Kühn, 1978). These proteins which bind to rhodopsin in a light-dependent manner have been identified as the G protein complex (G_{α}, M_r 39K; G_{β}, M_r 37K; G_{γ}, M_r 6K), rhodopsin kinase, and a protein of M_r 48,000 ("48K protein") (Kühn, 1981). The change in conformation at rhodopsin's cytoplasmic surface must provide binding sites for these proteins, sites which are not available prior to rhodopsin's exposure to light. The events following from the interaction of these proteins with rhodopsin probably account for the rapid responses to light by the rod cell (Vuong *et al.*, 1984).

IV. Light-Dependent Biochemistry of the Rod Cell

A. Cyclic GMP Is Important in Visual Transduction

One of the early events occurring in the rod cell following illumination is the hydrolysis of cGMP. Rod cells contain an unusually high level of cGMP in the dark: ~70 μM. The cGMP is hydrolyzed in response to light (Goridis *et al.*, 1974; Woodruff and Bownds, 1979; Kilbride and Ebrey, 1979; Goldberg *et al.*, 1983). A phosphodiesterase specific for cGMP is responsible for this rapid hydrolysis (Miki *et al.*, 1975; Yee and Liebman, 1978; Bitensky *et al.*, 1981). Hydrolysis of cGMP appears to be linked to visual transduction, since intracellular injection of cGMP mimics the electrophysiological dark state of the rod and increases the latency of the light response (Miller and Nichol, 1979). Injection of activated G protein or phosphodiesterase into rods also mimics the effect of light (Clack *et al.*, 1983). Therefore, it would appear that hydrolysis of cGMP is involved in closing Na^+ channels in the rod cell plasma membrane. The hypothesis that cGMP keeps the light-sensitive Na^+ channels open in the dark has recently received strong support (Fesenko *et al.*, 1985; Yau and Nakatani, 1985; Matthews *et al.*, 1985; Cobbs and Pugh, 1985). For discussion of this topic in detail see Farber and Shuster (this volume) and Stryer (1986). How do these events follow from the bleaching of rhodopsin?

It is the binding of G protein to bleached rhodopsin which leads to activation of the cGMP phosphodiesterase. In dark-adapted ROS the G protein consists of a three-subunit complex, $G_{\alpha\beta\gamma}$ (Kühn, 1981). The complex is peripherally bound to disk membranes and can be visualized on disk surfaces as 8- to 12-nm particles (Roof and Heuser, 1982). In the dark state the 39K α subunit contains bound GDP (Fung *et al.*, 1981). When this complex binds to photolyzed rhodopsin, G_α exchanges its bound GDP for a GTP, and G_α-GTP and $G_{\beta\gamma}$ are separately released from the membrane. The free Gα-GTP then binds to cGMP phosphodiesterase, activating it and thereby initiating the hydrolysis of cGMP. The $G_{\beta\gamma}$ complex, therefore, appears to facilitate G_α function. The finding that the G_α amino acid sequence shows significant homology to the *ras* oncogene strengthens the hypothesis that the G protein must have important cell regulatory functions (Lochrie *et al.*, 1985).

Phosphodiesterase (PDE) is a peripheral membrane protein consisting of three subunits: α (M_r 88K), β (M_r 84K), and γ (M_r 11K) (Baehr *et al.*, 1979). It has a very low catalytic activity in the dark state but upon activation becomes one of the most efficient enzymes known. PDE has been found to become activated as a result of removal of its γ subunit, which acts as an inhibitory subunit (Hurley *et al.*, 1981). In the rod cell, activation of phosphodiesterase occurs when the G protein G_α-GTP binds to PDE and removes the PDE γ inhibitory subunit (Yamazaki *et al.*, 1985). Once the cycle is activated, what restores it to the dark conditions?

A change is the conformation of rhodopsin's surface initiates the cycle, and it appears to be another such surface change which terminates it. The spontaneous relaxation of the surface to the dark conformation is a slow process (Kühn *et al.*, 1982) and is uninfluenced by regeneration of rhodopsin with 11-*cis*-retinal. Another process, turning off of the phosphodiesterase reaction, has been determined to require ATP (Liebman and Pugh, 1980). This ATP requirement has now been linked to the phosphorylation of rhodopsin (Sitaramayya and Liebman, 1983; Miller and Dratz, 1984). Not only does G protein bind to photolyzed rhodopsin—so does rhodopsin kinase (Kühn, 1978). The kinase uses ATP to incorporate as many as nine phosphate groups onto serines and threonines on rhodopsin's cytoplasmic surface (Wilden and Kühn, 1982). Most of the phosphorylated residues are located on rhodopsin's carboxyl-terminal region (Hargrave *et al.*, 1980; Thompson and Findlay, 1984), but one appears to be located at Ser-240 in the surface loop connecting helices V and VI (McDowell *et al.*, 1985). Addition of the new charged groups to rhodopsin's surface may change its ability to interact with other rod cell proteins, although this requires a direct demonstration. Thus, the phosphorylation of rhodopsin appears to be involved in deactivating photolyzed (activated) rhodopsin.

B. The Light-Activated Rhodopsin System Shows Homology with Hormone-Activated Adenylate Cyclase

Hormones which exert their effect through the adenylate cyclase system act in a manner analogous to the effect of light on rhodopsin which activates PDE (Shinozawa *et al.*, 1979; Stryer, 1983; Yamazaki *et al.*, 1985). The signal is light for the photoreceptor and a specific hormone for the hormone receptor. In each system the receptor causes an inactive G protein (which contains bound GDP) to be activated by binding GTP. The activated G-GTP complex then exerts its specific effect on its target enzyme(s), amplifying the signal and causing metabolic effects characteristic of light or hormone. The adenylate cyclase system contains two types of G protein (also called N protein); one stimulates and the other inhibits adenylate cyclase. Rod cell G protein bears both structural and functional relationships to the G proteins of the hormone system, and must be evolutionarily related. Components of the two systems are functionally interchangeable (Bitensky *et al.*, 1982). They show a further similarity in that the different G proteins are substrates for toxins which covalently modify them (Van Dop *et al.*, 1984). It is likely that the G proteins will show amino acid sequence homology (Gilman, 1984). These observations suggest that the cytoplasmic surface of rhodopsin which interacts with rod cell G protein may have similarities to the cytoplasmic surfaces of certain hormone receptors.

C. Ca^{2+} Fluxes Are Caused by Light

There has been considerable support for the suggestion that Ca^{2+} is a messenger in visual transduction (Hagins and Yoshikami, 1974). An increase in intracellular Ca^{2+} leads to hyperpolarization of the rod cell membrane. This has been demonstrated by increasing the extracellular Ca^{2+} (Yoshikami and Hagins, 1973; Brown and Pinto, 1974) by use of Ca^{2+} ionophores (Bastian and Fain, 1979) and by direct injection of Ca^{2+} into photoreceptors (Brown *et al.*, 1979). Many of the same experimenters also showed by a variety of techniques that decreasing the intracellular Ca^{2+} leads to rod cell membrane depolarization (Hagins and Yoshikami, 1977; Brown *et al.*, 1977).

Greater than 95% of the Ca^{2+} in rods is in the outer segment, where it is present in an amount equivalent to 1–2 Ca^{2+}/rhodopsin (Schröder and Fain, 1984). Upon illumination of the rod cell there is Ca^{2+} efflux to the extent of $\sim 10^4$ Ca^{2+}/rhodopsin bleached (Gold and Korenbrot, 1980; Schröder and Fain, 1984). Under steady light illumination over half of the rod cell Ca^{2+} is transported out of the rod within a few minutes. It is not yet clear what effects Ca^{2+} has in the rod cell, but a direct role for Ca^{2+} in blocking Na^+ conductance has been questioned (Yau and Nakatani, 1984). For a recent review see Korenbrot (1985).

D. *Inositol Phospholipids Are Implicated in Signal Transduction*

Signal transduction in a variety of cellular systems has been found to involve the turnover of inositol phospholipids (Streb *et al.*, 1983; Berridge, 1984). This leads to increased levels of intracellular Ca^{2+}, stimulation of the activity of a protein kinase, and often the release of arachidonate and an increase in the levels of cGMP (Nishizuka, 1983). A variety of cellular responses then ensue which are tissue specific. It is apparent that this particular stimulation and response system occurs in vertebrate and invertebrate photoreceptor cells and is of importance in their light-dependent biochemical processes (Fein *et al.*, 1984; Brown *et al.*, 1984; Ghalayini and Anderson, 1984; Kapoor and Chader, 1984). Although the complete mechanisms of this signal transduction pathway are far from being elucidated for any cellular system, a tentative working hypothesis may be developed as follows.

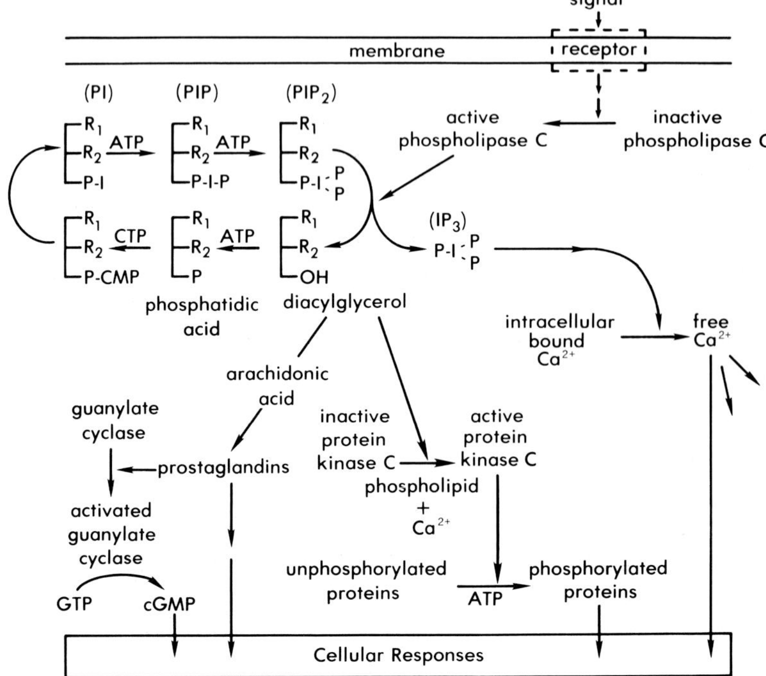

FIG. 10. Model for receptor-stimulated hydrolysis of phosphatidylinositol bisphosphate and its resultant cellular metabolic effects. See text for explanation.

In the phosphatidylinositol-dependent responses studied to date, a membrane-bound receptor protein is stimulated by its effector, which may be a hormone, neurotransmitter, growth factor, etc. (Fig. 10). This signal leads, via some intermediate mechanism, to activation of a phospholipase C which is specific for phosphatidylinositol 4,5-bisphosphate (PIP_2) (Berridge, 1983). Phospholipase C hydrolyzes PIP_2 yielding diacylglycerol and inositol 1,4,5-trisphosphate (IP_3). The diacyglycerol is rapidly metabolized by breakdown to form arachidonate and by phosphorylation to yield phosphatidic acid. Arachidonate is the precursor for a number of compounds, including some prostaglandins which can serve to activate guanylate cyclase (Graff *et al.*, 1978). The IP_3 formed by phospholipase C action appears to be involved via an unknown mechanism in mobilizing intracellular Ca^{2+}, thereby raising [Ca^{2+}]. Diacylglycerol causes the activation of protein kinase C by making the enzymes more sensitive to Ca^{2+} stimulation. Protein kinase C then phosphorylates its specific intracellular protein substrate(s) (Nishizuka, 1984). Ca^{2+} appears to act synergistically with protein kinase C to effect its cellular responses; however, cyclic nucleotides have no effect on the activity of protein kinase C. The resultant protein phosphorylation is undoubtedly of regulatory significance and influences enzymatic reactions, thus leading to the endpoint physiological events characteristic of the action of the signal (hormone, neurotransmitter, etc.) on the target cell.

E. Light Causes Turnover of Rod Cell Inositol Phospholipids

Phosphatidylinositol (PI) is a minor component of rod cell phospholipids, yet it is turned over at a faster rate than are other phospholipids (Anderson *et al.*, 1980). Details of the metabolism of PI have been partially elucidated for retina and the pathways appear to be basically similar to those of cytidine-dependent pathways in other tissues (Schmidt, 1983a). Light has been shown to enhance the turnover of PI in horizontal cells of *Xenopus* retina (Anderson and Hollyfield, 1981) and in rat retina (Schmidt, 1983a,b). Different cells in the retina respond differently to light but rod cells in the rat enhance both synthesis and hydrolysis of PI in response to light stimulus (Schmidt, 1983b). It has been recently shown that light specifically stimulates the hydrolysis of PIP_2 in rod outer segments (Ghalayini and Anderson, 1984). This suggests that a phospholipase C must become light activated and will cause the transient production of IP_3 and diacylglycerol in the rod cell. The mechanism of the phospholipase activation is unknown (however, one might speculate that some intermediary such as G protein might be involved in activation of the phospholipase; see Vandenberg and Montal, 1984). The production of diacylglycerol could lead to activation of the ubiquitious protein kinase C. Indeed, protein kinase C has recently been identified in rod outer segments and has been shown to phosphorylate as many as nine rod outer segment proteins (Kapoor and Chader, 1984). Whether diacylglycerol is further metabolized in rod cells to prostaglandins and other compounds, and whether such compounds affect enzymatic activities such as guanylate cyclase in the rod cell is currently unknown.

1,4,5-Inositol trisphosphate (IP_3), the other product of phospholipase C hydrolysis of PIP_2, has a pronounced effect on photoreceptor cells. Brown *et al.* (1984) have shown that the *Limulus* ventral eye can make IP_3 and that the amount of IP_3 is increased upon illumination. Furthermore, injection of IP_3 into the photoreceptor cell mimics the electrophysiological events observed for light (Brown *et al.*, 1984; Fein *et al.*, 1984). The injection of IP_3 into the *Limulus* eye leads to an increase in intracellular calcium concentration (Brown and Rubin, 1984). IP_3 may play a similar role in vertebrate rods since its injection into salamander rods induces a reversible hyperpolarization of the rod membrane (Waloga and Anderson, 1985). Further elucidation of the details of the light-activated phospholipase C cascade represent one of the recent important and potentially fruitful current developments in vision biochemistry.

Acknowledgment

During the preparation of this article the author was supported in part by research grants from the National Institutes of Health (EY 6225 and EY 6226), a Jules and Doris Stein Professorship from Research to Prevent Blindness, Inc., and an unrestricted departmental grant from Research to Prevent Blindness, Inc.

References

Adams, A. J., Tanaka, M., and Shichi, H. (1978). Concanavalin A binding to rod outer segment membranes: Usefulness for preparation of intact disks. *Exp. Eye Res.* **27,** 595–605.

Anderson, R. E., and Andrews, L. D. (1982). Biochemistry of retinal photoreceptor membranes in vertebrates and invertebrates. *In* "Visual Cells in Evolution," (J. A. Westfall, ed.), pp. 1–22. Raven, New York.

Anderson, R. E., and Hollyfield, J. G. (1981). Light stimulates the incorporation of inositol into phosphatidylinositol in the retina. *Biochim. Biophys. Acta* **665,** 619–622.

Anderson, R. E., Maude, M. B., and Kelleher, P. A. (1980). Metabolism of phosphatidylinositol in the frog retina. *Biochim. Biophys. Acta* **620,** 236–246.

Andrews, L. D. (1982). Freeze-fracture studies of vertebrate photoreceptor membranes. *In* "The Structure of the Eye" (J. G. Hollyfield, ed.), pp. 11–23. Elsevier, Amsterdam.

Andrews, L. D., and Cohen, A. I. (1980). Freeze-fracture observations of a specialized structure in the apical plasma membrane of the inner segments of photoreceptors of frogs and golfish. *Invest. Ophthalmol. Visual Sci.* **19** *(Suppl.),* 244.

Axelrod, D. (1983). Lateral motion of membrane proteins and biological function. *J. Membr. Biol.* **75,** 1–10.

Baehr, W., Devlin, M. J., and Applebury, M. L. (1979). Isolation and characterization of cGMP phosphodiesterase from bovine rod outer segments. *J. Biol. Chem.* **254,** 11669–11677.

Bastian, B. L., and Fain, G. L. (1979). Light adaptation in toad rods: Requirement for an internal messenger which is not calcium. *J. Physiol. (London)* **297,** 493–520.

Bastian, B. L., and Fain, G. L. (1982). The effects of calcium and background light on the sensitivity of toad rods. *J. Physiol. (London)* **330,** 307–329.

Berridge, M. J. (1983). Rapid accumulation of inositol trisphosphate reveals that agonists hydrolyse polyphosphoinositides instead of phosphatidylinositol. *Biochem. J.* **212,** 849–858.

Berridge, M. J. (1984). Inositol triphosphate and diacylglycerol as second messengers. *Biochem. J.* **220,** 345–360.

Biernbaum, M. S., Schobert, C. S., and Bownds, M. D. (1985). Frog rod outer segments with attached inner segment ellipsoids as an *in vitro* model for photoreceptors on the retina. *J. Gen. Physiol.* **85,** 83–105.

Bitensky, M. W., Wheeler, G. L., Yamazaki, A., Rasenick, M. M., and Stein, P. J. (1981). Cyclic nucleotide metabolism in vertebrate photoreceptors: A remarkable analogy and an unraveling enigma. *Curr. Top. Membr. Transp.* **15,** 237–271.

Bitensky, M. W., Wheeler, M. A., Rasenick, M. M., Yamazaki, A., Stein, P. J., Halliday, K. R., and Wheeler, G. L. (1982). Functional exchange of components between light-activated photoreceptor phosphodiesterase and hormone-activated adenylate cyclase systems. *Proc. Natl. Acad. Sci. U.S.A.* **79,** 3408–3412.

Bownds, D. (1967). Site of attachment of retinal in rhodopsin. *Nature (London)* **216,** 1178–1181.

Brett, M., and Findlay, J. B. C. (1983). Isolation and characterization of the CNBr peptides from the proteolytically derived N-terminal fragment of ovine opsin. *Biochem. J.* **211,** 661–670.

Bridges, C. D. B., Fong, S.-L., Liou, G. I., Alvarez, R. A., and Landers, R. A. (1983). Transport, utilization and metabolism of visual cycle retinoids in the retina and pigment epithelium. *Prog. Ret. Res.* **2,** 137–162.

Brown, J. E., and Rubin, L. J. (1984). A direct demonstration that inositol-trisphosphate induces an increase in intracellular calcium in *Limulus* photoreceptors. *Biochem. Biophys. Res. Commun.* **125,** 1137–1142.

Brown, J. E., Coles, J. A., and Pinto, L. H. (1977). Effects of injections of Ca^{++} and EGTA into the outer segment of the retinal rods of *Bufo marinus*. *J. Physiol. (London)* **269,** 707–722.

Brown, J. E., Rubin, L. J., Ghalayini, A. J., Tarver, A. P., Irvine, R. F., Berridge, M. J., and Anderson, R. E. (1984). *myo*-inositol polyphosphate may be a messenger for visual excitation in *Limulus* photoreceptors. *Nature (London)* **311,** 160–163.

Brown, P. K. (1972). Rhodopsin rotates in the visual receptor membrane. *Nature (London) New Biol.* **236,** 35–38.

Chabre, M. (1981). *In* "Membranes and Intercellular Communication" (R. Bellan *et al.*, eds.), pp. 251–265. Elsevier, Amsterdam.

Chaitin, M. H., and Bok, D. (1984). Immunoferritin localization of actin, myosin and calmodulin in photoreceptor outer segments. *J. Cell Biol.* **99,** 114a.

Chaitin, M. H., Schneider, B. G., Hall, M. O., and Papermaster, D. S. (1984). Actin in the photoreceptor connecting cilium: Immunochemical localization to the site of outer segment disk formation. *J. Cell Biol.* **99,** 239–247.

Clack, J. W., Oakley, B., and Stein, P. J. (1983). Injection of GTP-binding protein or cyclic GMP phosphodiesterase hyperpolarizes retinal rods. *Nature (London)* **305,** 50–52.

Clark, S. P., and Molday, R. S. (1979). Orientation of membrane glycoproteins in sealed rod outer segment disks. *Biochemistry* **18,** 5868–5873.

Cobbs, W. H., and Pugh, E. N., Jr. (1985). Cyclic GMP can increase rod outer-segment light-sensitive current 10-fold without delay of excitation. *Nature (London)* **313,** 585–587.

Cohen, A. L. (1972). Rods and cones. *In* "Physiology of Photoreceptor Organs" (M. G. F. Fuortes, ed.), pp. 63–110. Springer-Verlag, Berlin and New York.

Cone, R. A. (1972). Rotational diffusion of rhodopsin in the visual membrane. *Nature (London) New Biol.* **236,** 39–43.

Corless, J. M., McCaslin, D. R., and Scott, B. L. (1982). Two-dimensional rhodopsin crystals from disk membranes of frog retinal rod outer segments. *Proc. Natl. Acad. Sci. U.S.A.* **79,** 1116–1120.

Defore, D., and Bok, D. (1983). Rhodopsin chromophore exchanges among opsin molecules in the dark. *Invest. Ophthalmol. Visual Sci.* **24,** 1211–1226.

Del Priore, L. V., Lewis, A., Carley, W. W., and Webb, W. W. (1985). Fluorescence microscopy of the three dimensional organization of actin filaments in toad rod photoreceptors. Submitted.

Dratz, E. A. (1970). Vision. *In* "The Science of Photobiology" (K. C. Smith, ed.), pp. 241–279. Plenum, New York.

Dratz, E. A., and Hargrave, P. A. (1983). The structure of rhodopsin and the rod outer segment disk membrane. *Trends Biochem. Sci.* **8,** 128–131.

Dratz, E. A., Miljanich, G. P., Nemes, P. P., Gaw, J. E., and Schwartz, S. (1979). The structure of rhodopsin and its disposition in the rod outer segment disk membrane. *Photochem. Photobiol.* **29,** 661–670.

Drzymala, R. E., Weiner, H. L., Dearry, C. A., and Liebman, P. A. (1984). A barrier to lateral diffusion of porphyropsin in *Necturus* rod outer segment disks. *Biophys. J.* **45,** 683–692.

Dudley, P. A., Alligood, J. P., and O'Brien, P. J. (1979). Isolation of metabolically active truncated visual cells. *Invest. Ophthalmol. (Suppl).* **20,** 116.

Falk, G., and Fatt, P. (1969). Distinctive properties of the lamellar and disk-edge structures of the rod outer segment. *J. Ultrastruct. Res.* **28,** 41–60.

Fein, A., Payne, R., Corson, D. W., Berridge, M. J., and Irvine, R. F. (1984). Photoreceptor excitation and adaptation by inositol 1,4,5-trisphosphate. *Nature (London)* **311,** 157–160.

Fesenko, E. E., Kolesnikov, S. S., and Arkadiy, A. L. (1985). Induction by cyclic GMP of cationic conductance in plasma membrane of retinal rod outer segment. *Nature (London)* **313,** 310–313.

Fliesler, S. J., and Basinger, S. F. (1985). Tunicamycin blocks the incorporation of opsin into retinal rod outer segment membranes. *Proc. Natl. Acad. Sci. U.S.A.* **82,** 1116–1120.

Frye, L. D., and Edidin, M. (1970). The rapid intermixing of cell surface antigens after formation of mouse-human heterokaryons. *J. Cell Sci.* **7,** 319–335.

Fukuda, M. N., Papermaster, D. P., and Hargrave, P. A. (1979). Rhodopsin carbohydrate. Structure of small oligosaccharides attached at two sites near the NH_2 terminus. *J. Biol. Chem.* **254,** 8201–8207.

Fung, B. K.-K. (1983). Characterization of transducin from bovine retinal rod outer segments. Separation and reconstitution of the subunits. *J. Biol. Chem.* **258,** 10495–10502.

Fung, B. K.-K., and Hubbell, W. L. (1978). Organization of rhodopsin in photoreceptor membranes. Transmembranes organization of bovine rhodopsin: Evidence from proteolysis and lactoperoxidase-catalyzed iodination of native and reconstituted membranes. *Biochemistry* **17,** 4403–4410.

Fung, B. K.-K., Hurley, J. B., and Stryer, L. (1981). Flow of information in light-triggered cyclic nucleotide cascade of vision. *Proc. Natl. Acad. Sci. U.S.A.* **78,** 152–156.

Ghalayini, A., and Anderson, R. E. (1984). Phosphatidylinositol 4,5-bisphosphate: Light mediated breakdown in the vertebrate retina. *Biochem. Biophys. Res. Commun.* **124,** 503–506.

Gilman, A. G. (1984). G-proteins and dual control of adenylate cyclase. *Cell* **36,** 577–579.

Goldberg, N. D., Ames, A., Gander, J. E., and Walseth, T. F. (1983). Magnitude of increase of retinal cGMP metabolic flux determined by ^{18}O incorporation into nucleotide α-phosphoryls corresponds with intensity of photic stimulation. *J. Biol. Chem.* **258,** 9213–9219.

Goldman, B. M., and Blobel, G. (1981). *In vitro* biosynthesis, core glycosylation, and membrane integration of opsin. *J. Cell Biol.* **90,** 236–242.

Goridis, C., Virmaux, N., Cailla, H. L., and Delaage, M. A. (1974). Rapid, light-induced changes of retinal cyclic GMP levels. *FEBS Lett.* **49,** 167–169.

Graff, G., Stephenson, J. H., Glass, D. B., Haddox, M. K., and Goldberg, N. D. (1978). Activation of soluble splenic cell guanylate cyclase by prostaglandin endoperoxides and fatty acid hydroperoxides. *J. Biol. Chem.* **253,** 7662–7676.

Greenblatt, R. E. (1983). Adapting lights and lowered extracellular free calcium desensitize toad photoreceptors by differing mechanisms. *J. Physiol. (London)* **336,** 579–605.

Hagins, W. A., and Yoshikami, S. (1974). A role for Ca^{++} in excitation or retinal rods and cones. *Exp. Eye Res.* **18,** 299–305.

Hagins, W. A., and Yoshikami, S. (1977). Intracellular transmission of visual excitation in photoreceptors: Electrical effects of chelating agents introduced into rods by vesicle fusion. *In* "Vertebrate Photoreception" (H. B. Barlow and P. Fatt, eds.), pp. 97–139. Academic Press, New York.

Hamm, H. E., and Bownds, M. D. (1985). The protein complement of frog retinal rod outer segments. Submitted.

Hargrave, P. A. (1977). The amino-terminal glycopeptide of bovine rhodopsin. *Biochim. Biophys. Acta* **492,** 83–94.

Hargrave, P. A. (1982). Rhodopsin chemistry, structure and topography. *Prog. Ret. Res.* **1,** 1–51.

Hargrave, P. A., and Fong, S.-L. (1977). The amino- and carboxyl-terminal sequence of bovine rhodopsin. *J. Supramol. Struct.* **6,** 559–570.

Hargrave, P. A., Fong. S.-L., McDowell, J. H., Mas, M. T., Curtis, D. R., Wang, J. K., Juszczak, E., and Smith, D. P. (1980). The partial primary structure of bovine rhodopsin and its topography in the retinal rod cell disc membrane. *Neurochem. Int.* **1,** 231–244.

Hargrave, P. A., McDowell, J. H., Curtis, D. R., Wang, J. K., Juszczak, E., Fong, S.-L., Rao, J. K. M., and Argos, P. (1983). The structure of bovine rhodopsin. *Biophys. Struct. Mech.* **9,** 235–244.

Hargrave, P. A., McDowell, J. H., Feldmann, R. J., Atkinson, P. H., Rao, J. K. M., and Argos, P. (1984). Rhodopsin's protein and carbohydrate structure: Selected aspects. *Vision Res.* **24,** 1487–1499.

Honig, B., Dinur, U., Nakanishi, K., Balogh-Nair, V., Gawinowicz, M. A., Arnaboldi, M., and Motto, M. G. (1979). An external point-charge model for wavelength regulation in visual pigments. *J. Am. Chem. Soc.* **101,** 7084–7086.

Hurley, J. B., Barry, B., and Ebrey, T. (1981). Isolation of an inhibitory protein for the cyclic GMP phosphodiesterase of bovine rod outer segments. *Biochim. Biophys. Acta* **675,** 359–365.

Irvine, R. F., Brown, K. D., and Berridge, M. J. (1984). Specificity of inositol trisphosphate-induced calcium release from permeabilized Swiss-mouse 3T3 cells. *Biochem. J.* **222,** 269–272.

Kapoor, C. L., and Chader, G. J. (1984). Endogenous phosphorylation of retinal photoreceptor outer segment proteins by calcium phospholipid-dependent protein kinase. *Biochem. Biophys. Res. Commun.* **122,** 1397–1403.

Kilbride, P., and Ebrey, T. G. (1979). Light-initiated changes of cyclic guanosine monophosphate levels in the frog retina measured with quick-freezing techniques. *J. Gen. Physiol.* **74,** 415–426.

Korenbrot, J. I. (1985). Signal mechanisms of phototransduction in retinal rod. *Crit. Rev. Biochem.* **17,** 223–256.

Kühn, H. (1974). Light-dependent phosphorylation of rhodopsin in living frogs. *Nature (London)* **250,** 588–590.

Kühn, H. (1978). Light-regulated binding of rhodopsin kinase and other proteins to cattle photoreceptor membranes. *Biochemistry* **17,** 4389–4395.

Kühn, H. (1981). Interactions of rod cell proteins with the disk membrane: Influence of light, ionic strength, and nucleotides. *Curr. Top. Membr. Transp.* **15,** 171–201.

Kühn, H., Cook, J. H., and Dreyer, W. J. (1973). Phosphorylation of rhodopsin in photoreceptor membranes: A dark reaction after illumination. *Biochemistry* **12,** 2495–2502.

Kühn, H., Mommertz, O., and Hargrave, P. A. (1982). Light-dependent conformational change at rhodopsin's cytoplasmic surface detected by increased susceptibility to proteolysis. *Biochim. Biophys. Acta* **679,** 95–100.

Kuwabara, T. (1965). Microtubules in the retina. *In* "Structure of the Eye." Schattauer, Stuttgart.

Liang, C.-J., Yamashita, K., Muellenberg, C. G., Shichi, H., and Kobata, A. (1979). Structure of the carbohydrate moieties of bovine rhodopsin. *J. Biol. Chem.* **254,** 6414–6418.

Liebman, P. A., and Entine, G. (1974). Lateral diffusion of visual pigment in photoreceptor disk membranes. *Science* **185,** 457–459.

Liebman, P. A., and Pugn, E. N. (1980). ATP mediates rapid reversal of cyclic GMP phosphodiesterase activation in visual receptor membranes. *Nature (London)* **287,** 734–736.

Liebman, P. A., and Sitaramayya, A. (1984). Role of G protein–receptor interaction in amplified phosphodiesterase activation of retinal rods. *Adv. Cyclic Nucleotide Res.* **17,** 215–225.

Liu, R. S. H., Asato, A. E., Denny, M., and Mead, D. (1984). The nature of restrictions in the binding site of rhodopsin. A model study. *J. Am. Chem. Soc.* **106,** 8298–8300.

Lochrie, M. A., Hurley, J. B., and Simon, M. I. (1985). Sequence of alpha subunit of photoreceptor G protein: Homologies between transducin, *ras* and elongation factors. *Science* **228,** 96–99.

McDowell, J. H., Curtis, D. R., Abu Bakar, U., and Hargrave, P. A. (1985). Phosphorylation of rhodopsin: Localization of phosphorylated residues in the helix V–helix VI connecting loop. *Invest. Ophthalmol. Visual Sci. (Suppl.)* **26,** 291.

Mathies, R. A., Smith, S. O., and Palings, I. (1985). Determination of retinal chromophore structure in rhodopsins. *In* "Biological Applications of Raman Spectroscopy" (T. G. Spiro, ed.), Wiley, New York, in press.

Matthews, H. R., Torre, V., and Lamb, T. D. (1985). Effects on the photoresponse of calcium buffers and cyclic GMP incorporated into the cytoplasm of retinal rods. *Nature (London)* **313,** 582–585.

Miki, N., Baraban, J. M., Keirns, J. J., Boyce, J. J., and Bitensky, M. W. (1975). Purification and properties of the light-activated cyclic nucleotide phosphodiesterase of rod outer segments. *J. Biol. Chem.* **250,** 6320–6327.

Miljanich, G. P., Sklar, L. A., White, D. L., and Dratz, E. A. (1979). Disaturated and dipolyunsaturated phospholipids in the bovine retinal rod outer segment disk membrane. *Biochim. Biophys. Acta* **552,** 294–306.

Miller, J. L., and Dratz, E. A. (1984). Phosphorylation sites near the carboxyl-terminus of rhodopsin regulate light-activated phosphodiesterase. *Vision Res.* **24,** 1509–1521.

Miller, W. H., and Nicol, G. D. (1979). Evidence that cyclic GMP regulates membrane potential in rod photoreceptors. *Nature (London)* **280,** 64–66.

Molday, R. S., and Molday, L. L. (1979). Identification and characterization of multiple forms of rhodopsin and minor proteins in frog and bovine rod outer segment disc membranes. *J. Biol. Chem.* **254,** 4653–4660.

Nathans, J., and Hogness, D. S. (1983). Isolation, sequence analysis and intron-exon arrangement of the gene encoding bovine rhodopsin. *Cell* **34,** 807–814.

Nathans, J., and Hogness, D. S. (1984). Isolation and nucleotide sequence of the gene encoding human rhodopsin. *Proc. Natl. Acad. Sci. U.S.A.* **81,** 4851–4855.

Nir, I., and Papermaster, D. S. (1983). Differential distribution of opsin in the plasma membrane of frog photoreceptors: An immunochemical study. *Invest. Ophthalmol. Visual Sci.* **24,** 868–878.

Nir, I., Cohen, D., and Papermaster, D. S. (1984). Immunocytochemical localization of opsin in the cell membrane of developing rat retinal photoreceptors. *J. Cell Biol.* **98,** 1788–1795.

Nishizuka, Y. (1984a). The role of protein kinase C in cell surface signal transduction and tumour promotion. *Nature (London)* **308,** 693–698.

Nishizuka, Y. (1984b). Protein kinases in signal transduction. *Trends Biochem. Sci.* **9,** 163–166.

O'Brien, P. J. (1978). Rhodopsin: A light-sensitive membrane glycoprotein. *In* "Receptors and

Recognition'' (P. Cuatrecasas and M. F. Greaves, eds.), Ser. A, Vol. 6, pp. 109–150. Chapman & Hall, London.

Olive, J. (1980). The structural organization of mammalian retinal disc membrane. *Int. Rev. Cytol.* **64,** 107–169.

O'Tousa, J. E., Baehr, W., Martin, R. L., Hirsh, J., Pak, W. L., and Applebury, M. L. (1985). The *Drosophila* nina*E* gene encodes an opsin. *Cell* **40,** 839–850.

Ovchinnikov, Y. A. (1982). Rhodopsin and bacteriorhodopsin: Structure–function relationships. *FEBS Lett.* **148,** 179–191.

Ovchinnikov, Y. A., Abdulaev, N. G., Feigina, M. Y., Artamonov, I. D., Zolotarev, A. S., Kostina, M. B., Bogachuk, A. S., Moroshnikov, A. I., Martinov, V. I., and Kudelin, A. B. (1982). The complete amino acid sequence of visual rhodopsin. *Bioorg. Khim.* **8,** 1011–1014.

Papermaster, D. S., and Dreyer, W. J. (1974). Rhodopsin content in the outer segment membranes of bovine and frog retinal rods. *Biochemistry* **13,** 2438–2444.

Papermaster, D. S., and Schneider, B. G. (1982). Membrane biogenesis in photoreceptors. *In* ''Cellular Aspects of the Eye'' (D. S. McDevitt, ed.), pp. 475–531. Academic Press, New York.

Papermaster, D. S., Schneider, B. G., Zorn, M. A., and Kraehenbuhl, J. P. (1978). Immunocytochemical localization of a large intrinsic membrane protein to the incisures and margins of frog rod outer segment disks. *J. Cell Biol.* **78,** 415–425.

Papermaster, D. S., Reilly, P., and Schneider, B. G. (1982). Cone lamellae and red and green rod outer segment disks contain a large intinsic membrane protein on their margins: An ultrastructural immunocytochemical study of frog retinas. *Vision Res.* **22,** 1417–1428.

Peters, K.-R., Palade, G. E., Schneider, B. G., and Papermaster, D. S. (1983). Fine structure of a periciliary ridge complex of frog retinal rod cells revealed by ultrahigh resolution scanning electron microscopy. *J. Cell Biol.* **96,** 265–276.

Pfister, C., Dorey, C., Vadot, E., Mirshahi, M., Deterre, P., and Chabre, M. (1984). Identité de la protéine dite ''48K'' qui interagit avec la rhodopsine illuminée dans les bâtonnets rétiniens et de l' ''antigène S rétinien'' inducteur de l'uvéo-retinte autoimmune expérimentale. *C.R. Acad. Sci. Paris* **299,** 261–265.

Plantner, J. J., Poncz, L., and Kean, E. L. (1980). Effect of tunicamycin on the glycosylation of rhodopsin. *Arch. Biochem. Biophys.* **201,** 527–532.

Pober, J. S., and Bitensky, M. W. (1979). Light regulated enzymes of vertebrate retinal rods. *Adv. Cyclic Nucleotide Res.* **11,** 265–301.

Poo, M., and Cone, R. A. (1974). Lateral diffusion of rhodopsin in the photoreceptor membrane. *Nature (London)* **247,** 438–441.

Roof, D. J., and Applebury, M. L. (1984). Cytoskeletal specializations of the vertebrate rod outer segment. *J. Cell Biol.* **99,** 114a.

Roof, D. J., and Heuser, J. E. (1982). Surfaces of rod photorecptor disk membranes: Integral membrane components. *J. Cell Biol.* **95,** 487–500.

Rosenkranz, J. (1977). New aspects of the ultrastructure of frog rod outer segments. *Int. Rev. Cytol.* **50,** 26–158.

Schaeffer, J. M. (1983). Preparation and properties of dispersed rat retinal cells. *In* ''Methods in Enzymology'' (P. M. Conn, ed.), Vol. 103, pp. 362–368. Academic Press, New York.

Schechter, I., Burstein, Y., Zemell, R., Ziv, E., Kantor, F., and Papermaster, D. S. (1979). Messenger RNA of opsin from bovine retina: Isolation and partial sequence of the *in vitro* translation product. *Proc. Natl. Acad. Sci. U.S.A.* **76,** 2654–2658.

Schmidt, S. Y. (1983a). Phosphatidylinositol synthesis and phosphorylation are enhanced by light in rat retinas. *J. Biol. Chem.* **258,** 6863–6868.

Schmidt, S. Y. (1983b). Light enhances the turnover of phosphatidylinositol in rat retinas. *J. Neurochem.* **40,** 1630–1638.

Schnetkamp, P. P. M., Klompmakers, A. A., and Daemen, F. J. M. (1979). The isolation of stable cattle rod outer segments with an intact plasma membrane. *Biochim. Biophys. Acta* **552,** 379–389.

Schröder, W. H., and Fain, G. L. (1984). Light-dependent calcium release from photoreceptors measured by laser micro-mass analysis. *Nature (London)* **309,** 268–270.

Shinozawa, T., Sen, I., Wheeler, G., and Bitensky, M. W. (1979). Predictive value of the analogy between hormone-sensitive adenylate cyclase and light-sensitive photoreceptor cyclic GMP phosphodiesterase: A specific role for light-sensitive GTPase as a component in the activation sequence. *J. Supramol. Struct.* **10,** 185–190.

Shuster, T. A., and Farber, D. B. (1984). Phosphorylation in sealed rod outer segments: Effects of cyclic nucleotides. *Biochemistry* **23,** 515–521.

Singer, S. J., and Nicolson, G. L. (1972). The fluid-mosaic model of the structure of membranes. *Science* **175,** 720–731.

Singh, A. K., Balogh-Nair, V., and Nakanishi, K. (1984). Photoaffinity labeling of bovine rhodopsin. *Tetrahedron* **40,** 493–500.

Sitaramayya, A., and Liebman, P. A. (1983a). Phosphorylation of rhodopsin and quenching of cyclic GMP phosphodiesterase activation by ATP at weak bleaches. *J. Biol. Chem.* **258,** 12106–12109.

Sitaramayya, A., and Liebman, P. (1983b). Mechanism of ATP quench of phosphodiesterase activation in rod disc membranes. *J. Biol. Chem.* **258,** 1205–1209.

Stone, W. L., Farnsworth, C. C., and Dratz, E. A. (1979). A reinvestigation of the fatty acid content of bovine, rat and frog retinal rod outer segments. *Exp. Eye Res.* **28,** 387–397.

Streb, H., Irvine, R. F., Berridge, M. J., and Schulz, I. (1983). Release of Ca^{2+} from a non-mitochondrial intracellular store in pancreatic acinar cells by inositol-1,4,5-trisphosphate. *Nature (London)* **306,** 67–69.

Stryer, L. (1983). Transducin and the cyclic GMP phosphodiesterase: Amplifier proteins in vision. *Cold Spring Harbor Symp. Quant. Biol.* **48,** 841–852.

Stryer, L. (1986). Cyclic GMP cascade of vision. *Annu. Rev. Neurosci.* **9,** in press.

Thomas, D. D., and Stryer, L. (1982). The transverse location of the retinal chromophore of rhodopsin in the rod outer segment disc membranes. *J. Mol. Biol.* **154,** 145–157.

Thompson, P., and Findlay, J. B. C. (1984). Phosphorylation of ovine rhodopsin. Identification of the phosphorylated sites. *Biochem. J.* **220,** 773–780.

Usukura, J., and Yamada, E. (1981). Molecular organization of the rod outer segment. A deep-etching study with rapid freezing using unfixed frog retina. *Biomed. Res.* **2,** 177–193.

Vandenberg, C. A., and Montal, M. (1984). Light-regulated biochemical events in invertebrate photoreceptors. 2. Light-regulated phosphorylation of rhodopsin and phosphoinositides in squid photoreceptor membranes. *Biochemistry* **23,** 2347–2352.

VanDop, C., Yamanaka, G., Steinberg, F., Sekura, R. D., Manclark, C. R., Stryer, L., and Bourne, H. R. (1984). ADP-ribosylation activation of transducin by pertussis toxin blocks the light-stimulated hydrolysis of GTP and cGMP in retinal photoreceptors. *J. Biol. Chem.* **259,** 23–26.

Vuong, T. M., Chabre, M., and Stryer, L. (1984). Millisecond activation of transducin in the cyclic nucleotide cascade of vision. *Nature (London)* **311,** 659–661.

Waloga, G., and Anderson, R. E. (1985). Effects of inositol-1,4,5-trisphosphate injections into salamander rods. *Biochem. Biophys. Res. Commun.* **126,** 59–62.

Wey, C.-L., Cone, R. A., and Edidin, M. A. (1981). Lateral diffusion of rhodopsin in photoreceptor cells measured by fluorescence photobleaching and recovery. *Biophys. J.* **33,** 225–232.

Wilden, U., and Kühn, H. (1982). Light-dependent phosphorylation of rhodopsin: The number of phosphorylation sites. *Biochemistry* **21,** 3014–3022.

Woodruff, M. L., and Bownds, M. D. (1979). Amplitude, kinetics and reversibility of a light-induced decrease in guanosine 3′,5′-cyclic monophosphate in frog photoreceptor membranes. *J. Gen. Physiol.* **73,** 629–653.

Yamazaki, A., Halliday, K. R., George, J. S., Nagao, S., Kuo, C.-H., Ailsworth, K. S., and Bitensky, M. W. (1985). Homology between light-activated photoreceptor phosphodiesterase and hormone-activated adenylate cyclase systems. *Adv. Cyclic Nucleotide Protein Phosphorylation Res.* **19,** 113–124.

Yau, K.-W., and Nakatani, K. (1984). Electrogenic Na–Ca exchange in retinal rod outer segment. *Nature (London)* **311,** 661–663.

Yau, K. -W., and Nakatani, K. (1985). Light-induced regulation of cytoplasmic free calcium in retinal rod outer segment. *Nature (London)* **313,** 579–582.

Yee, R., and Liebman, P. A. (1978). Light-activated phosphodiesterase of the rod outer segment. Kinetics and parameters of activation and deactivation. *J. Biol. Chem.* **253,** 8902–8909.

Yoshikami, S., and Hagins, W. A. (1971). Light, calcium and the photocurrent of rods and cones. *Biophys. Soc. Abstr.* **15,** 47a.

Young, R. W. (1971). The renewal of rod and cone outer segments in the *Rhesus* monkey. *J. Cell Biol.* **49,** 303–318.

Zimmerman, W. F., Lion, F., Daemen, F. J. M., and Bonting, S. L. (1975). Biochemical aspects of the visual process. Distribution of stereo specific retinol dehydrogenase activities in subcellular fractions of bovine retina and pigment epithelium. *Exp. Eye Res.* **21,** 325–332.

Zuckerman, R., Buzdygon, B., and Liebman, P. (1984). Characterization of the 48 kilodalton protein of retinal rod outer segments as a light-dependent ATP-binding protein. *Invest. Ophthalmol. Visual Sci. (Suppl.)* **25,** 112.

Zuker, C. S., Cowman, A. F., and Rubin, G. M. (1985). Isolation and structure of a rhodopsin gene from *D. melanogaster. Cell* **40,** 851–858.

CYCLIC NUCLEOTIDES IN RETINAL FUNCTION AND DEGENERATION

DEBORA B. FARBER AND TERRENCE A. SHUSTER

Jules Stein Eye Institute
UCLA School of Medicine
Los Angeles, California

I. Introduction

Considerable evidence has accumulated suggesting that cyclic nucleotides play an important role in the metabolism and function of the retina. Furthermore, abnormalities in the regulation of cyclic nucleotide levels have been shown to be associated with several inherited or drug-induced degenerations affecting the retinal photoreceptor cells.

Cyclic GMP and cyclic AMP are compartmentalized differently within the retinal neurons and participate in the modulation of different biochemical and physiological processes. The cyclic GMP system is most important in rod visual cells and its principal effector is light; cyclic AMP metabolism is minimal in rods. In contrast to rods, the cyclic AMP system seems to predominate in cone visual cells and is responsive to light; minimal levels of cyclic GMP are present in cones. Cyclic AMP also acts as a second messenger in many of the metabolic reactions that take place in the neurons of the inner retina.

We have divided this review into three main sections. In the first we will discuss the involvement of cyclic nucleotides in the normal functioning of retinas which have a majority of rod photoreceptors. We will also consider the cyclic nucleotide metabolism of the inner retinal layers which is common to both rod- and cone-dominated tissues. The second section will have a discussion of cyclic nucleotide involvement in retinas in which cone visual cells are predominant, and in the last section we will review abnormalities in cyclic nucleotide metabolism which have been associated with degenerative diseases affecting the retinas of several species of animals. We will also include in this section a discussion of studies in which the use of drugs that interfere with cyclic nucleotide metabolism has promoted retinal degenerations similar to those observed in the inherited disorders.

II. Cyclic Nucleotide Metabolism in Rod-Dominant Retinas

A. *The Rod Photoreceptor*

The photoreceptor layer in rod-dominant retinas (see Fig. 2 in Farber and Adler, this volume) varies considerably in size from one species to another. In some species, this layer can constitute up to 40% of the total retinal mass. The structure of rods as well as many of their molecular properties is described in the article by Paul Hargrave in this volume, and thus only those details necessary for discussion of our subject will be mentioned here. The outer segments of rods contain stacks of disks formed by lipid bilayers which provide a matrix for protein molecules (Fig. 1). These disks are contained by but are not continuous with the plasma membrane of the photoreceptors; they are, however, connected to the plasma membrane by filamentous structures (Roof and Heuser, 1982). The visual pigment rhodopsin is the major protein of the rod outer segment. Rhodopsin is an intrinsic glycoprotein which spans the disk membrane thickness crossing the membrane interface several times (Hubbell *et al.*, 1977; Hubbell and Fung, 1978). The lipid composition of the disk membranes makes them relatively fluid, allowing rhodopsin and other components of the outer segments to rotate and

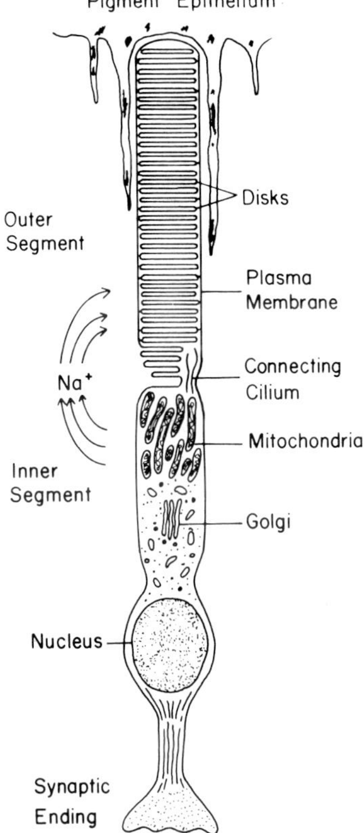

FIG. 1. Diagram of a vertebrate rod photoreceptor cell.

diffuse in the membranes (Poo and Cone, 1974) and perhaps facilitating the reactions that occur during visual transduction. A modified cilium connects the outer segments to the inner segment, a compartment rich in mitochondria. In the inner segment the cell synthesizes proteins, produces the energy-containing nucleoside triphosphates, and assembles the disk membranes. The rest of the photoreceptor cell consists of a cell soma and a synaptic ending.

B. The Phototransduction Process of Rods and the Search for a Second Messenger of the Light Signal

When the photoreceptor cell is in the dark, there is a continuous current, the "dark current," consisting of Na^+ extruded from the cell body and returning across the outer segment (Fig. 1). The initial step of the visual process is the absorption of light by rhodopsin which causes the isomerization of the chromophore, 11-*cis*-retinal, to the all-*trans* configuration. This process results in a reduction in the permeability of the plasma membrane to Na^+, which, in vertebrates, causes hyperpolarization of the photoreceptor cell membrane (Werblin, 1974).

The link between photon capture in the disk membrane and the change in Na^+ permeability and electrical potential of the plasma membrane is not known. To explain this reaction, thought to be transmitted across the cytoplasm, one must assume that light causes the release or production of a "second messenger" at the disk membrane, which then diffuses to the plasma membrane and reduces its conductance for Na^+ (Baylor and Fuortes, 1970; Hagins, 1972; Cone, 1973). The identity of this internal messenger has puzzled many investigators and has been the subject of considerable research in the last decade. Recently, Fein and Szuts (1982) have discussed a set of criteria which must be met by the proposed transmitter before it can be considered as genuine. These criteria are (1) the transmitter must be present in the rod outer segments in sufficient quantities, (2) light should affect the transmitter concentration or location, (3) agents which interfere with the metabolism of the putative transmitter should affect light-induced responses, (4) changes in the concentration of the transmitter must mimic the effect of light in the absence of light, and (5) the metabolizing enzymes of the transmitter must be present in the system to enable the transmitter to function.

It is evident that the transmitter must be soluble in the aqueous millieu of the cytoplasm that surrounds the disks and that many molecules of messenger must be released per photon absorbed by rhodopsin so that enough Na^+ channels are closed in the plasma membrane to produce a measurable electric response. This phenomenon implies a tremendous amplification of the initial single molecular event. Furthermore, the effect of the second messenger has to be fast, on the order of milliseconds, since this is the time frame in which the change in conductance occurs.

Investigations into the nature of the internal transmitter of the light signal have reduced the list of major candidates to calcium ion and cyclic GMP, and a new third candidate, inositol triphosphate, an end product of phosphatidyl inositol metabolism. Protons, which were considered for a transmitter role some time ago, have lost popularity.

Yoshikami and Hagins (1971) were the first to observe that an elevation of extracellular Ca^{2+} mimicked the effect of light in rod photoreceptors; that is, Ca^{2+} suppressed the dark current. They assumed that the increased extracellular Ca^{2+} led to Ca^{2+} influx into the cell, and on this basis they proposed a model in which Ca^{2+}, sequestered within the disks in the dark-adapted state, is released to the cytoplasm following the absorption of light by rhodopsin. Ca^{2+} then binds to the Na^+ channels, thereby inhibiting Na^+ entrance across the outer segment membrane. The experimental observations which do or do not support this hypothesis are confusing and in many instances contradictory. Recent reviews of the subject include those of Uhl and Abrahamson (1981), Fein and Szuts (1982), and Kaupp and Schnetkamp (1982).

The only criterion which Ca^{2+} satisfies as a putative transmitter of the light signal is that the rod disks contain a high enough concentration of Ca^{2+} to sustain release even for saturating light flashes (Szuts and Cone, 1977). Although it has been demonstrated that illumination increases the Ca^{2+} concentration in the subretinal space around the rod outer segments (Gold and Korenbrot, 1980; Yoshikami *et al.*, 1980; Schroeder and Fain, 1984), actual release of Ca^{2+} from disks within milliseconds of light exposure, the most direct proof of the Ca^{2+} theory, has never been confirmed (Hemminki, 1975; Weller *et al.*, 1975a; Sorbi and Cavaggioni, 1975; Szuts and Cone, 1977; Smith *et al.*, 1977; Kaupp and Junge, 1977; Liebman, 1978). Experiments by Yau and Nakatani (1985) suggest that upon illumination there is only a transient efflux of Ca^{2+} from the rod and that the internal Ca^{2+} levels actually fall and recover only in the dark. Also, recent work by Fesenko *et al.* (1985), discussed later in this article, showed that Ca^{2+} has no effect on the cationic conductance of "inside-out" patches from the plasma membrane of rod outer segments. Thus, there is increasing evidence that the Ca^{2+} hypothesis may not be operative in rod photoreceptors; in other words, Ca^{2+} may not directly control their light-dependent conductance.

An alternate hypothesis has gained support in the last few years due to the detection of several light-activated enzyme systems directly or indirectly related to cyclic GMP metabolism in rod photoreceptors. This hypothesis considers cyclic GMP as the second messenger of the light signal. It has been proposed that cyclic GMP interacts with the plasma membrane maintaining the Na^+ channels open. Therefore, cyclic GMP would act as a "negative" transmitter, the conductance of the plasma membrane changing in response to a decrease in cyclic GMP concentration. In the next sections, we discuss observations which are important in establishing the probable role of cyclic GMP as a component of visual transduction.

Of considerable interest is the function that inositol triphosphate (IP_3) may have in rod photoreceptors. A light-induced turnover of IP_3 has been observed in retinas of rats (Schmidt, 1983) and of toads (Anderson and Hollyfield, 1981), and a light-induced hydrolysis of phosphatidylinositol 4,5-biphosphate has been reported to occur in frog retina (Ghalayini and Anderson, 1984). Furthermore, intracellular injections of IP_3 into *Limulus* ventral photoreceptors indicated that IP_3 causes a transient depolarization of their plasma membrane, with characteristics similar to those of the light-induced membrane depolarization (Brown *et al.*, 1984; Fein *et al.*, 1984). Notice that these experiments were carried out using the visual cells of an invertebrate. In invertebrates, photoreceptor cells respond to light by depolarizing whereas in vertebrates there is a hyperpolarization of the plasma membrane. Recent work with vertebrate retinas (Waloga and Anderson, 1985) demonstrated that pressure injections of IP_3 into outer segments of dark-adapted rats caused hyperpolarizing responses and that these responses were attenuated after light adaptation of the cells. Furthermore, IP_3 injection decreased light-induced receptor potentials. Thus, these initial studies present a new question: Are there branched pathways involving more than one transmitter of the light signal in the transduction mechanism of rod photoreceptors? This is a promising area of future investigation.

C. Cyclic GMP Metabolism in Rods

1. Localization of Cyclic GMP in Rod Photoreceptors

Rod-dominant retinas have the highest concentration of cyclic GMP of any tissue studied, about 10-fold that in cerebellum and 100 times higher than in any other region of the central nervous system (Farber and Lolley, 1974, 1977a; Goridis *et al.*, 1974, 1977; Krishna *et al.*, 1976; Ferrendelli, 1978). In these retinas, cyclic GMP is concentrated in the photoreceptors, as concluded from different types of investigations. Studies of retinal development showed that the levels of the cyclic nucleotide increase as the visual cells differentiate and the outer segments grow and mature (Farber and Lolley, 1977a). In animals affected with inherited degenerations of the retina, all photoreceptors are lost as a consequence of the disease, and the retina is left with the inner layers apposed to the pigment epithelium. The cyclic GMP concentration in retinas of affected mice (Farber and Lolley, 1974, 1977a), rats (Lolley and Farber, 1976), and dogs (Aguirre *et al.*, 1978; Woodford *et al.*, 1982) is reduced drastically following the death of the visual cells. Furthermore, in human and monkey retinas which contain mixed populations of rods and cones, the cyclic GMP levels are found to vary by area and are highest in the rod-rich zones and lowest in the cone-rich macula (Newsome *et al.*, 1980; Farber *et al.*, 1985) as shown in Fig. 2. In addition, microdissection of the different retinal layers followed by analyses of cyclic GMP concentration in each layer indicated that approximately 90% of the total retinal cyclic GMP is found in the rod photoreceptor cells (Farber and Lolley, 1974), with a great enrichment in rod outer segments (Orr *et al.*, 1976; Berger *et al.*, 1980). Consistent with these data, isolated rod outer segments have very high levels of cyclic GMP (Fletcher and Chader, 1976). In frog rod outer segments, the cyclic GMP concentration is in the order of 10^{-5} *M* (Woodruff and Bownds, 1979).

FIG. 2. Three-dimensional computerized representation of the distribution of rod photoreceptors (a), cone photoreceptors (b), and cyclic GMP levels (c) in human retina. (From Farber *et al.*, 1985.)

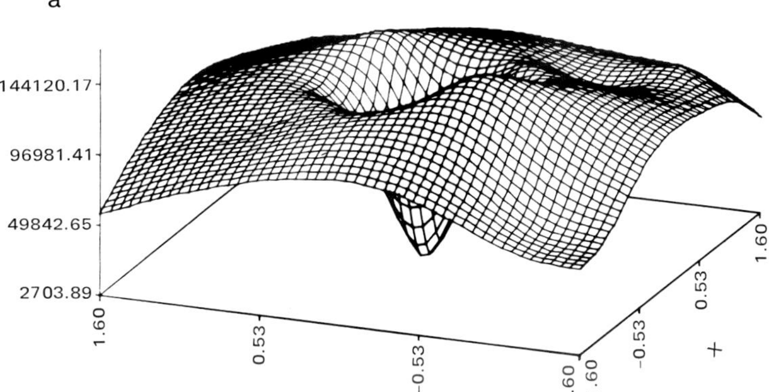

a
144120.17
96981.41
49842.65
2703.89
1.60
0.53
−0.53
−1.60
Y
−1.60
−0.53
0.53
1.60
X

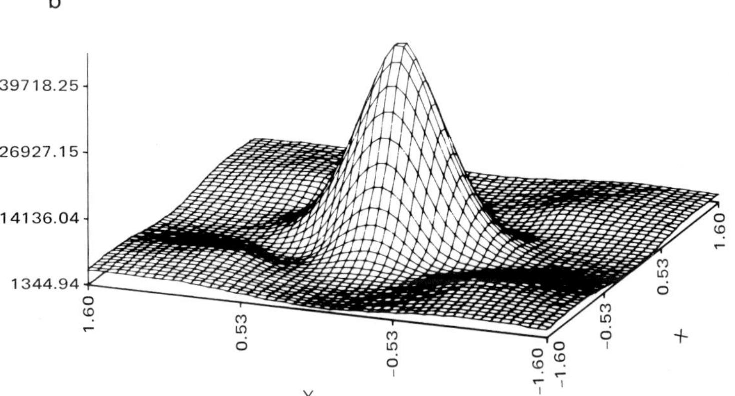

b
39718.25
26927.15
14136.04
1344.94
1.60
0.53
−0.53
−1.60
Y
−1.60
−0.53
0.53
1.60
X

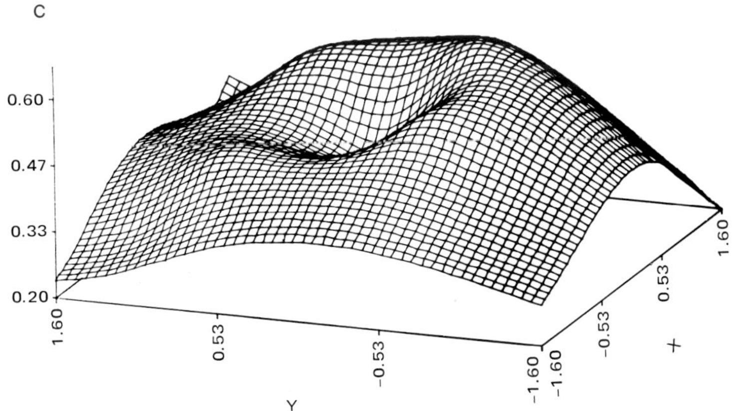

c
0.60
0.47
0.33
0.20
1.60
0.53
−0.53
−1.60
Y
−1.60
−0.53
0.53
1.60
X

2. Modulation of Cyclic GMP Concentration by Light

Light decreases the cyclic GMP content of rod-dominant retinas, whole photoreceptor cells, and also isolated rod outer segments (Goridis *et al.*, 1974, 1977; Fletcher and Chader, 1976; Brodie and Bownds, 1976; Ferrendelli and Cohen, 1976; Krishna *et al.*, 1976; Orr *et al.*, 1976; Farber and Lolley, 1977a; Woodruff *et al.*, 1977; Cohen *et al.*, 1978; Mitzel *et al.*, 1978; de Azeredo *et al.*, 1978; Woodruff and Bownds, 1979). When the inner layers of the retina are exposed to light following microdissection, no change in their cyclic GMP level is observed (Orr *et al.*, 1976). Thus, the effect of light on retinal cyclic GMP concentration is apparently restricted to the visual cells.

The time course for the cyclic GMP decrease is not always the same for the different types of retinal preparations studied. Several seconds of exposure to light are necessary to reduce the cyclic GMP levels of whole retinas (Goridis *et al.*, 1977; Mitzel *et al.*, 1978; Kilbride and Ebrey, 1979; Govardovskii and Berman, 1981); however, the intensity of the light used in the experiments can modify the rate of cyclic GMP decrease. Thus, it takes a full minute before a significant reduction in cyclic GMP levels is observed following a strong flash of light; whereas a weaker flash, near the physiological working range of the rods, causes a fall in cyclic GMP concentration in only 3 sec and an almost maximal response after 30 sec. This rate of change of cyclic GMP content in whole retinas, however, is not fast enough. The change must occur in the order of milliseconds to have a determining effect in excitation. It might be argued that not all of the cyclic GMP present in the retinal photoreceptors is responsive to light. For instance, if the light-sensitive cyclic GMP pool is compartmentalized and, in addition, if some cyclic GMP is bound to enzyme systems and not available for light-initiated degradation, measurements of changes in its concentration in whole retina may not reflect what actually occurs in a reduced area of the visual cells.

Furthermore, even when studying rod outer segment preparations, differences in the degree of dark adaptation, the time taken to obtain the isolated organelles, their integrity, the buffers used during the preparation (removal of bicarbonate/CO_2 reduces the dark cGMP levels to those observed on illumination; Meyertholen *et al.*, 1980), the ions present or absent in the buffers (effect of Ca^{2+}, described below), the redox conditions (addition of diamide, a specific glutathione oxidant, reduces the dark cGMP levels to those observed in the light; Winkler *et al.*, 1984), the temperature at which the rod outer segments are handled, in other words, all the environmental conditions, could have tremendous effects on the rate of cyclic GMP hydrolysis in response to light. This may explain why Krishna *et al.* (1976) and Salceda *et al.* (1982) did not detect any change in cyclic GMP levels until rod outer segments had been illuminated for several minutes, whereas Woodruff *et al.* (1977) and Woodruff and Bownds

(1979) found that the half-time for the light-induced cyclic GMP decrease is approximately 125 msec. These investigators reported that with light exposures which bleach less than 100 rhodopsin molecules per outer segment, at least 10^4–10^5 molecules of cyclic GMP are hydrolyzed for each rhodopsin molecule bleached. With continuous illumination, the decrease in cyclic GMP concentration becomes larger as illumination increases and varies linearly with the logarithm of light intensity at levels that bleach between 5 and 5000 rhodopsin molecules/outer segment per second. Over this same range of light intensities the ionic permeability of rod outer segments is suppressed, as assayed *in vitro* by Brodie and Bownds (1976).

Woodruff and Bownds (1979) carried out their experiments in buffers containing low concentrations of Ca^{2+}, and in these conditions, Cohen *et al.* (1978) had shown that the cyclic GMP content of dark-adapted retinas as well as that of rod outer segments could be increased 10- to 20-fold. Kilbride (1980) suggested that only because of this large increase in cyclic GMP levels, induced by a reduction of Ca^{2+} concentration, could the effect of light on cyclic GMP levels be measured. However, Polans *et al.* (1981) demonstrated that although the dark levels of cyclic GMP in outer segments could be decreased by increasing the concentration of Ca^{2+} to above 10^{-9} *M*, the resultant cyclic GMP concentration does not vary between 2×10^{-9} and 10^{-3} *M* Ca^{2+}. Moreover, illumination reduces the cyclic GMP levels by about 50% within 1 sec from the onset of the light stimulus, regardless of the Ca^{2+} concentration. In support of these findings, Yee and Liebman (1978) showed that in fragmented rod outer segments the proton release accompanying cyclic GMP hydrolysis could be detected within 100 msec following a bright flash. Therefore, the reduction of cyclic GMP levels by light seems to occur rapidly in outer segments and represents a large amplification of the light signal, since the bleaching of one rhodopsin molecule can activate the hydrolysis of many thousands of cyclic GMP molecules.

The time course for the recovery of cyclic GMP content to dark-adapted levels after light exposures varies according to both the intensity and the length of the light exposure. Recovery is slower when the light intensity is greater and the exposure is longer (Woodruff and Bownds, 1979; Kilbride and Ebrey, 1979).

Recently, Goldberg *et al.* (1983) performed experiments with intact rabbit retinas in which they determined the cyclic GMP metabolic flux in response to light by monitoring ^{18}O incorporation into guanine α phosphoryls. These investigators found that at low light levels, cyclic GMP concentration in intact retina remains unchanged with a continuous 20-sec pulse of light, but the rate of cyclic GMP flux increases linearly with increases in light intensity. Goldberg's group postulated that to achieve this, cyclic GMP synthesis must be stimulated as cyclic GMP hydrolysis increases in response to light. (We will discuss these reactions in detail in the next sections.) These observations are very intriguing; however, the majority of studies described to date have indicated that light causes a

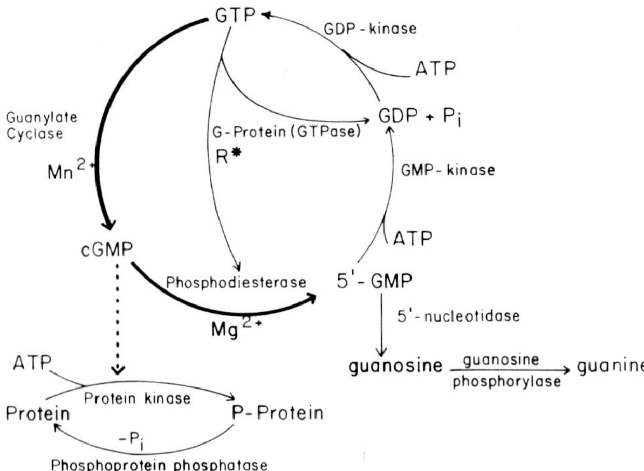

FIG. 3. Major enzymes of cyclic GMP metabolism. Dark arrows indicate the synthesis and the degradation reactions. The dotted arrow points to reactions that could be stimulated by cyclic GMP. R*, Photolyzed rhodopsin.

decrease in cyclic GMP levels. Differences in methodology, conditions of the tissue, and illumination may account for the discrepancies. Evidently, this is an important issue that requires clarification.

3. Cyclic GMP Synthesis: Guanylate Cyclase

All the enzymes necessary for cyclic GMP metabolism have been studied in retinal tissues. Figure 3 depicts the reactions in which they are involved, together with their substrates, known cofactors, and metabolites. We will discuss in detail cyclic GMP synthesis in this section, and degradation and physiological effects in the following sections. The enzymes that participate in the regeneration of the cyclic GMP precursor, GTP, and in the degradation of the cyclic GMP metabolite, 5′-GMP, have been studied by Berger *et al.* (1980) and by Kavipurapu *et al.* (1982) and will not be considered here.

Developmental studies of rodent retina have shown a close correspondence between differentiation of the photoreceptor cells, the growth of their outer segments, and the postnatal increases in the activity of guanylate cyclase (Farber and Lolley, 1976). In addition, the retinas of photoreceptorless adult *rd* mice exhibit minimal guanylate cyclase activity, suggesting that the enzyme is compartmentalized in rods (Farber and Lolley, 1976). Indeed, about 90% of the total guanylate cyclase activity in retinas of various species has been found in rod outer segments (Virmaux *et al.*, 1976; Zimmerman *et al.*, 1976; Berger *et al.*, 1980). Within these organelles, the enzyme is apparently localized mainly in the ciliary axonemes, where it has a specific activity 50-fold higher than in unfractionated rods (Fleishman and Denisevich, 1979; Fleishman *et al.*, 1980; Khoury and Farber, 1981; Toibana *et al.*, 1982). Interestingly, a high level of guanylate cyclase activity seems to be a common feature of ciliated cells, such as those of the lung, sperm, and rods (outer segments can be considered modified cilia).

Guanylate cyclase has been found in soluble and particulate forms in many tissues, including retina. However, the enzyme from rod outer segments appears to be entirely membrane associated or bound to cellular structures, since it is not extracted by low or high ionic strength buffers after repeated freezing and thawing (Goridis *et al.*, 1976). Detergent treatment also does not solubilize the enzyme but causes a loss of activity (Bensinger *et al.*, 1974a; Krishna *et al.*, 1976; Troyer *et al.*, 1978). This guanylate cyclase requires divalent cations for its activity, which becomes maximal with Mn^{2+} when present in a 1 : 1 ratio with GTP (Krishnan *et al.*, 1978). Ca^{2+} seems to have a biphasic effect on the enzyme. At 2–3 m*M*, Ca^{2+} stimulates the cyclase of bovine and human rod outer segments whereas it inhibits the activity at higher concentrations (Pannbacker, 1973). These studies were carried out using Mg^{2+} as cofactor, which supports only 10% of the guanylate cyclase activity obtained when Mn^{2+} is the cofactor. Only when using Mg^{2+} is the effect of Ca^{2+} evident. Lolley and Racz (1982) have reported also that the activity of guanylate cyclase in intact retinas incubated in the dark increased when Ca^{2+} concentrations were below 10^{-6} *M* Ca^{2+} and the enzyme appears to be fully active at Ca^{2+} levels lower than 10^{-9} *M*. With Mn^{2+} in the reaction mixture, Ba^{2+}, Mg^{2+}, and Co^{2+} inhibited the enzyme when tested in the range of 0.1–0.5 m*M* (Krishnan *et al.*, 1978).

The effect of nucleoside triphosphates other than GTP on the guanylate cyclase activity of rod outer segments has also been investigated. ATP at 1 m*M* inhibits the enzyme by 30%, whereas at concentrations between 0.01 and 0.1 m*M*, ATP appears to stimulate the activity by a mechanism that is unresolved (Krishnan *et al.*, 1978). UTP and ITP are also inhibitors of this guanylate cyclase. Other agents known to activate guanylate cyclase from brain and other tissues, such as sodium azide, hydroxylamine (Troyer *et al.*, 1978), prostaglandin F_2, carbamylcholine, glutamic acid, insulin, and cholera toxin do not have any effect on rod outer segment guanylate cyclase (Bitensky *et al.*, 1975, 1978).

The initial reports of guanylate cyclase activity in rod outer segments suggested that the enzyme was inhibited by light (Pannbacker, 1973, 1974; Bensinger *et al.*, 1974b; Krishna *et al.*, 1976). However, these reports were followed by contradictory results. Since the activity of cyclic GMP phosphodiesterase in rod outer segments is very high, accurate guanylate cyclase measurements depend upon the ability of phosphodiesterase inhibitors to block the degradation of cyclic GMP during the time of assay. Inhibitors of cyclic GMP phosphodiesterase are not 100% effective and, in addition, this enzyme is activated by light in the presence of

GTP, the guanylate cyclase substrate. Thus, the assay of guanylate cyclase activity reflects degradation as well as synthesis of cyclic GMP during the time of incubation and it is difficult to distinguish between light inhibition of guanylate cyclase and light activation of phosphodiesterase. Using a variety of conditions, Goridis *et al.* (1973, 1975) and Bitensky *et al.* (1978) apparently demonstrated that guanylate cyclase activity is unchanged by illumination in rod outer segment homogenates. However, Goldberg *et al.* (1983) suggested recently that light increases guanylate cyclase activity and, furthermore, Toibana *et al.* (1982), using histochemical methods, have shown that, *in situ,* light and dark adaptation may affect the activity and distribution of the cyclic GMP synthetic enzyme.

4. Cyclic GMP Degradation: Phosphodiesterase

Several classes of phosphodiesterases have been observed in retina. Kinetic analysis of microdissected samples, and of developing normal retinas and those affected with photoreceptor degeneration, showed two cyclic nucleotide phosphodiesterases in the inner layers and at least one in the visual cells (Lolley and Farber, 1975; Farber and Lolley, 1976; Schmidt and Lolley, 1973; Aguirre *et al.,* 1978; Pannbacker and Lovett, 1977). Because of its light regulation and possible involvement in the visual process, the latter phosphodiesterase is the enzyme that has received most of the attention. It has been isolated, purified, and characterized by several laboratories. In contrast, phosphodiesterases from the inner layers of the retina have not been thoroughly studied. They may be similar to the enzymes of brain; at least, they have comparable K_m values for cyclic AMP and cyclic GMP.

The cyclic nucleotide phosphodiesterase from rod photoreceptors has been localized specifically by histochemical and immunocytochemical methods in the outer segments (Robb, 1974, 1978; Ueno *et al.,* 1984; Hurwitz *et al.,* 1984; Farber and Bok, 1984), as shown in Fig. 4. Although this phosphodiesterase can hydrolyze cyclic AMP and cyclic GMP *in vitro,* it displays a marked preference for cyclic GMP. For example, at 10^{-7} *M* substrate concentration, the hydrolysis of cyclic GMP is 23 times greater than that of cyclic AMP (Bitensky *et al.,* 1973). Phosphodiesterase preparations from retinas of different species have K_m values for cyclic GMP in the 10^{-4}–10^{-5} *M* range, whereas the K_m values for cyclic AMP are in the 10^{-3} *M* range. These K_m values are identical for the dark- or light-activated enzyme. However, V_{max} values vary with illumination, suggesting that light does not change the affinity of phosphodiesterase for cyclic GMP but increases the number of enzyme molecules available for the hydrolysis of cyclic GMP.

Cyclic GMP phosphodiesterase requires Mg^{2+} or Mn^{2+} for maximal activity, with Mg^{2+} being the preferred cation. Unlike the cyclic GMP phosphodiesterase of brain, the rod outer segment enzyme is not activated by Ca^{2+}, even in the

FIG. 4. Immunocytochemical localization of cyclic GMP phosphodiesterase in rod outer segments. (From Farber and Bok, 1984.)

presence of calmodulin (Bitensky *et al.*, 1975). In fact, Ca^{2+} at 1.0 m*M* concentration causes 90% inhibition of the enzyme. EGTA and calcium ionophores (X537A) do not change the activity of the rod outer segment phosphodiesterase (Chader *et al.*, 1981).

a. Isolation and Purification of Cyclic GMP Phosphodiesterase from Rod Outer Segments. Phosphodiesterase is bound to the disk membranes but can be eluted with low ionic strength buffers. Miki *et al.* (1975) found that the omission of Mg^{2+} and the use of EDTA facilitate this extraction. Since Mg^{2+} favors the binding of the enzyme to the membranes, many soluble proteins can be washed off the disks before extracting the phosphodiesterase with buffer containing no Mg^{2+}.

The complete purification of the cyclic GMP phosphodiesterase from different species has been achieved utilizing various chromatographic techniques. The enzyme extracted from frog rod outer segments was bound to agarose–polyhistidine and then eluted with imidazole buffer (Miki *et al.*, 1975). This enzyme binds irreversibly to or is denatured on DEAE–cellulose and activity cannot be eluted with buffers of pH from 4.5 to 10 (Yamazaki *et al.*, 1982b). The frog phosphodiesterase accounts for approximately 0.5% of the disk membrane protein; it has an isoelectric point of 5.7 and a sedimentation coefficient of 12.4 S, which corresponds to an apparent molecular weight of 240,000. According to Miki *et al.* (1975) the solubilized enzyme is composed of two subunits of 120,000 and 110,000 molecular weight and has a K_m for cyclic GMP of 70 μ*M*, which is apparently lower than that of the enzyme bound to the disk membranes. The solubilized enzyme also has a lower pH optimum (7.5) for activity than the disk-bound phosphodiesterase (8.0). Miki *et al.* (1975) calculate a catalytic constant of 48,000 molecules of cyclic GMP hydrolyzed/min/mol of enzyme, based on a specific activity of 185 μmol of cyclic GMP hydrolyzed/min/mg of protein (using 2 m*M* cyclic GMP as substrate) and a molecular weight of 240,000.

The cyclic GMP phosphodiesterase from mammalian rod outer segments has characteristics which are different from those of the frog enzyme. The mammalian phosphodiesterase has been purified by several laboratories (Coquil *et al.*, 1975; Baehr *et al.*, 1979; Hurley and Ebrey, 1979; Fung and Stryer, 1980; and Kohnken *et al.*, 1981a), using combinations of DEAE–cellulose and Sephadex G-100 or G-200 chromatography, or hexylagarose or DEAE–cellulose followed by HPLC gel filtration. The protein has a molecular weight of 170,000–185,000 (depending on the report cited) and is composed of three subunits, two separating as a doublet on SDS–polyacrylamide gel electrophoresis (α, β; molecular weights 88,000 and 84,000, respectively,) and the third (γ, MW 13,000) having an inhibitory function. An isoelectric pH of 5.0–5.5 has been determined for the bovine phosphodiesterase. It has been estimated that the rhodopsin : phosphodiesterase ratio is approximately 150 : 1 in cow and 900 : 1 in frog rod outer segments. It is interesting to note that Takemoto *et al.* (1984) have compared the peptide maps of the subunits (in the doublet) from bovine, frog, chicken, and dog cyclic GMP phosphodiesterases and reported that the enzyme has been conserved through evolution and that all these species have remarkably similar 88,000 and 84,000 MW subunits. [The reason for the different molecular weight of the subunits from frog phosphodiesterase isolated by Miki *et al.* (1975) is unclear.] Several questions come to mind regarding the 88,000 and 84,000 MW proteins: Why does the rod cyclic GMP phosphodiesterase have these two subunits? Do they have different functions or does each subunit contain a catalytic site? Do both bind cyclic GMP?

b. Activation of Cyclic GMP Phosphodiesterase. 1. Activation of the purified enzyme. Removal of the cyclic GMP phosphodiesterase from the disk membranes substantially reduces its activity. However, there are various ways of activating the purified enzyme. Polycations such as protamine, polylysine, and polyarginine increase the V_{max} without changing the K_m for cyclic GMP. A similar effect is caused by histones and by heparin, a polyanion (Bitensky *et al.*, 1975). Nonionic detergents, such as Triton X-100 (0.01%; Chader *et al.*, 1974b) and Lubrol (Pannbacker *et al.*, 1972), or ions, such as fluoride (Sitaramayya *et al.*, 1977), also have been shown to increase the hydrolysis of cyclic GMP. A partial proteolysis of the enzyme with trypsin is more effective in activating the enzyme than protamine, although if trypsin treatment is prolonged, the activity is again reduced to basal levels. This activation is consistent with the dissociation of an inhibitory component of the enzyme complex by trypsin. In fact, Hurley and Stryer (1982) have shown that trypsin treatment degrades the low molecular weight subunit of phosphodiesterase (γ) and that the trypsin-activated enzyme can be turned off by addition of the purified γ subunit. Light and nucleoside triphosphates, which do activate the enzyme on the disk membranes, cannot increase the activity of the solubilized phosphodiesterase (Manthorpe and McConnell, 1975), but this can be achieved with the addition of the purified regulatory components described next.

2. *Activation in the rod outer segments.* The activation of cyclic GMP phosphodiesterase from rod outer segments by light is a complex process requiring the participation of several different components of the organelle. We will consider each component separately.

Rhodopsin requirement. Photon absorption by rhodopsin yielding activated rhodopsin (R*) is the first step in the cascade of events that leads to hydrolysis of cyclic GMP. Keirns *et al.* (1975) showed that the most efficient wavelength of light for activation of phosphodiesterase (Fig. 5) corresponded to the wavelength of maximum absorption for rhodopsin (500 nm). All-*trans*-retinal does not mimic the effect of light, indicating that the protein moiety of rhodopsin is the essential

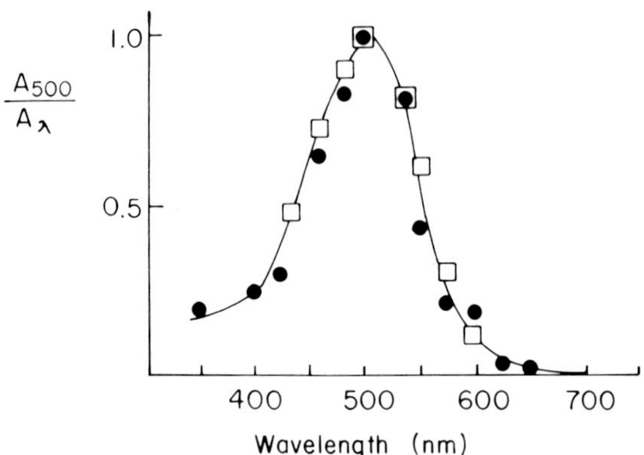

FIG. 5. Action spectra for activation of cyclic GMP phosphodiesterase (●) and G protein (□). Maximal activity for both enzymes was observed when rod outer segments were exposed to monochromatic light of 500 nm. Solid curve, absorption spectrum of rhodopsin in 1% hexadecyltrimethylammonium bromide. (Adapted from Keirns *et al.*, 1975; and Wheeler *et al.*, 1977.)

agent in the activation process. Furthermore, these authors estimated that half-maximal activation of the enzyme in frog rod outer segments is achieved during a light exposure that bleaches only 1 in 2000 rhodopsin molecules. Estimates by Yee and Liebman (1978) were even more dramatic: half-maximal activation was obtained with bleaches of 1 in 80,000 molecules of rhodopsin. These investigators also calculated that 500 phosphodiesterase molecules are activated by bleaching one molecule of rhodopsin in frog rod outer segments. Liebman and Pugh (1982) estimated later that about 2000 phosphodiesterase molecules are gradually activated by one bleached rhodopsin molecule, yielding a final enzyme velocity of 2.5×10^6 cyclic GMP hydrolyzed/bleached rhodopsin/sec. This multiplier mechanism provides extraordinary gain for the system.

Nucleotide requirement. The activation of phosphodiesterase by light requires the presence of a nucleoside triphosphate (Chader *et al.*, 1974c; Bitensky *et al.*, 1975; Manthorpe and McConnell, 1975). GTP is the most effective, causing maximal activation of phosphodiesterase at concentrations in the 10^{-7}–10^{-8} M range (Wheeler and Bitensky, 1977). The nonhydrolyzable imido analog of GTP, GMP-PNP, can substitute for GTP and irreversibly activate the enzyme, suggesting that GTP is an allosteric effector rather than an energy or phosphate donor. In fact, it has been shown that phosphodiesterase is not activated by phosphorylation since it does not appear to incorporate phosphate in darkness or upon exposure to light (Chader *et al.*, 1976). The basis for the nucleotide requirement is discussed in the next section.

G protein requirement. In 1977 Wheeler *et al.* reported the presence in rod outer segments of a GTPase which showed remarkable sensitivity to light. The action spectrum of this GTPase corresponded perfectly to the absorption spectrum of rhodopsin (Fig. 5). Half-maximal activation of the enzyme was observed when 1 in 2000–2500 rhodopsin molecules were photoisomerized (Bitensky *et al.*, 1978; Caretta *et al.*, 1979a). The apparent K_m of GTPase for GTP was 0.5 μM with a V_{max} of about 500 pmol of GTP hydrolyzed/min/mg protein. The ionic requirements of the enzyme were not specific; the hydrolysis of GTP occurred in the presence or absence of Mg^{2+} and with or without Ca^{2+} in the medium (Bignetti *et al.*, 1978). Wheeler *et al.* (1977) estimated that a single bleached rhodopsin molecule could activate both GTPase and phosphodiesterase. Furthermore, they found that the activation of phosphodiesterase by GTP showed a time-dependent decay which correlated well with the light-stimulated GTPase activity. However, repeated addition of GTP to rod outer segment membranes could elicit sequential increases in the activity of cyclic GMP phosphodiesterase. At a concentration of 0.2 μM, GMP-PNP completely inhibited the light-activated GTPase and irreversibly activated the rod outer segment phosphodiesterase. These observations, together with the analogous kinetics of phosphodiesterase and light-activated GTPase, led Wheeler and Bitensky (1977) to conclude that the allosteric site at which GTP activates phosphodiesterase may correspond to the catalytic site of the light-activated GTPase.

Several laboratories have studied the properties of the light-activated GTPase (Robinson and Haggins, 1979a; Kühn, 1978, 1980b; Godchaux and Zimmerman, 1979; Bitensky *et al.*, 1978; Bignetti *et al.*, 1978; Fung *et al.*, 1981). The enzyme can be easily removed from the unilluminated rod outer segment membranes, together with cyclic GMP phosphodiesterase, with aqueous buffers either at low or very high ionic strength (5 mM Tris buffer or 1.0 M NH_4Cl), but not at moderate ionic strength. This suggests that both the GTPase and the phosphodiesterase are peripherally bound membrane proteins. Upon illumination, GTPase activity disappears from the supernatant and is again found on the rod outer segment membranes. This light-induced binding of GTPase to rod outer segment membranes is completely inhibited if GTP is present during illumination and is quickly reversed if GTP is added in the dark after illumination (Kühn, 1980a, 1982). Taking advantage of these properties, it is relatively easy to purify the enzyme. Light-activated GTPase has a native molecular weight of 80,000 and is composed of three subunits, α, β, and γ, of 37,000, 35,000, and 5000 molecular weight, respectively (Kühn, 1980a; Baehr *et al.*, 1982).

Shinozawa *et al.* (1979) suggested that the GTPase could be an intermediary between bleached rhodopsin and the cyclic GMP-hydrolyzing enzyme, and Kohnken *et al.* (1981a) showed that the GTPase complex is an absolute requirement for the activation of phosphodiesterase by light. This enzymatic system has received different names in the literature: "G protein," GTP-binding protein, or transducin, and its components are also referred to differently depending upon the laboratory involved. Thus, it is common to find references to G_α, $G_{\beta\gamma}$, T_α, $T_{\beta\gamma}$, or even "G protein" and "H protein" (or "helper protein") as Shinozawa *et al.* refer to the α and $\beta\gamma$ subunits.

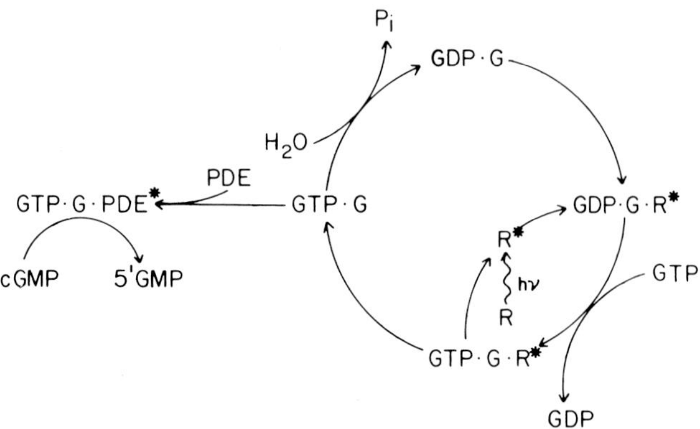

FIG. 6. Proposed mechanism by which photolyzed rhodopsin catalyzes the formation of the GTP-G complex which in turn activates cyclic GMP phosphodiesterase. (Adapted from Fung *et al.*, 1981.)

3. Proposed mechanism of action. A mechanism proposed by Fung *et al.* (1981) for the activation of cyclic GMP phosphodiesterase is shown in Fig. 6. According to this model, GDP is tightly bound to the G protein in the dark. On illumination, photolyzed rhodopsin (R*) forms a complex with GDP-G and catalyzes the exchange of GTP for GDP. The new complex, GTP-G-R*, dissociates generating free GTP-G, which then activates cyclic GMP phosphodiesterase, and R*, which is free to bind again to another GDP-G. Thus, one single bleached rhodopsin molecule can catalyze the exchange of GTP for GDP on many molecules of G protein, which in turn can activate many molecules of phosphodiesterase. This system is restored to the unactivated state by means of the built-in GTPase activity. The GTP in GTP-G is slowly hydrolyzed and the resulting GDP-G complex no longer activates the phosphodiesterase.

The GTP-G complex obtained upon illumination of rod outer segments can be separated from the membranes and is capable of activating purified phosphodiesterase (Fung *et al.*, 1981; Uchida *et al.*, 1981). In addition, Bennett (1982) has shown that high concentrations of GTP or its nonhydrolyzable imido analog GMP-PNP can force the formation of the GTP-G complex in the dark and activate the phosphodiesterase without photolyzed rhodopsin. The α subunit of the G protein binds to GTP and is the only subunit required for the activation process. In fact, in solution, G_α-GTP rapidly dissociates from $G_{\beta\gamma}$. Yamazaki *et al.* (1983) have suggested that GTP-G is responsible for the removal of the inhibitory γ subunit from phosphodiesterase and that the rest of the phosphodiesterase molecule remains catalytically active until the γ subunit is released by the G protein. However, this has not been conclusively demonstrated.

Light-scattering data obtained by Kühn *et al.* (1981) have confirmed the basic mechanism discussed above. The GDP-G complex has a high affinity for R* but not for R, and therefore a "binding signal" is obtained and detected with light-scattering when a 1 : 1 stoichiometric complex GDP-G-R* is formed. The complex is stable in the absence of GTP, but if GTP is present, rapid exchange of GTP for GDP occurs and, since the resulting GTP-G-R* has low affinity for R*, it dissociates into GTP-G and R*, originating the "dissociation signal," also detected spectrophotometrically. From this type of study, Kühn has estimated an amplification factor (the number of times that R* is recycled) of about 100, which is in the range of the amplification factor calculated by Stryer's group for the GDP/GMP-PNP exchange (70–500; Fung and Stryer, 1980; Fung *et al.*, 1981) and for the phosphodiesterase activation (factor of about 500; Yee and Liebman, 1978). Also, from the light-scattering changes, Kühn *et al.* (1981) have determined that all the GTP-associated reactions occur within 100 msec following the absorption of light; thus, they occur within the time frame of the electrophysiological response of the rod photoreceptor.

A sequence of photoproducts is formed when rhodopsin is bleached by light (Hargrave, this volume). Within the first millisecond metarhodopsin I (MI, 480 nm) and metarhodopsin II (MII, 380 nm) are produced, and the further decay into MIII (470 nm) is much slower, taking a few minutes. Bennett *et al.* (1982) and Emeis *et al.* (1982) have reported that the form of R* that binds to the GDP-G complex is MII. Likewise, Fukada and Yoshizawa (1981) and Fukada *et al.* (1981) have evidence to suggest that the actual activation of cyclic GMP phosphodiesterase is effected by MII.

c. Deactivation of Cyclic GMP Phosphodiesterase. The concentrations of GTP found in rod outer segments exposed to light can support full activation of phosphodiesterase. According to de Azeredo *et al.* (1978), *in vivo* GTP levels are 400 μM in dark-adapted organelles and 170 μM after 2 min of illumination. Biernbaum and Bownds (1979), working with isolated rod outer segments, found that the light-induced decrease in GTP can be as large as 70% with a half-life of 7 sec, and that this decrease is a linear function of the logarithm of continuous light intensity at levels which bleach between 5×10^2 and 5×10^6 rhodopsin molecules/outer segment/sec. Robinson and Hagins (1979b) obtained similar results. If after light exposure the endogenous GTP is not replenished, the GTP will finally be depleted with the subsequent deactivation of phosphodiesterase and "turnoff" of the photoreceptor cell. However, this mechanism of deactivation is too slow and does not explain the rate of dark adaptation that has been observed for rod photoreceptors.

Liebman and Pugh (1979) have described a more rapid, ATP-mediated, rod outer segment phosphodiesterase shutoff mechanism. They observed that when measuring phosphodiesterase activity using 1 mM cyclic GMP as substrate, if both ATP (0.5 mM) and GTP (0.25 mM) were present in the reaction medium, individual flashes of light only caused the turnover of approximately 100 μM cyclic GMP. If only GTP was present in the reaction mixture, all of the cyclic GMP was hydrolyzed. ATP was consumed in this shutoff process and if the nonhydrolyzable ATP analog, AMP-PNP was used, no quenching of phosphodiesterase was observed. The authors concluded that ATP was involved in the phosphorylation of bleached rhodopsin and that phosphorylated rhodopsin would no longer support activation of the G protein, resulting in loss of phosphodiesterase activity.

Phosphorylation of rhodopsin is the major phosphorylation event in the rod outer segment. After the initial descriptions of this light-activated reaction (Kühn and Dreyer, 1972; Bownds *et al.*, 1972), investigators concluded that phosphorylation of rhodopsin occurs when the visual pigment undergoes the light-induced conformational change (Frank and Bensinger, 1974; Frank and Buzney, 1975; McDowell and Kühn, 1977) and apparently takes place at the metarhodopsin II state (Yamamoto and Shichi, 1983). Rhodopsin kinase, the enzyme that catalyzes the transfer of phosphoryl groups, preferably from ATP, to rhodopsin (Shichi and Somers, 1978), is potentially active in the dark and is not stimulated by light (McDowell and Kühn, 1977; Frank and Buzney, 1975). Paulsen and Bentrop (1984) have recently studied the dephosphorylation of rhodopsin in the blowfly, and found that regeneration of the visual pigment greatly enhanced its dephosphorylation. Earlier investigators of vertebrate rhodopsin phosphorylation had shown that following regeneration of rhodopsin and dephosphorylation, the phosphorylation of the visual pigment is again stimulated by light (Miller *et al.*, 1977) and that conditions designed to minimize recycling and regeneration of rhodopsin are able to suppress the reactivation of the rhodopsin phosphorylation (McDowell and Kühn, 1977). While there has been some debate regarding the total amount of phosphate incorporated into rhodopsin in response to light, the current estimate by Wilden and Kühn (1982) is that under optimal conditions, an average of six to seven and up to nine phosphates can be attached per rhodopsin molecule. Seven phosphorylation sites have been localized to the region at the carboxyl-terminus of the molecule (Hargrave *et al.*, 1980).

Initial estimates for the speed of rhodopsin phosphorylation by Kühn and Bader (1976) indicated the half-time of the light-induced phosphorylation to be about 2 min in frog retinas incubated at 21°C and 1–2 min in bovine retinas kept at 36°C. These authors and others (Weller *et al.*, 1975b; Kühn *et al.*, 1977) suggested that the slow rates of phosphate incorporation into the visual pigment preclude a role for rhodopsin phosphorylation in the transduction process and suggested, instead, that rhodopsin phosphorylation/dephosphorylation may be an important component in the control of light/dark adaptation. Recently, however, Sitaramayya and Liebman (1983b) demonstrated that the phosphorylation of rhodopsin can occur very rapidly. In fact, reaction kinetics under weak bleaches seemed to be rapid enough to cause cGMP phosphodiesterase quenching. Aton and Litman (1981) showed that the degree of inhibition of the activation of phosphodiesterase is related to the number of sites that become phosphorylated in the rhodopsin molecule. Three phosphates incorporated per molecule of rhodopsin caused 50% inhibition. Miller and Dratz (1984) found that at low to intermediate levels of bleach, increasing the phosphorylation of rhodopsin from four to six phosphates inhibits the V_{max} of PDE and controls the lifetime of the excited rhodopsin. Lisman *et al.* (1985) suggested that the multiple phosphorylation of rhodopsin in *Limulus* represents a very reliable off switch for the visual pigment, one necessary to reduce the probability that dark responses occur by spontaneous formation of active pigment by a reverse reaction.

Sitaramayya and Liebman (1983a) have studied reconstituted systems in which they recombined some or all of the components of the cyclic GMP cascade, that is, purified G protein, phosphodiesterase, rhodopsin kinase, and rod outer segment membranes stripped of these enzymes. They found that in the absence of rhodopsin kinase, ATP did not turn off the phosphodiesterase or reduce the sensitivity of the system. In addition, when they enzymatically cleaved the C-terminal region of rhodopsin, ATP did not change the phosphodiesterase activity. These results are consistent with rhodopsin phosphorylation being involved in the deactivation of phosphodiesterase. However, phosphorylated, regenerated rhodopsin binds to the GDP-G complex and seems to be able to allow the exchange of GDP for GTP. Therefore, in addition to rhodopsin kinase, some other component must be involved in mediation of the ATP effect in the shutoff mechanism. This component may be a protein of molecular weight 48,000 (48K protein) which may compete for light-activated rhodopsin with the G protein when rhodopsin is phosphorylated. The 48K protein binds to the rod outer segment membranes upon illumination, and ATP or GTP strengthens the binding (Pfister *et al.*, 1983). Kühn (1984) comments that in experiments not published as yet, they have evidence that the 48K protein strongly binds to membrane disks with phosphorylated R*. Thus, the 48K protein is a good candidate for the role of terminating the action of R* in conjunction with the phosphorylation mechanism. This is an area that needs to be explored.

d. Other Inhibitors of Cyclic GMP Phosphodiesterase. The methylxanthines, which are potent inhibitors of phosphodiesterase in most tissues, do not completely block the activity of the enzyme from rod outer segments. IBMX (isobutylmethylxanthine) is, however, a better inhibitor than theophylline, caffeine, papaverine, SQ 20,009, dipyridamol (Persantine), and the dibutyryl and 8-Br derivatives of cyclic AMP or of cyclic GMP (Pannbacker *et al.*, 1972; Chader *et al.*, 1974a,b; Hollyfield *et al.*, 1982).

In 1976, Dumler and Etingof reported the presence of a protein inhibitor of phosphodiesterase in bovine retina and rod outer segment preparations. They determined that this inhibitor was not tissue specific, had a molecular weight of

38,000, was trypsin sensitive, could withstand boiling, and appeared to affect the hydrolysis of cyclic AMP more than that of cyclic GMP. Berman and Usova (1978) found that this inhibitor was localized in rod outer segments and was not observable in any other retinal layer. Other heat-stable, trypsin-sensitive protein inhibitors have been described by Liu and Wong (1979) and Yamazaki *et al.* (1982a). The latter authors indicated that their inhibitor not only prevented the action of phosphodiesterase but also stimulated the binding of cyclic GMP to noncatalytic sites on the phosphodiesterase. This inhibitor apparently must be released from the phosphodiesterase for activation of the enzyme by the GTP-G complex to occur. Perhaps all of these preparations actually contain the γ subunit of cyclic GMP phosphodiesterase, which has been purified by Hurley (1980) and is similar to the above-mentioned protein inhibitors in its resistance to heat and its degradation by trypsin.

Other proteins present in rod outer segments may also have a regulatory role in phosphodiesterase activity. The G protein involved in the light-induced activation process inhibits phosphodiesterase activity by 50% when added in a 1 : 1 ratio to phosphodiesterase from bovine rod outer segments (Baehr *et al.*, 1979). Kohnken *et al.* (1981b) described a similar observation. In addition, it has been suggested that other inhibitory factors may exist which are lost during the purification of the rod outer segments (Robinson *et al.*, 1980) or during the extraction of phosphodiesterase from the disk membranes (Bitensky *et al.*, 1975).

e. A Model for the Regulation of Cyclic GMP Phosphodiesterase Action. Figure 7 is a diagram illustrating the different components, described in the preceding sections, which may regulate the hydrolysis of cyclic GMP. In this model, photolyzed rhodopsin has undergone a conformational change that allows it to interact with the α subunit of the G-GDP complex. Upon exchange of GDP for GTP, G-GTP removes the inhibitory subunit of phosphodiesterase, thus enabling the enzyme to express its full activity. The system may be inactivated by independent mechanisms acting in concert. (1) Rhodopsin kinase phosphorylates the C-terminal region of rhodopsin. (2) This phosphorylated R* strongly binds the 48K protein which prevents the activation of more molecules of G protein. (3) The G protein slowly hydrolyzes the bound GTP converting the G-GTP into G-GDP which releases the γ subunit of phosphodiesterase, with the consequent turnoff of the enzyme.

5. Cyclic GMP Modulation of Protein Phosphorylation

Cyclic nucleotides are known to express their actions primarily through the activation of protein kinases which phosphorylate specific proteins (Greengard, 1976). These phosphoproteins, in turn, may be involved in the regulation of functions that are fundamental to the cell. For example, in neurons, phosphopro-

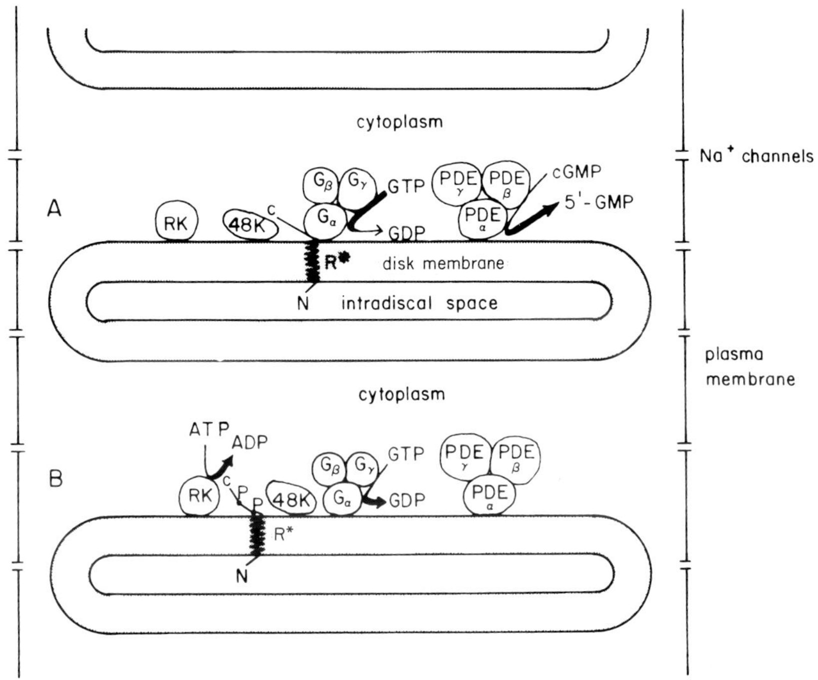

FIG. 7. Diagram illustrating the components of the rod outer segment involved in the activation (A) and deactivation (B) of cyclic GMP phosphodiesterase. R*, Photolyzed rhodopsin with its N-terminus in the intradiscal space and its carboxy-terminus (C) in the cytoplasmic space. G_{α}, G_{β}, and G_{γ}: subunits of G protein. PDE_{α}, PDE_{β}, and PDE_{γ}: subunits of phosphodiesterase. RK, Rhodopsin kinase; 48K, 48,000-Da protein; P, phosphate groups. The darker section of the arrows indicate the part of the reaction that prevails.

teins have been implicated in the control of membrane permeability to certain ions (Nathanson, 1977). The high levels of cyclic GMP in dark-adapted rod outer segments and the modulation of cyclic GMP levels by light suggested that a cyclic GMP-dependent protein kinase could be involved in the phosphorylation of specific proteins of the rod outer segment which in turn could modulate its membrane permeability.

In 1977, Lolley *et al.* (1977a) described the phosphorylation of a soluble protein from bovine rod outer segments which had a molecular weight of 30,000. The phosphorylation reaction was stimulated by physiological concentrations of cyclic GMP, although cyclic AMP was effective at lower concentrations. The enzyme that catalyzed this process had the characteristics of a type II cyclic AMP-dependent protein kinase (according to the classification of Corbin *et al.*, 1975), similar to the cyclic AMP-dependent kinases present in most tissues (Farber *et al.*, 1979). It was proposed, however, that since in rod outer segments the levels of cyclic GMP are severalfold higher than those of cyclic AMP, this protein kinase could be modulated by the light-induced changes in cyclic GMP concentrations.

Two years later, Polans *et al.* (1979) showed that two membrane associated proteins from frog rod outer segments (molecular weights 12,000 and 13,000) were phosphorylated in a cyclic GMP-dependent manner. At concentrations of 10^{-5}–10^{-3} *M*, cyclic GMP stimulated the incorporation of phosphate into these proteins by fivefold. The 12,000 and 13,000 molecular weight proteins were also observed in intact retinas of frog where they were dephosphorylated when the retinas were illuminated and were rephosphorylated when illumination ceased. However, the dephosphorylation process was not observed in the isolated outer segments (Polans *et al.*, 1979). In addition, drugs that influence the concentration of cyclic GMP also altered the levels of phosphorylation of the two membrane proteins from frog retina. Cyclic AMP enhanced the phosphorylation of these proteins to the same extent as cyclic GMP in intact retina, but did not have any effect in isolated rod outer segments (Hermolin *et al.*, 1982). Calcium inhibited the phosphate incorporation into the 12,000 and 13,000 molecular weight proteins.

The data on the cyclic nucleotide-dependent phosphorylation in bovine and frog photoreceptors provided an attractive mechanism to support the involvement of cyclic GMP in the regulation of the light-sensitive pathways. In 1978, Farber and Lolley speculated that during the dark phase, high levels of cyclic GMP would stimulate the phosphorylation of soluble or membrane-bound proteins. These phosphoproteins would keep the sodium channels open and thus maintain the sodium permeability of the rod outer segment plasma membrane. When light decreased the levels of cyclic GMP, the cyclic nucleotide-dependent protein kinase would not be as active and a phosphoprotein phosphatase would dephosphorylate the phosphoproteins. Upon removal of the phosphate groups, the proteins would no longer maintain the permeability of the plasma membrane to sodium. However, a mechanism supporting the involvement of phosphoproteins other than rhodopsin in the phototransduction process of the photoreceptor cells has not been supported by experimental data. As mentioned above, the kinase identified in bovine rod outer segments and further characterized by Lee *et al.* (1981) is actually a type II cyclic AMP protein kinase which is present in large quantities in many types of cells. There is conflicting evidence for a cyclic GMP-dependent protein kinase localized in outer segments. Its absence from these organelles is suggested by experiments in which cyclic GMP binding sites were not found on a protein with a molecular weight corresponding to that of cyclic GMP-dependent protein kinase (Yamazaki *et al.*, 1980) and because cyclic GMP-dependent protein kinase activity and the 30,000 molecular weight proposed substrate did not copurify with rod outer segments which maintained an intact, sealed plasma membrane during purification (Shuster and Farber, 1984).

The 30,000 molecular weight protein has been suggested by DeVries and Ferrendelli (1983) to be a component of rod inner segments. The involvement of cyclic GMP-dependent phosphorylation/dephosphorylation as a controlling mechanism in transduction is currently thought to be less likely than a few years ago.

It is interesting to note that levels of cyclic GMP normally found in dark-adapted rod outer segments can have substantial inhibitory effects upon *in vitro* rhodopsin phosphorylation (Hermolin *et al.*, 1982; Shuster and Farber, 1984). Thus, cyclic GMP may participate in the control of rhodopsin phosphorylation *in vivo,* although the mechanism and the importance of this process are currently unknown.

6. Effects of Cyclic GMP on Membrane Permeability and Other Physiological Processes

In 1976, Brodie and Bownds measured changes in the swelling of isolated rod outer segments as a function of light in the presence of phosphodiesterase inhibitors. They found that an increase in cyclic GMP levels augmented the dark permeability of the tissue, but it did not have a substantial influence on the amount of illumination required for the supression of the permeability. In other words, the increase in cyclic GMP seemed to modulate the amplitude of the response to light without affecting the sensitivity of the system. Caretta *et al.* (1979b) also tested the effect of cyclic GMP on membrane permeability. They found that the addition of cyclic GMP increased the efflux of Na^+ and K^+ from perfused, broken outer segments independent of Ca^{2+} and the ionic strength of the perfusing medium. Cavaggioni and Sorbi (1981) observed that cyclic GMP caused the release of Ca^{2+} from specific binding sites on the disk membrane. On the other hand, Ca^{2+}-dependent changes in cyclic GMP levels were not correlated with opening and closing of the light-dependent permeability in toad visual cells, as shown by Woodruff and Fain (1982). A possible interaction between Ca^{2+} and cyclic GMP in the control of transduction might help explain conflicting data regarding effects of Ca^{2+} and cyclic GMP, and this has been suggested by several researchers. A discussion of the subject, particularly with respect to cyclic GMP modulation of free Ca^{2+} levels by activation of the disk Ca^{2+} pump, has been presented by Lipton and Dowling (1981). These authors also discuss possible effects of Ca^{2+} on cyclic GMP levels.

Lipton and collaborators (1977a,b) have shown that when the retina was superfused with cyclic GMP, dibutyryl cyclic GMP, or IBMX, the electrical effects recorded from rods closely matched those observed when extracellular Ca^{2+} levels were lowered; that is, there was depolarization of the plasma membrane, an increase in response amplitude, and some changes in waveform. Application of high Ca^{2+} to the retina blocked the effect of cyclic GMP. Thus, increased cyclic GMP and lowered Ca^{2+} produced similar alterations in the electrical activity of the rods. However, the extent of the change in cyclic GMP concentration inside the rod was not known with this approach. A more direct method was to inject cyclic GMP intracellularly into the rod outer segments of the isolated toad retina (Nicol and Miller, 1978; Miller and Nicol, 1979; Waloga and Brown, 1979). Excess cyclic GMP in dark-adapted preparations increased both the latency and amplitude of the response to light and depolarized the rod membrane to a potential close to the equilibrium potential for sodium. The extent of this depolarization increased with the amount of cyclic GMP injected, and it was abolished if the injection of cyclic GMP was preceded by a strong flash of light. A flash of light after the injection accelerated the return of the cell to normal potential. Intracellular injection of 5′-GMP or cyclic AMP did not mimic the effects of injection of cyclic GMP. Thus, elevating the levels of cyclic GMP appears to produce a situation in the rod analogous to that observed naturally in the dark.

Recently, interesting experiments were carried out by Fesenko *et al.* (1985). These authors studied the effects of possible conductance modulators on excised "inside-out" patches from the plasma membrane of rod outer segments. They found that cyclic GMP in the micromolar range (30 μM), acting from the inner side of the membrane, markedly increased the cationic conductance of the patches in a reversible manner; that is, the conductance went back to baseline when cyclic GMP was washed out. In contrast, Ca^{2+} as well as cyclic AMP and 2′,3′-cyclic GMP were ineffective. Since the cyclic GMP-induced conductance increase occurred in the absence of nucleoside triphosphates, it must not be mediated by protein phosphorylation. Fesenko *et al.* (1985) concluded that cyclic GMP acts directly on channels in the membrane and suggested that, *in vivo*, the cyclic GMP-sensitive conductance may be responsible for the generation of the rod photoresponse.

Although this is the first known report of the direct action of cyclic GMP on the conductance of the rod plasma membrane, the complete scheme of how the photoreceptor cell responds to light is still not clear. Matthews *et al.* (1985) and Cobbs and Pugh (1985) have reported that exogenous cyclic GMP does not slow the rising phase and, in addition, it prolongs the duration of the photoresponse. These findings are not consistent with the idea that binding of cyclic GMP to the membrane keeps its channels open.

D. *Cyclic AMP Metabolism in Rod-Dominant Retinas*

1. Localization and Factors That Modulate Cyclic AMP Concentration

Rod-dominant retinas have considerably lower concentrations of cyclic AMP than of cyclic GMP (Farber *et al.*, 1981). Cyclic AMP is uniformly distributed throughout the layers of the retina with the exception of the rod outer segments which contain the lowest levels (Orr *et al.*, 1976). In contrast to cyclic GMP, light does not affect the cyclic AMP concentration in the outer segments (Fletcher and Chader, 1976; Krishna *et al.*, 1976; Orr *et al.*, 1976), but does reduce the cyclic AMP levels in the outer nuclear layer and in the outer plexiform layer (Orr *et al.*, 1976). De Vries *et al.* (1978) observed a light-induced 40% decrease in the cyclic AMP concentration of dark-adapted mouse retina, which could have occurred in the same retinal layers. Since photoreceptor terminals release neurotransmitter in the dark (Byzov and Trifonov, 1968; Dowling and Ripps, 1973; Cervetto and Piccolino, 1974), higher dark-cyclic AMP levels present in the layer that contains the photoreceptor terminals and the postsynaptic elements could indicate cyclic AMP participation in synaptic transmission. However, it has not been established which cell type in the outer plexiform layer accumulates cyclic AMP.

As in other neural tissues, anoxia or ischemia elevates the cyclic AMP levels in rabbit (Orr *et al.*, 1976) and mouse (Mitzel *et al.*, 1978) retinas. These cyclic AMP changes occur predominantly in the inner layers. Brown and Makman (1972) reported that depolarizing agents such as ouabain and high K^+ also elevate the cyclic AMP content of bovine retina. Although Ca^{2+} modulates cyclic GMP levels, it does not affect cyclic AMP concentrations (Cohen *et al.*, 1978).

Of the putative neurotransmitters present in the retina (see articles by Marc, this volume, and by Iuvone, Part II), dopamine (discussed below) and neuropeptides such as glucagon and VIP (Kuwamaya *et al.*, 1982; Watling and Dowling, 1983; Koh and Chader, 1984; Koh *et al.*, 1984) appear to be involved in the modulation of cyclic AMP concentrations.

2. Cyclic AMP Synthesis

Developmental studies and work with dystrophic retinas which are devoid of photoreceptor cells have indicated that adenylate cyclase, the enzyme that synthesizes cyclic AMP from ATP, is primarily localized in the inner layers of the retina (Farber and Lolley, 1977b; Lolley *et al.*, 1974). In addition, treatment of neonatal rats with glutamate, which selectively destroys the inner retina leaving the photoreceptor cells intact, decreases adenylate cyclase activity (Makman *et al.*, 1975). Consistent with these data, Clement-Cormier and Redburn (1978) showed that the P_2 subcellular fraction of rabbit retina, which is enriched in synaptosomes from amacrine cells, and to a lesser extent, from bipolar and horizontal cells, has a higher specific activity than the P_1 fraction, which contains synaptosomes mainly from photoreceptor cells.

Dopamine-stimulated adenylate cyclase has been found in many types of retinas, including primates (Makman *et al.*, 1975), fishes (Watling *et al.*, 1980), birds (Schwarcz and Coyle, 1976), and invertebrates (Makman *et al.*, 1975). In mice and rats, dopamine-stimulated adenylate cyclase activity increases markedly during the early postnatal period, as retinal maturation occurs (Lolley *et al.*, 1974; Makman *et al.*, 1975). In chick embryonic retina, a sharp threefold increase in cyclic AMP levels is observed between the sixteenth and eighteenth embryonic day and remains constant thereafter (De Mello, 1978). There is a high degree of specificity for agonists at retinal dopamine receptors. The effects of some agonists and of a few dopamine antagonists on adenylate cyclase activity have been discussed in a previous review (Farber, 1982) and are also considered in the article by Michael Iuvone (Part II).

d-LSD (lysergic acid diethylamide) has also been shown to stimulate retinal adenylate cyclase activity (Spano *et al.*, 1976). This stimulation is abolished by neuroleptics and seems to occur at the same site of action of dopamine, since *d*-LSD inhibits dopamine activation of adenylate cyclase in a dose-dependent manner (Schorderet and Magistretti, 1980).

Light deprivation and degeneration of the photoreceptor cells can affect normal cyclic AMP synthesis in retina. The dopamine- or LSD-stimulated cyclic AMP formation is approximately twice as high in rats which are kept in the dark 65 hr compared to rats kept under a light cycle of 12 hr/day (Trabucchi *et al.*, 1976). Since it has been shown that cyclic AMP synthesis increases when dopamine is depleted from nerve terminals by reserpine injection, and retinal dopamine is released mainly as a consequence of light stimulation, Trabucchi *et al.* (1976) speculate that light deprivation is a sort of functional denervation. In addition, the deafferentiation from the photoreceptor cells of pre- or postdifferentiated neurons of the inner retina can modulate the long-term metabolism of cyclic AMP, as seen in studies of retinal degenerations (Farber and Lolley, 1977b). For example, in the *rd* mouse retina, the degeneration of the photoreceptor cells occurs during the period when the neurons from the inner retina are maturing and forming their synapses (Blanks *et al.*, 1974). In the *rd* retina, the content of cyclic AMP and the activity of adenylate cyclase increase above that of control retina during development, and both remain elevated in the adult retina after the photoreceptors have degenerated. In contrast, in the retinal degeneration of RCS rats, the photoreceptor cells are mature and the synaptic connections with the inner neurons have been established before the visual cells become abnormal. In this retina, the content of cyclic AMP and the activity of adenylate cyclase are normal at all times during development (Farber and Lolley, 1977b).

3. Physiological Effects of Cyclic AMP

Cyclic AMP has been implicated in the regulation of several metabolic processes in the retina. For example, it activates phosphorylase *b* kinase and inactivates glycogen synthase I, thereby increasing retinal glycogenolysis and glycolysis (Orr *et al.*, 1976). Cyclic AMP also increases dopamine formation through activation of tyrosine hydroxylase (Tunicliff, 1977). Although the precise mechanism of this activation is not clear, Iuvone *et al.* (1982) have provided evidence that cyclic AMP stimulates a cyclic AMP-dependent protein kinase which phosphorylates tyrosine hydroxylase. This enzyme is activated in the phosphorylation process. Other retinal proteins have also been shown to be phosphorylated in a cyclic AMP-dependent manner. In addition to the 30,000 molecular weight protein studied by Farber and Lolley (1979), Hesketh *et al.* (1978) reported that the phosphorylation of three proteins from a crude actomyosin fraction, with molecular weights 22,000, 33,000, and 200,000, was markedly increased by cyclic AMP. The 200,000 MW protein has been identified as the heavy chain of retinal myosin. Retinal synapsin I (about 80,000 MW) is also a good substrate of cyclic AMP-dependent protein kinase (J. M. Bronstein, C. G. Wasterlain, and D. B. Farber, unpublished observation).

A striking effect of cyclic AMP has been observed in the teleost photoreceptor and pigment epithelial cells, which undergo dramatic movement within the retina in response to diurnal light changes. These photomechanical movements have been discussed in the article by Beth Burnside and Allen Dearry (this volume). Cyclic AMP also appears to have an effect on the shedding of photoreceptor outer segments and on the subsequent phagocytosis of shed membranes by the retinal pigment epithelium. This process is discussed in this volume by Joseph Besharse.

Wiglusz (1973, 1975a,b) has shown that cyclic AMP and drugs which elevate its concentration in the retina increase the amplitude of the b wave of the electroretinogram (ERG) and stimulate active sodium transport, while reduction of cyclic AMP levels by stimulation of phosphodiesterase decreases both parameters. Since ouabain inhibits the bioelectrical response to light as well as the active sodium transport, it seems that these actions of cyclic AMP are effective only in the presence of active Na^+,K^+-ATPase.

Studies by Yeh *et al.* (1983) suggest an effect of cyclic AMP in the process of differentiation of retinal neurons. They showed that the 8-Br derivative of cyclic AMP induced the ability of cultured embryonic retinal neurons to release acetylcholine at synapses in response to an excitatory stimulation and that this occurred before the time that the cells normally respond to the stimuli. IBMX produced the same effect. Thus, cyclic AMP may accelerate presynaptic events linking neuronal depolarization with acetylcholine release.

Cyclic AMP has been shown to have a marked effect on the pigmentation of cultured retinal pigment epithelial cells. Redfern *et al.* (1976) reported that dibutyryl cyclic AMP, added with IBMX to the medium, increases melanin synthesis by 40% after 13 days of culture. In addition, the cells appear more differentiated, with more mature pigment granules. Dibutyryl cyclic AMP also seems to favor the elongation of microvilli in pigment epithelial cells, which may be important in the phagocytosis of shed outer segments. Similar data were reported by Takeuchi and Kajishima (1976), who found that cyclic AMP as well as theophylline prevented the degradation of pigment granules while enhancing their synthesis. In contrast to these observations, Newsome *et al.* (1974) showed that in similar conditions, dibutyryl cyclic AMP decreased the final number of pigment epithelial cells per culture and considerably reduced visible pigmentation of the cells. Although the basis of these differences is not understood, it may be possible that since Newsome *et al.* (1974) had not included IBMX in their incubation medium, the specific pools of cyclic AMP involved in the process in question could have been reduced by the action of a phosphodiesterase.

Another important effect of cyclic AMP is that of regulation of ion and fluid transport across the pigment epithelium. As mentioned by Farber and Adler (this volume), pigment epithelial cells are responsible for the exchange of nutrients, metabolites, ions, and fluid between the blood supply provided by the choroid and the outer retina. Miller and Farber (1984) have shown that elevation of the cellular cyclic AMP levels by the exogenous application of cyclic AMP, phosphodiesterase inhibitors, or adenylate cyclase stimulators significantly alters the mechanisms of active and passive transport that are located in the apical and basal membranes of the pigment epithelium. This results in an increase in the transepithelial active transport of Na^+ and K^+ and in an inhibition of the active Cl^- transport. Cyclic AMP apparently does this in a specific way, by increasing the secretion of NaCl and the absorption of K^+. The unidirectional fluxes of Na^+ and Cl^- in the retina to choroid (absorptive) direction and K^+ in the choroid to retina (secretory) direction remain relatively unchanged with the elevation of cyclic AMP levels. In addition, Hughes *et al.* (1984) have reported that cyclic AMP causes a significant decrease in net fluid transport in the direction of retina to choroid. This may be a mechanism for the control of fluid volume in the subretinal space and therefore of critical importance for the normal functioning of the retina.

III. Cyclic Nucleotides in Cone-Dominant Retinas

In contrast to rod photoreceptors, cell biological and biochemical studies of cone visual cells have not been as extensive. It is only in the last decade that investigations on the cyclic nucleotide metabolism of retinas dominated by cones have begun. In 1978, Farber and Lolley reported that the cone-dominant retina of the thirteen-lined ground squirrel had cyclic AMP levels which were considerably higher than those of cyclic GMP (91.0 vs 11.1 pmol/mg protein). The cyclic AMP/cyclic GMP ratio (8.2) was 35–40 times higher than that of rod-dominant retinas. Similarly, the Western fence lizard, which also has a cone-dominant retina, had higher levels of cyclic AMP than of cyclic GMP, with a cyclic AMP/cyclic GMP ratio of 2.2 (Farber *et al.*, 1980). However, in contrast

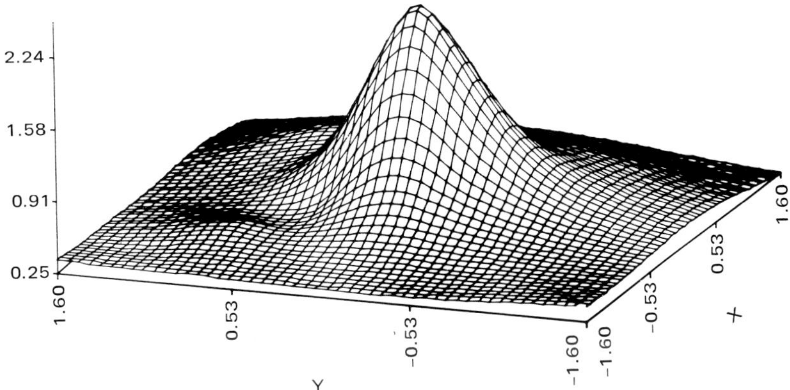

FIG. 8. Three-dimensional computerized representation of the distribution of cyclic AMP levels in human retina. Compare with the distribution of cyclic GMP shown in Fig. 2. (From Farber *et al.*, 1985.)

to these results, De Vries *et al.* (1979) found cyclic GMP concentrations slightly greater than cyclic AMP concentrations in ground squirrel retina. Possible explanations for this discrepancy may be related to the manner in which the tissue samples were obtained. Since cone outer segments are ensheathed by retinal pigment epithelium processes, it is almost impossible to isolate the retina without contamination by pigment epithelial cells. Attempts to minimize pigment epithelium contamination can result in loss of outer segment material. De Vries *et al.* (1979) suggested that the samples of Farber *et al.* may have experienced such a loss. On the other hand, Farber *et al.* (1981) found that freezing reduces the levels of cyclic AMP but not of cyclic GMP in ground squirrel retina. All of the samples of De Vries *et al.* (1979) were extracted from freeze-dried tissues and therefore may have contained lowered cyclic AMP levels. Recent studies by Farber *et al.* (1985) on human retina indicate that higher levels of cyclic AMP than of cyclic GMP are present in the macula, a retinal area enriched in cone photoreceptors (Fig. 8).

In cone-dominant retinas, the metabolism of cyclic AMP seems to be affected by light, whereas that of cyclic GMP apparently remains unchanged. For example, the content of cyclic AMP in dark-adapted ground squirrel retina is reduced by approximately 55% by exposure to light, and a 30% reduction is observed in the Western fence lizard retina after a similar exposure (Farber *et al.*, 1981). Burnside *et al.* (1982) have also seen this in teleost retina. However, De Vries *et al.* (1979) were unable to demonstrate the light effect in their ground squirrel preparations. This also may be due to the fact that all of their data were obtained using freeze-dried microdissected retinal layers and that freezing reduces the levels of cyclic AMP in ground squirrel retina independent of dark or light

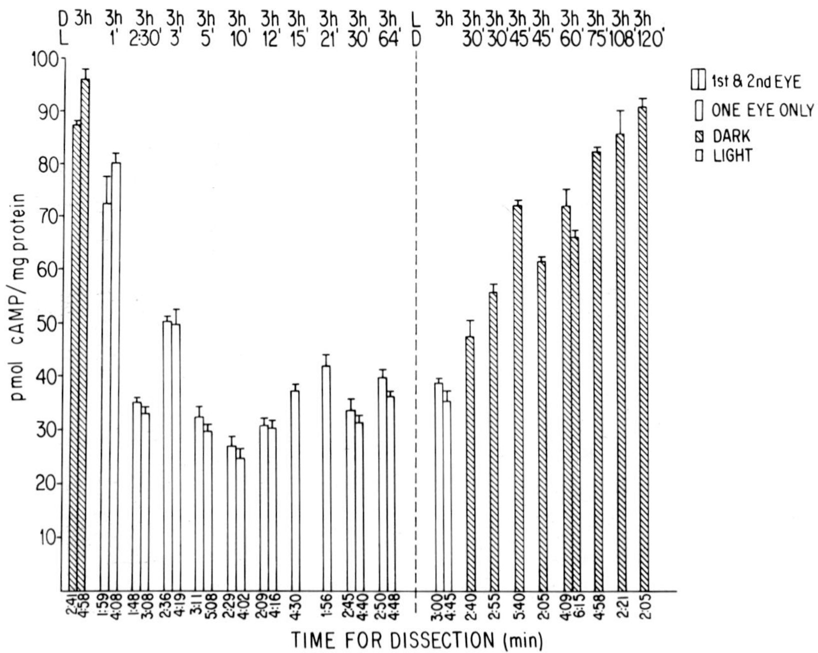

FIG. 9. Time course of reduction (by light exposure) and recovery (in darkness) of cyclic AMP levels in ground squirrel retina *in vivo*. Animals were dark- or light-adapted for 3 hr and then exposed to light or kept in darkness for the indicated times. Cyclic AMP was extracted immediately after enucleation and dissection of the retinas. The time elapsed from sacrifice to cyclic AMP extraction for each specimen is indicated on the abscissa. (From Farber *et al.*, 1982.)

conditions (Farber *et al.* 1981). De Vries *et al.* (1982) suggested ischemia as another possibility to explain Farber *et al.*'s higher levels of cyclic AMP in the dark than in the light. They suggested that retinas could be dissected faster in the light than under dim red light or infrared illumination and that the longer ischemic condition in the dark would increase the dark levels of cyclic AMP. However, in experiments in which a comparison was made of the cyclic AMP levels of the two retinas from the same animal, dissected sequentially after enucleation under identical conditions of illumination, no statistical difference in cyclic AMP levels between the two specimens was observed (Farber *et al.*, 1982). Furthermore, the time course of light reduction and dark recovery of cyclic AMP content of ground squirrel retina *in vivo*, shown in Fig. 9 (more than 50% reduction after 3 min of exposure to light of a dark-adapted animal; almost 2 hr of darkness to recover from light to dark levels), is another indication that light modulates cyclic AMP metabolism in ground squirrel retina (Farber *et al.*, 1982).

As mentioned previously, studies of retinal degenerations in different species, which we will discuss below, have yielded considerable information on biochemical mechanisms and structure–function relations in photoreceptor cells. For example, the content of any particular component of the visual cells can be estimated by measuring the difference between normal and photoreceptorless retinas. In the case of the ground squirrel, iodoacetic acid injection has been successfully employed to destroy retinal cones (Farber *et al.*, 1981). The cyclic AMP level in the dark-adapted retina of the iodoacetate-treated animal is reduced by approximately 46% and that of cyclic GMP by 94%. Thus, both cyclic nucleotides may be located in the cones, with cyclic AMP almost equally distributed between the photoreceptors and the neurons of the inner retina and cyclic GMP mostly present in the visual cells. The absolute loss of cyclic AMP after iodoacetic acid treatment (from 91.0 to 49.2 pmol/mg protein) is about four times greater than that of cyclic GMP (from 11.1 to less than 1.0 pmol/mg protein). Therefore, a reasonable estimate is that cones contain four times more cyclic AMP than cyclic GMP (Farber *et al.*, 1981).

Although preliminary as yet, it is interesting to note other differences and similarities between rod- and cone-dominant retinas regarding cyclic nucleotide metabolism. Biochemical studies have indicated that the cyclic GMP phosphodiesterase characteristic of rod outer segments is not present in detectable concentrations in ground squirrel retina (Farber and Souza, 1982). Furthermore, immunocytochemical methods confirm the apparent absence of this enzyme from cone photoreceptors (Farber and Bok, 1985). On the other hand, it seems that the visual pigment of the all-cone lizard retina may undergo light-induced phosphorylation in a fashion similar to the phosphorylation of rhodopsin (Walter *et al.*, 1985). In addition, an intriguing observation has been made by Haynes and Yau (1985), using perfused inside-out patches of catfish cone membranes. These authors found a cyclic GMP sensitive conductance in the outer segments and have suggested that cyclic GMP is the internal transmitter for phototransduction in cones. In agreement with this idea, Cobbs *et al.* (1985) have indicated that cyclic GMP increases photocurrent and light sensitivity of retinal cones. Evidently this is an area of research where many questions remain to be answered.

IV. Abnormalities in Cyclic Nucleotide Metabolism in Retinal Degenerations

Retinal degenerations leading to blindness occur in several species of animals (Table I). In some of these disorders, the photoreceptor cells are selectively affected and the rest of the retinal neurons remain intact and functional. In other cases, most of the retinal cells become involved. In diseases such as retinitis pigmentosa (RP), which is a grouping of inherited human disorders, cells of the inner retina become abnormal following the degeneration of the visual cells

TABLE I
GENETIC RETINAL DEGENERATIONS

Early onset		
Fast	Slow	Late onset
rd mouse	*rds* mouse	*pcd* mouse
Irish setter dog	Norwegian elkhound dog	"Nervous" mouse
Collie dog		RCS rat
		Poodle dog

(Merin and Auerbach, 1976; Berson, 1976). The time during postnatal life at which the degenerative process begins has been used to classify these diseases as early or late onset. Within the early onset disorders, one more distinction has been made, related to the length of time it takes for the photoreceptor cells to degenerate and be removed from the retina. Thus, there are cases of early onset, fast degeneration, and others of early onset, slow degeneration. A substantial amount of work has been done on animal models affected with these genetic diseases, with the hope of uncovering a basic biochemical lesion associated with the cause of photoreceptor death. As yet, no mechanism leading to the loss of the visual cells has been established for any particular disease. However, cyclic GMP metabolism has been found to be abnormal in the early onset, fast degeneration type of disorders. We will discuss in the next few pages the data documenting these observations.

A. *Early Onset, Fast Degeneration Inherited Diseases*

1. Retinal Degeneration (*rd*) Mouse

The *rd* mutation is carried as an autosomal recessive characteristic (Sidman and Green, 1965; LaVail and Sidman, 1974). The *rd* gene, located in linkage group XVII of chromosome number 5, is expressed specifically in the retinal visual cells when they differentiate during postnatal life. Affected photoreceptors begin to form outer segments and a synaptic ribbon complex (LaVail and Sidman, 1974; Blanks *et al.*, 1974), but they never reach maturity. During the short period in which the visual cells are viable (Noell, 1965), however, the *rd* retina is responsive to high-intensity illumination. The pigment epithelium is not affected in this inherited disease (LaVail and Sidman, 1974).

The first sign of pathology, observed at the ultrastructural level on postnatal day 8, is a swelling in the mitochondria of the inner segments (Sonohara and Shiose, 1968). After day 10, cell death occurs rapidly and, by day 20, virtually all of the rod visual cells have degenerated (Noell, 1965). The few cells that

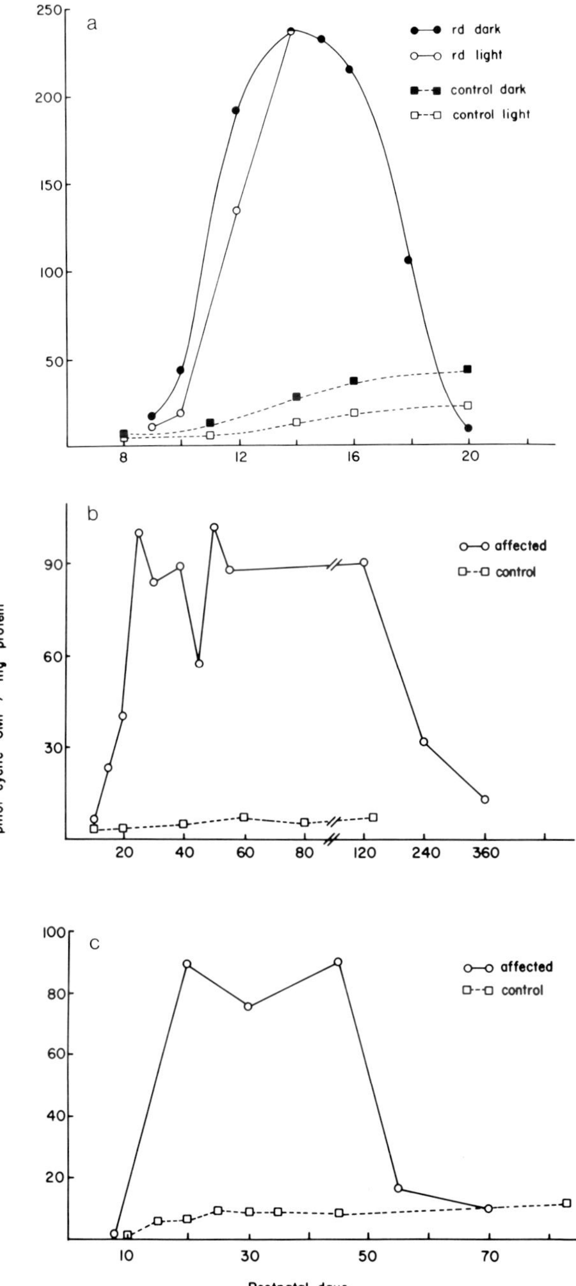

FIG. 10. Developmental pattern of retinal cyclic GMP content in animals affected with inherited diseases characterized by an early onset, fast degeneration of the photoreceptor cells: (a) *rd* mouse; (b) Irish setter dog, (c) Collie dog. (Adapted from Farber and Lolley, 1977; Aguirre *et al.*, 1982; and Woodford *et al.*, 1982.)

remain longer in the photoreceptor layer appear to be cones (Carter-Dawson *et al.*, 1978). The loss of total protein from the retina reflects the loss of rods (Farber and Lolley, 1973).

An abnormality in cyclic GMP metabolism has been detected in the *rd* retina 2 days before the photoreceptor cells begin to degenerate (Farber and Lolley, 1974). Cyclic GMP levels begin to rise above normal by postnatal day 6 and peak by day 14 (Fig. 10a). This accumulation of cyclic GMP occurs exclusively in the photoreceptor cells, as determined by measurements carried out using microdissected samples of the photoreceptor and inner layers of the *rd* retina (Farber and Lolley, 1974). The elevated levels of cyclic GMP are not the result of a hyperactive guanylate cyclase, since this enzyme behaves normally in the *rd* retina from the onset of differentiation, but result from an apparent deficiency in the activity of cyclic GMP phosphodiesterase (Farber and Lolley, 1976). In *rd* retina which has been freeze-dried, the K_m of phosphodiesterase for cyclic GMP is similar to that of photoreceptor cells of control retina, although its apparent V_{max} is quite below normal (Farber and Lolley, 1977a). *In vivo,* light activates the enzyme of *rd* photoreceptors slightly. This is indicated by a small reduction in cyclic GMP content caused by light adaptation during the short period while the *rd* visual cells have intact outer segments (Farber and Lolley, 1977a). The abnormality in cyclic GMP phosphodiesterase activity has not been observed in other body tissues of the *rd* mouse, such as blood components (including plasma, serum, lymphocytes, and erythrocytes), muscle, brain, and skin (Lolley and Farber, 1980). The high levels of cyclic GMP accumulated by the *rd* photoreceptors do not decrease immediately following the degeneration of the outer segments, suggesting that some of the light-insensitive cyclic GMP is localized in the inner segment or other parts of the cell.

The fact that the cyclic GMP phosphodiesterase is present but barely activated by light in *rd* photoreceptors indicates that there may be a lesion in the phosphodiesterase molecule itself or in any of the components required for its activation. *In vitro* studies of rhodopsin phosphorylation in developing retinas of control and *rd* mice by Shuster and Farber (1986) have demonstrated that this reaction is also substantially reduced in the young *rd* retinas, when rhodopsin is still present (Fig. 11). It is possible that *in vivo* rhodopsin is permanently phosphorylated in the degenerative retina, in which case the incorporation of phosphate *in vitro* would not be detectable. Furthermore, phosphorylated rhodopsin would not support cyclic GMP phosphodiesterase activation (Sitaramayya and Liebman, 1983b), a situation which is consistent with the findings described above.

At this time, there is still no conclusive evidence regarding the site of the abnormality in the *rd* disease, and the actual genetic defect may be quite complex. The lesion may not stem from an aberrant phosphodiesterase but rather from a biochemical system acting prior to the activation of the enzyme by light.

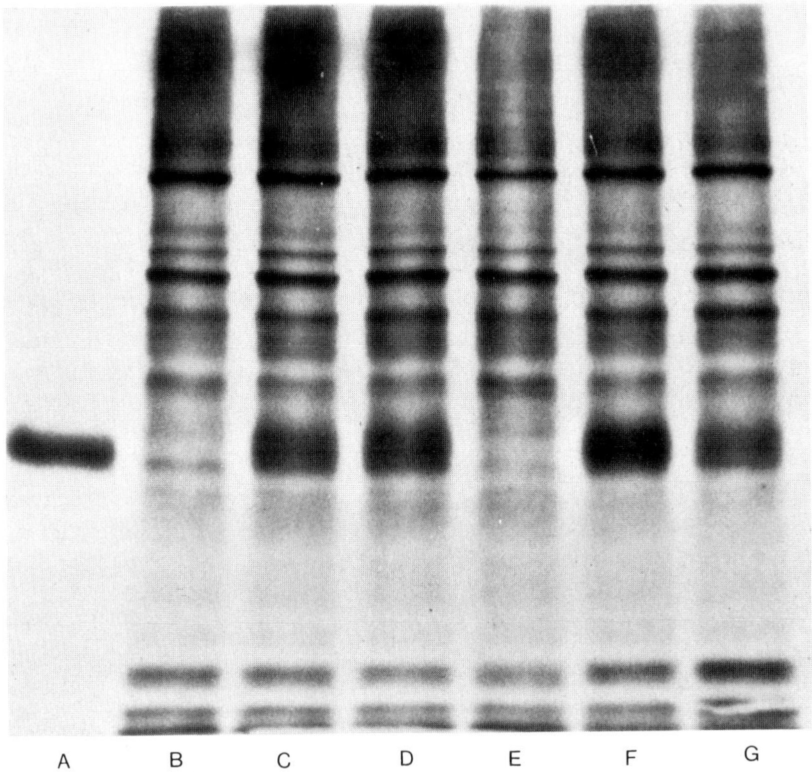

FIG. 11. Rhodopsin in clearly phosphorylated in normal but not in *rd* retina. A, Rhodopsin (R); B and E, homozygous *rd* retina (rdle); C and F, heterozygous *rd* and C57BL retina (rd/+); D and G, C57BL retina (+/+). B–D, Retinas from 12-day-old animals; E–G, retinas from 13-day-old animals. (From Shuster and Farber, 1986.)

Thus, the abnormality could reside in the rhodopsin molecule or its packing in the disk membrane, in rhodopsin kinase, in phosphorhodopsin phosphatase, in the G protein as well as in the cyclic GMP phosphodiesterase, or else in other sites unrelated to photoreception. Wherever the lesion is, the accumulation of cyclic GMP in rod visual cells is an early and perhaps causative factor for the death of the *rd* photoreceptors. The mechanism by which cyclic GMP could act in this process is still unresolved, however. If cyclic GMP is associated with visual transduction regulating the ion permeability of the photoreceptors, an accumulation of cyclic GMP could increase ion entry into the cell, perhaps leading to osmotic imbalance and eventual death. Alternatively, cyclic GMP may regulate other processes of the photoreceptors. An abnormal accumulation of cyclic GMP in the *rd* visual cells could disrupt these processes and lead to degeneration of the cells.

2. Irish Setter Dog

Irish setters also carry an autosomal recessive mutation that causes an early onset retinal degeneration in homozygous animals (Aguirre and Rubin, 1975). The morphology of the retina appears normal until postnatal day 13. However, following their initial differentiation, the photoreceptor cells stop developing and, as a result, have diminutive inner segments and form only small and disorganized outer segments. These abnormalities become evident by the third postnatal week, and thereafter, loss of rod outer segments and subsequent visual cell death occurs. Rods are affected preferentially and by 18–20 weeks, the retina has only a small number of abnormal cones in the photoreceptor layer. The inner retinal layers remain unchanged. Thus, this disease is similar to that of the *rd* mouse in its pattern of visual cell degeneration.

There are also biochemical similarities between the Irish setter and *rd* mouse disorders. In the affected dog, retinal cyclic GMP levels are higher than control levels prior to postnatal day 10 (Chader *et al.*, 1980), a time when photoreceptor outer segments are in their earliest stages of formation and when morphological signs of pathology are not yet apparent (Fig. 10b). By 3 weeks of age, cyclic GMP levels have increased about 10-fold above levels in control retinas (Aguirre *et al.*, 1982). This abnormality is apparently restricted to the photoreceptors since cyclic GMP concentrations in liver, brain, or the pigment epithelium–choroid complex are not altered (Aguirre *et al.*, 1982). Cyclic AMP levels remain normal in affected retinas throughout development.

The Irish setter retina also has a deficiency in rod outer segment cyclic GMP phosphodiesterase activity (Aguirre *et al.*, 1978). By postnatal day 9, the activity of this enzyme is already lower than in the retina of a 9-day-old normal dog. Liu *et al.* (1979) have reported that cyclic GMP phosphodiesterase of the Irish setter retina is calmodulin dependent throughout life. However, the levels of calmodulin present in this retina are lower than normal and probably not sufficient to cause the activation of the enzyme. Research in this area has not continued any further; therefore, it is difficult to explain the observations of Liu *et al.* Nevertheless, these abnormalities result in greatly increased levels of cyclic GMP and may lead to photoreceptor degeneration.

3. Collie Dog

The inherited degeneration of the collie retina is also characterized by the abnormal development of the photoreceptor cells. Outer segments have disappeared from the retina by postnatal week 6 and the visual cells are dying (Woodford *et al.*, 1982). The collie retina also has a deficiency in cyclic GMP phosphodiesterase activity which results in elevated levels of cyclic GMP (Fig. 10c). Unlike the Irish setter, however, no calmodulin dependency is observed in this retina (Woodford *et al.*, 1982). Thus, even though the three early onset, fast degenerations described have in common elevated levels of cyclic GMP, different genetic lesions may be present in each particular case.

B. Early Onset, Slow Degeneration Inherited Diseases

The *rds* mouse (retinal degeneration slow) is a recently described model of retinal degeneration (van Nie *et al.*, 1978; Sanyal and Jansen, 1981). In this mutant, the photoreceptor cells degenerate at a slower rate than in the *rd* mouse; however, their outer segments never develop.

Although cyclic GMP phosphodiesterase activity is lower than control activity in the *rds* retina throughout life (Sanyal *et al.*, 1984), cyclic GMP levels are also lower than normal (Cohen, 1983; Sanyal *et al.*, 1984). This suggests that a reduced synthesis or an increased utilization of cyclic GMP may be involved in the control of cyclic GMP levels in the *rds* retina. The *rds* photoreceptor cells degenerate in spite of the fact that cyclic GMP levels are not elevated. An issue that remains to be investigated is whether concentrations of cyclic GMP substantially lower than normal support specific functions of the visual cells; if not, the lowered levels may cause photoreceptor degeneration.

C. Late Onset Inherited Diseases

Retinal degeneration occurs in several strains of rats, such as the Royal College of Surgeons (RCS), the Campbell, and the Hunter. In the RCS rat, the genetic lesion is expressed in the pigment epithelium rather than in the photoreceptor cells. This gene appears to block the ability of the pigment epithelium to phagocytize the shed outer segments (Bok and Hall, 1971; Herron *et al.*, 1969; La Vail *et al.*, 1972). As a consequence of this, debris accumulates between the retina and the pigment epithelium (Dowling and Sidman, 1962), ultimately causing the death of the photoreceptor cells.

An abnormality in cyclic GMP metabolism has also been found in the RCS retina, but this occurs only after debris accumulates in the intercellular space between photoreceptors and pigment epithelium (Lolley and Farber, 1975). Both synthesis and hydrolysis of cyclic GMP are altered. The light-activated cyclic GMP phosphodiesterase appears to increase its affinity for cyclic GMP as a consequence of the presence of a heat-denaturable, nondialyzable protein in the debris (Lolley and Farber, 1975, 1976). *In vivo,* the levels of cyclic GMP in the RCS retina are decreased by light prior to debris accumulation, but thereafter, both in dark and light, cyclic GMP levels are lower than control (Farber and Lolley, 1977a). Like the *rds* mouse model, the RCS rat retinal degeneration is another case in which reduced levels of cyclic GMP may not support the viability of the photoreceptor cells. This is the only biochemical defect described so far which occurs prior to pathological changes in the visual cells.

Dewar *et al.* (1975a,b) have studied phosphodiesterase activity in two strains of rats which bear dystrophic retinas, the albino Campbell and the pigmented Hunter. Using cyclic AMP as substrate, they found that the albino retina is deficient in phosphodiesterase activity before histological signs of degeneration are evident, whereas the Hunter retina shows reduced enzyme activity secondary to the degeneration of the photoreceptors. There is, however, no evidence suggesting the cause of death of the visual cells in these rat retinas. This is an open field for investigation.

D. *Drug-Induced Degenerations*

1. Studies on Normal *Xenopus* Retinas

To test whether the accumulation of cyclic GMP could be a causative factor in the degeneration of photoreceptor cells observed in the early onset, fast degeneration models, the effects of phosphodiesterase inhibitors and cyclic nucleotide derivatives on normal photoreceptor viability were investigated. The embryonic retina of *Xenopus laevis* differentiates and develops in culture (Hollyfield *et al.*, 1975), changing from a neuroepithelium to a well-organized and layered structure containing rods and cones, all of which have typical outer segments and synaptic contacts with horizontal and bipolar neurons. The differentiation process also occurs normally in incubation media to which low concentrations ($<10^{-5}$ *M*) of IBMX or the inhibitor Squibb 65,422 have been added (Lolley *et al.*, 1977b). When the concentration of inhibitor in the cultures is raised from 10^{-5} to 10^{-3} *M*, however, the levels of cyclic GMP increase considerably (four- to eightfold) within the retinas and the selective degeneration of rod photoreceptors occurs. Higher concentrations of IBMX or of the Squibb inhibitor result in cell death throughout the retina (Lolley *et al.*, 1977b).

To avoid the use of phosphodiesterase inhibitors which may modulate other enzymes or cellular functions, Hollyfield *et al.* (1982) assessed the effect of high levels of cyclic AMP or cyclic GMP by growing the eye rudiments in medium containing either the cyclic nucleotides or their dibutyryl or 8-bromo derivatives. Neither cyclic AMP nor cyclic GMP caused disruption of retinal morphology, probably because they did not enter the cells or because they were rapidly hydrolyzed by the phosphodiesterases of the normal retinas once they had crossed the cell membranes. After 3 days in culture, 8-bromo cyclic AMP (1 m*M*) produced widespread cell death in all retinal layers, whereas 8-bromo cyclic GMP (>1 m*M*) prevented the elaboration of photoreceptor outer segments but did not cause cell death. High levels of dibutyryl cyclic GMP (8 m*M*) blocked differentiation in immature retinas but did not cause cell death. The selective degeneration of photoreceptors occurred only when dibutyryl cyclic GMP was presented to partially differentiated visual cells. In contrast, dibutyryl cyclic AMP caused degeneration of all retinal neurons, regardless of stage of development. Dibutyryl cyclic GMP and IBMX at levels which are nontoxic individually acted synergistically to cause selective photoreceptor cell death. These results indicate that elevated levels of cyclic AMP or cyclic GMP are toxic to neurons of the retina, but cyclic GMP can initiate the specific destruction of normal photoreceptor cells.

2. Studies on Human Retinas

Ulshafer *et al.* (1980) adapted the medium and culture conditions used in the experiments described above for *Xenopus* eye rudiments and repeated them on fresh human retinas. These authors found that IBMX and dibutyryl cyclic GMP caused selective degeneration of rod photoreceptors and had only slight effects on cone morphology, causing the cones to round up. The inner retinal layers remained unaltered. One of the toxic actions of the phosphodiesterase inhibitors and of the cyclic nucleotide derivatives may occur at the level of protein synthesis. Human retinas incubated with 4×10^{-3} *M* IBMX, a dose which does not alter rod morphology during a 4-hr incubation, inhibited [^{3}H]leucine incorporation into protein by 60% after 4 hr of exposure, exclusively in rod visual cells. The dibutyryl derivatives of cyclic AMP and cyclic GMP had a similar effect, although the inhibition of protein synthesis was only about half that observed with IBMX. These studies support the hypothesis that elevated levels of cyclic GMP may be involved in some forms of human retinal degenerations, similar to what has been suggested for the early onset animal models described before.

3. Iodoacetic Acid Effects on Ground Squirrel Retina

Intravenous injections of sodium iodoacetate produce a characteristic visual cell degeneration in several species of animals (Schubert and Bornschein, 1951; Noell, 1952). The remaining cells of the retina are not damaged unless high doses of iodoacetate are used. As in the inherited diseases of *rd* mice, and Irish setter and collie dogs, in all the drug-induced disruptions of retinal metabolism that have been studied, rod photoreceptors seem to be more vulnerable than cones and are the cells that die first. From all of these investigations, however, it cannot be determined whether cone visual cells degenerate as a consequence of the death of rods. This issue has been addressed in the work of Farber *et al.* (1983), in which the selective degeneration of photoreceptors of the cone-dominant retina of ground squirrel was induced by intracardial injection of iodoacetate. The following is a summary of the sequence of events observed.

Degenerating cones were first detected 1 day after iodoacetate treatment, and all cone photoreceptors were reduced to compact, dense masses by day 4. Cellular debris was removed, starting at the synaptic terminals, by macrophages which entered the photoreceptor layer from the inner nuclear layer as early as day 3. After completion of this process the macrophages retreated, leaving the pigment epithelium in contact with the intact neurons of the remaining retinal layers.

The cyclic nucleotide content of ground squrrel retina at different stages of cone cell damage and debris clearance resembles the cyclic nucleotide changes observed during postnatal development in the early onset, fast degeneration inherited diseases (Fig. 12). The main difference is that, whereas in those disorders only cyclic GMP levels become elevated prior to the pathological morphology of the cells, in the iodoacetate-treated retina both cyclic AMP and cyclic GMP reach levels much higher than normal before any signs of degeneration are detected. This may result from a nonspecific inhibition by iodoacetate of all types of phosphodiesterases present in the ground squirrel retina, since iodoacetic acid is a potent inhibitor of sulfhydryl-containing enzymes. In addition, this inhibition may also explain the loss of photosensitivity of retinal cyclic AMP that is observed very soon after injection.

Thus, elevated levels of cyclic nucleotides may be involved in the degeneration of cone visual cells, as they are in rods. It is still necessary to determine which, cyclic AMP or cyclic GMP, independently or in a combined action, is responsible for the death of the photoreceptors in the cone-dominant retina of the ground squirrel.

4. Degeneration in a Pure-Cone Retina *in Vitro*

Recently, it has been found that pure-cone retinas of adult Western fence lizards can be maintained well for 2 days *in vitro* (Williams *et al.*, 1986a). This preparation has been used to assess the damaging effects of high levels of cyclic nucleotides on cones more directly than in the *in vivo* situation with the ground squirrel, described above.

Williams *et al.* (1986b) found that 10^{-3} *M* IBMX, but not 10^{-5} *M*, elevated levels of retinal cyclic AMP and cyclic GMP (by 5 to 10-fold after 2–10 hr), decreased retinal protein synthesis by approximately 50% and induced degeneration of the cones. Dibutyryl cyclic AMP, 10^{-2} *M*, also inhibited protein synthesis and had a deleterious effect on cone morphology (albeit milder than 10^{-3} *M* IBMX). In comparison, 10^{-2} *M* dibutyryl cyclic GMP did not significantly inhibit protein synthesis or affect morphology.

At these concentrations of IBMX and dibutyryl cyclic AMP, inner retinal cells were also affected, but to a much lesser extent than the cones. Although the

Fig. 12. Time course of changes in cyclic nucleotide levels after intracardial iodoacetic acid injection in ground squirrel. (a) Cyclic AMP, (b) cyclic GMP. (From Farber *et al.*, 1983.)

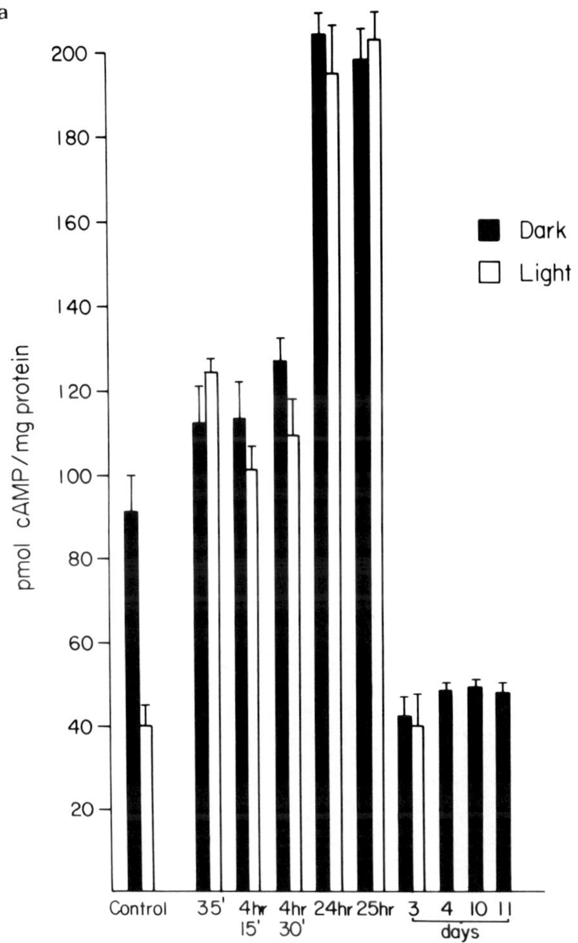
a
pmol cAMP/mg protein
200
180
160
140
120
100
80
60
40
20
Dark
Light
Control
35'
4hr 15'
4hr 30'
24hr
25hr
3
4
10
11
days
Time after injection

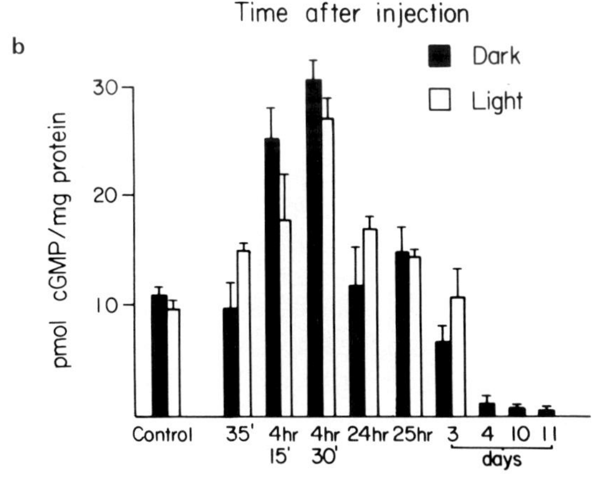
b
pmol cGMP/mg protein
30
20
10
Dark
Light
Control
35'
4hr 15'
4hr 30'
24hr
25hr
3
4
10
11
days
Time after injection

cones showed greater resilience than the developing rods in *Xenopus* (cf. concentrations above), this study shows that the photoreceptors are still the most susceptible retinal cells to high levels of cyclic nucleotides, even in a pure-cone retina. It also indicates that cyclic AMP may be more toxic than cyclic GMP to cones.

V. Future Prospects

Throughout this review, an attempt was made to point out some of the issues which need to be addressed regarding second messengers and retinal metabolism. We would like to briefly restate some of the major questions concerning the photoreceptor cells.

First of all, the identity of the second messenger participating in the transduction of rod outer segments needs clarification. Currently, cyclic GMP is probably the major candidate, particularly since so much of the rod outer segment metabolic machinery is directed toward cyclic GMP light-stimulated hydrolysis. Furthermore, recent data suggest cyclic GMP may directly control the membrane pores conducting the dark current. On the other hand, inositol metabolites are emerging as possible transduction mediators and calcium is present in large quantities in the disks, presumably for some reason. It may be that multiple regulatory pathways for the light response exist, and each of these second messenger candidates may participate somehow in modulating or controlling the action of the other candidates. For example, a model has been proposed in which cyclic GMP or calcium could modulate the levels of each other (Fatt, 1982). In addition, some or all of these messenger candidates may participate in dark and light adaptation. Moreover, an as-yet-unidentified second messenger may also be at work during transduction and adaptation. Research in cone photoreceptors has recently begun, so that many questions, including those just posted for rod photoreceptors regarding the second messengers, have to be answered in this area. Generally, much less is known about cone outer segment metabolism since these outer segments have proven difficult to purify.

Questions regarding the role cyclic nucleotides play in the development and degeneration of photoreceptor cells are plentiful. Do the high levels of cyclic GMP actually cause photoreceptor death? Or are the elevated levels interfering with fundamental metabolic processes which, when impaired, cannot support any longer the life of the cell? Elucidation of these issues would contribute to our knowledge of human retinal degenerative diseases and also afford us a better understanding of visual cell differentiation and development.

Finally, much more needs to be known about the fine control of outer segment metabolic processes. For example, we do not completely understand how all the various nucleotides in the outer segments participate in the modulation of rhodopsin phosphorylation, quenching of phosphodiesterase activity, and other activities, and how the various proteins interact to control transduction and adaptation. Clearly, there are many exciting discoveries yet to be made regarding the roles second messengers play in photoreceptor biology.

Acknowledgments

Our appreciation to Drs. David Williams and Alfred Walter for their constructive criticism of the manuscript. Special thanks to Ms. Ethel Mason for help with the manuscript preparation. Our deepest appreciation to Dr. John Flannery and Ms. Beth Jones for their invaluable help with computer programming. We also want to acknowledge the dedicated assistance of Lee J. Stone and the support from the National Eye Institute (Grants EY 02651 and EY 00331) and from the National Retinitis Pigmentosa Foundation Fighting Blindness, Baltimore, Maryland.

References

Aguirre, G. D., and Rubin, L. F. (1975). Pathology of hemeralopia in the Alaskan malamute dog. *J. Am. Vet. Med. Assoc.* **166,** 257–259.

Aguirre, G., Farber, D., Lolley, R., Fletcher, R. T., and Chader, G. J. (1978). Rod–cone dysplasia in Irish Setters: A defect in cyclic GMP metabolism in visual cells. *Science* **201,** 1133–1134.

Aguirre, G., Farber, D., Lolley, R., O'Brien, P., Alligood, J., Fletcher, R., and Chader, G. (1982). Retinal degenerations in the dog. III. Abnormal cyclic nucleotide metabolism in rod–cone dysplasia. *Exp. Eye. Res.* **35,** 625–642.

Anderson, R. E., and Hollyfield, J. G. (1981). Light stimulates the incorporation of inositol into phosphatidylinositol in the retina. *Biochim. Biophys. Acta* **665,** 619–622.

Aton, G. B., and Litman, B. J. (1981). The ability of rhodopsin to activate ROS phosphodiesterase is reduced by phosphorylation. *Invest. Ophthalmol.* **20** (Suppl.), 208.

Baehr, W., Devlin, M. J., and Applebury, M. L. (1979). Isolation and characterization of cGMP phosphodiesterase from bovine rod outer segments. *J. Biol. Chem.* **254,** 11669–11677.

Baehr, W., Morita, E. A., Swanson, R. J., and Applebury, M. L. (1982). Characterization of bovine rod outer segment G-protein. *J. Biol. Chem.* **257,** 6452–6460.

Baylor, D. A., and Fuortes, M. G. F. (1970). Electrical responses of single cones in the retina of the turtle. *J. Physiol. (London)* **207,** 77–92.

Bennett, N. (1982). Light-induced interaction between rhodopsin and the GTP-binding protein. Relation with phosphodiesterase activation. *Eur. J. Biochem.* **123,** 133–139.

Bennett, N., Michel-Vilaz, M., and Kuhn, H. (1982). Light-induced interactions between rhodopsin and the GTP-binding protein. Metarhodopsin II is the major photoproduct involved. *Eur. J. Biochem.* **127,** 97–103.

Bensinger, R. E., Fletcher, R. T., and Chader, G. J. (1974a). "Piggyback" chromatography: Assay for guanylate cyclase in retina and other neural tissue. *J. Neurochem.* **22,** 1131–1134.

Bensinger, R. E., Fletcher, R. T., and Chader, G. J. (1974b). Guanylate cyclase: Inhibition by light in retinal photoreceptors. *Science* **183,** 86–87.

Berger, S. J., De Vries, G. W., Carter, J. G., Schulz, D. W., Passoneau, P. W., Lowry, O. H., and Ferrendelli, J. A. (1980). The distribution of the components of the cyclic GMP cycle in retina. *J. Biol. Chem.* **255,** 3128–3133.

Berman, A. L., and Usova, A. A. (1978). Protein inhibitor of the retinal cyclic nucleotide phosphodiesterase: Its localization in the outer segment of a photoreceptor (in Russian). *Biokheimiia* **43,** 486–490.

Berson, E. L. (1976). Retinitis pigmentosa and allied retinal diseases: Electrophysiological findings. *Trans. Am. Acad. Ophthalmol. Otolaryngol.* **81,** 659–666.

Biernbaum, M. S., and Bownds, M. D. (1979). Influence of light and calcium on guanosine 5′-triphosphate in isolated frog rod outer segments. *J. Gen. Physiol.* **74,** 649–669.

Bignetti, E., Cavaggioni, A., and Sorbi, R. T. (1978). Light-activated hydrolysis of GTP and cyclic GMP in the rod outer segments. *J. Physiol. (London)* **279,** 55–69.

Bitensky, M. W., Miki, N., Marcus, F. R., and Keirns, J. J. (1973). The role of cyclic nucleotides in visual excitation. *Life Sci.* **13,** 1451–1472.

Bitensky, I. W., Miki, N., and Kierns, J. J. (1975). Activation of photoreceptor disk membrane phosphodiesterase by light and ATP. *Adv. Cyclic Nucleotide Res.* **5,** 213–240.

Bitensky, M. W., Wheeler, G. L., Aloni, B., Vetury, S., and Matuo, Y. (1978). Light- and GTP-activated photoreceptor phosphodiesterase regulation by a light-activated GTPase and identification of rhodopsin as the phosphodiesterase binding site. *Adv. Cyclic Nucleotide Res.* **9,** 553–572.

Blanks, J. C., Adinolfi, A. M., and Lolley, R. N. (1974). Photoreceptor degeneration and synaptogenesis in retinal-degenerative (*rd*) mice. *J. Comp. Neurol.* **156,** 95–106.

Bok, D., and Hall, M. O. (1971). The role of the pigment epithelium in the etiology of inherited retinal dystrophy in the rat. *J. Cell Biol.* **49,** 664–682.

Bownds, D., Dawes, L., Miller, L., and Stahlman, M. (1972). Phosphorylation of frog photoreceptor membranes induced by light. *Nature (London) New Biol.* **237,** 125–127.

Brodie, A. E., and Bownds, D. (1976). Biochemical correlates of adaptation processes in isolated frog photoreceptor membranes. *J. Gen. Physiol.* **63,** 1–11.

Brown, J. E., Rubin, L. J., Ghalayini, A. J., Tarver, A. P., Irvine, R. F., Berridge, M. J., and Anderson, R. E. (1984). *myo*-inositol polyphosphate may be a messenger for visual excitation in *Limulus* photoreceptors. *Nature (London)* **311,** 1160–1163.

Brown, J. H., and Makman, M. H. (1972). Stimulation by dopamine of adenylate cyclase in retinal homogenates and of adenosine-3′:5′-cyclic monophosphate formation in intact retina. *Proc. Natl. Acad. Sci. U.S.A.* **69,** 539–543.

Burnside, B., Evans, M., Fletcher, R., and Chader, G. (1982). Induction of dark-adaptive retinomotor movement (cell elongation) in teleost retinal cones by cyclic adenosine 3′,5′-monophosphate. *J. Gen. Physiol.* **79,** 759–774.

Byzov, A. L., and Trifonov, Y. A. (1968). The response to electric stimulation of horizontal cells in the carp retina. *Vision Res.* **8,** 817–822.

Caretta, A., Cavaggioni, A., and Sorbi, R. T. (1979a). Phosphodiesterase and GTPase in rod outer segments. Kinetics *in vitro*. *Biochim. Biophys. Acta* **583,** 1–13.

Caretta, A., Cavaggioni, A., and Sorbi, R. T. (1979b). Cyclic GMP and the permeability of the discs of the frog photoreceptors. *J. Physiol. (London)* **295,** 171–178.

Carter-Dawson, L. D., LaVail, M. M., and Sidman, R. L. (1978). Differential effect of the *rd* mutation on rods and cones in the mouse retina. *Invest. Ophthalmol. Visual Sci.* **7,** 489–498.

Cavaggioni, A., and Sorbi, R. T. (1981). Cyclic GMP affects calcium movements in the disc membrane of photoreceptors. *Proc. Natl. Acad. Sci. U.S.A.* **78,** 3964–3968.

Cervetto, L., and Piccolino, M. (1974). Synaptic transmission between photoreceptors and horizontal cells in the turtle retina. *Science* **183,** 417–419.

Chader, G., Johnson, M., Fletcher, R., and Bensinger, R. (1974a). Cyclic nucleotide phosphodiesterase of the bovine retina: Activity, subcellular distribution and kinetic parameters. *J. Neurophysiol.* **22,** 93–99.

Chader, G., Fletcher, R., Johnson, M., and Bensinger, R. (1974b). Rod outer segment phosphodiesterase: Factors affecting the hydrolysis of cyclic AMP and cyclic GMP. *Exp. Eye Res.* **18,** 509–515.

Chader, G. J., Herz, L. R., and Fletcher, R. T. (1974c). Light activation of phosphodiesterase activity in retinal rod outer segments. *Biochim. Biophys. Acta* **347,** 491–493.

Chader, G. J., Fletcher, R. R., O'Brien, P. J., and Krishna, G. (1976). Differential phosphorylation by GTP and ATP in isolated rod outer segments of the retina. *Biochemistry* **15,** 1615–1620.

Chader, G., Liu, Y., O'Brien, P., Fletcher, R., Kirshna, G., Aguirre, G., Farber, D., and Lolley, R. (1980). Cyclic GMP phosphodiesterase activator: Involvement in a hereditary retinal degeneration. *In* "Neurochemistry of the Retina" (N. G. Bazan and R. N. Lolley, eds.), pp. 441–458. Pergamon, Oxford.

Chader, G., Liu, Y., Fletcher, R., Aguirre, G., Santos-Anderson, R., and T'so, M. (1981). Cyclic GMP phosphodiesterase and calmodulin in early-onset inherited retinal degenerations. *Curr. Top. Membr. Transp.* **15,** 133–156.

Clement-Cormier, Y. C., and Redburn, D. A. (1978). Dopamine-sensitive adenylate cyclase in retina-subcellular distribution. *Biochem. Pharmacol.* **27,** 2281–2282.

Cobbs, W. H., and Pugh, E. N., Jr. (1985). Cyclic GMP can increase rod outer segment light-sensitive current 10-fold without delay of excitation. *Nature (London)* **313,** 585–587.

Cobbs, W. H., Barkdoll, A. E., III, and Pugh, E. N., Jr. (1985). Cyclic GMP increases photocurrent and light sensitivity of retinal cones. *Nature (London)* **317,** 64–66.

Cohen, A. (1983). Some cytological and initial biochemical observations on photoreceptors in retinas of *rds* mice. *Invest. Ophthalmol. Visual Sci.* **24,** 832–843.

Cohen, A. I., Hall, I. A., and Ferrendelli, J. A. (1978). Calcium and cyclic nucleotide regulation in incubated mouse retina. *J. Gen. Physiol.* **71,** 595–612.

Cone, R. A. (1973). The internal transmitter model for visual excitation: Some quantitative implications. *In* "Biochemistry and Physiology of Visual Pigments" (H. Langer, ed.), pp. 275–282. Springer-Verlag, Berlin and New York.

Coquil, J. F., Virmaux, N., Mandel, P., and Goridis, C. (1975). Cyclic nucleotide phosphodiesterase of retinal photoreceptors. Partial purification and some properties of the enzyme. *Biochim. Biophys. Acta* **403,** 425–437.

Corbin, J. D., Keely, S., and Park, C. R. (1975). Distribution and dissociation of cyclic adenosine 3′:5′-monophosphate-dependent protein kinases in adipose, cardiac, and other tissues. *J. Biol. Chem.* **250,** 218–225.

de Azeredo, F. A., Lust, W. D., and Passonneau, J. V. (1978). Guanine nucleotide concentrations *in vivo* in outer segments of dark and light adapted frog retina. *Biochem. Biophys. Res. Commun.* **85,** 293–300.

De Mello, F. G. (1978). The ontogeny of dopamine-dependent increase of adenosine 3′,5′-cyclic monophosphate in the chick retina. *J. Neurochem.* **31,** 1049–1053.

De Vries, G. W., and Ferrendelli, J. A. (1983). Localization of an endogenous substrate for cAMP-stimulated protein phosphorylation in retina. *Exp. Eye Res.* **36,** 505–515.

De Vries, G. W., Cohen, A. I., Hall, I. A., and Ferrendelli, J. A. (1978). Cyclic nucleotide levels in normal and biologically fractionated mouse retina: Effects of light and dark adaptation. *J. Neurochem.* **31,** 1345–1351.

De Vries, G. W., Cohen, A. I., Lowry, O. H., and Ferrendelli, J. A. (1979). Cyclic nucleotides in the cone-dominant ground squirrel retina. *Exp. Eye Res.* **29,** 315–321.

De Vries, G. W., Cohen, A. I., Lowry, O. H., and Ferrendelli, J. A. (1982). Cyclic nucleotide levels in light- and dark-adapted ground squirrel whole eyes. *Vision Res.* **22,** 1237–1240.

Dewar, A. J., Barron, G., and Richmond, J. (1975a). Adenosine 3′:5′-cyclic monophosphate phosphodiesterase activity in the dystrophic rat retina. *Biochem. Soc. Trans.* **3,** 265–268.

Dewar, A. J., Barron, G., and Richmond, J. (1975b). Retinal cyclic-AMP phosphodiesterase activity in two strains of dystrophic rat. *Exp. Eye Res.* **21,** 299–306.

Dowling, J. E., and Ripps, H. (1973). Effect of magnesium on horizontal cell activity in the skate retina. *Nature (London)* **242,** 101–103.

Dowling, J. E., and Sidman, R. L. (1962). Inherited retinal dystrophy in the rat. *J. Cell Biol.* **14,** 73–109.

Dumler, I. L., and Etingof, R. N. (1976). Protein inhibitor of cyclic adenosine 3′,5′-monophosphate phosphodiesterase in retina. *Biochim. Biophys. Acta* **429,** 474–478.

Emeis, D., Kuhn, H., Reichert, J., and Hofmann, K. (1982). Complex formation between metarhodopsin II and GTP-binding protein in bovine photoreceptor membranes leads to a shift of the photoproduct equilibrium. *FEBS Lett.* **143,** 29–34.

Farber, D. B. (1982). The role of cyclic nucleotide metabolism in the eye. *In* "Handbook of Experimental Pharmacology" (J. W. Kebabian and J. A. Nathanson, eds.), pp. 465–524. Springer-Verlag, Berlin and New York.

Farber, D. B., and Bok, D. (1984). cGMP-phosphodiesterase of retinal photoreceptors: Specific localization in rods demonstrated by immunocytochemistry. *J. Cell Biol.* **99,** 64a.

Farber, D. B., and Bok, D. (1985). Light-activated cGMP-phosphodiesterase is absent from cone photoreceptors. *Invest. Ophthalmol. Visual Sci. (Suppl.)* **26,** 334.

Farber, D. B., and Lolley, R. N. (1973). Proteins in the degenerative retina of C3H mice: Deficiency of a cyclic-nucleotide phosphodiesterase and opsin. J. Neurochem. **21,** 817–828.

Farber, D. B., and Lolley, R. N. (1974). Cyclic guanosine monophosphate: Elevations in degenerating photoreceptor cells of the C3H mouse retina. *Science* **186,** 449–451.

Farber, D. B., and Lolley, R. N. (1976). Enzymatic basis for cyclic GMP accumulation in degenerative photoreceptor cells of mouse retina. *J. Cyclic Nucleotide Res.* **2,** 139–148.

Farber, D. B., and Lolley, R. N. (1977a). Light-induced reduction in cyclic GMP of retinal photoreceptor cells *in vivo:* Abnormalities in the degenerative diseases of RCS rats and *rd* mice. *J. Neurochem.* **28,** 1089–1095.

Farber, D. B., and Lolley, R. N. (1977b). Influence of visual cell maturation or degeneration on cyclic AMP content of retinal neurons. *J. Neurochem.* **29,** 167–170.

Farber, D. B., and Lolley, R. N. (1978). Cyclic-AMP and cyclic-GMP content of cone-dominant retinas of ground squirrel. *ARVO Abstr. Suppl. Invest. Ophthalmol. Visual Sci.* 255.

Farber, D. B., and Lolley, R. N. (1979). Phosphoproteins as proposed modulators of visual function. *Adv. Exp. Med. Biol.* **116,** 103–115.

Farber, D. B., and Souza, D. W. (1982). Characterization of phosphodiesterases from ground squirrel retina. *Am. Soc. Neurochem.* **13,** 83.

Farber, D. B., Brown, B. M., and Lolley, R. N. (1978). Cyclic GMP: Proposed role in visual cell function. *Vision Res.* **18,** 497–499.

Farber, D. B., Brown, B. M., and Lolley, R. N. (1979). Cyclic nucleotide-dependent protein kinase and the phosphorylation of endogenous proteins of retinal rod outer segments. *Biochemistry* **18,** 370–378.

Farber, D. B., Chase, D. G., and Lolley, R. N. (1980). Cyclic nucleotides in rod- and cone-dominant retinas. *In* "Neurochemistry of the Retina" (N. G. Bazan and R. N. Lolley, eds.), pp. 327–336. Pergamon, Oxford.

Farber, D. B., Souza, D. W., Chase, D. G., and Lolley, R. N. (1981). Cyclic nucleotides of cone-dominant retinas: Reduction of cyclic AMP levels by light and by cone degeneration. *Invest. Ophthalmol. Visual Sci.* **20,** 24–31.

Farber, D. B., Souza, D. W., and Lolley, R. N. (1982). Cyclic AMP content of a cone-dominant retina: Time course of light reduction and dark recovery, *in vivo*. *Soc. Neurosci. Abstr.* **8,** 342.

Farber, D. B., Souza, D. W., and Chase, D. G. (1983). Cone visual cell degeneration in ground squirrel retina: Disruption of morphology and cyclic nucleotide metabolism by iodoacetic acid. *Invest. Ophthalmol. Visual Sci.* **24,** 1236–1249.

Farber, D. B., Flannery, J., Lolley, R. N., and Bok, D. (1985). Distribution patterns of photoreceptors, protein and cyclic nucleotides in the human retina. *Invest. Ophthalmol. Visual Sci.* **26,** 1558–1568.

Fatt, P. (1982). An extended Ca^{2+}-hypothesis of visual transduction with the role for cyclic GMP. *FEBS. Lett.* 149, 159–166.

Fein, A., and Szuts, E. Z. (1982). "Photoreceptors: Their Role in Vision." Cambridge Univ. Press, London and New York.

Fein, A., Payne, R., Carson, D. W., Berridge, M. J., and Irvine, R. F. (1984). Photoreceptor excitation and adaptation by inositol 1,4,5-triphosphate. *Nature (London)* **311,** 157–160.

Ferrendelli, J. A. (1978). Distribution and regulation of cyclic GMP in the central nervous system. *Adv. Cyclic Nucleotide Res.* **9,** 453–464.

Ferrendelli, J. A., and Cohen, A. I. (1976). The effects of light and dark adaptation on the levels of cyclic nucleotides in retinas of mice heterozygous for a gene for photoreceptor dystrophy. *Biochem. Biophys. Res. Commun.* **74,** 421–427.

Fesenko, E. E., Kolesnikov, S. S., and Lyubarsky, A. L. (1985). Induction by cyclic GMP of cationic conductance in plasma membrane of retinal rod outer segment. *Nature (London)* **313,** 310–313.

Fleishman, D., and Denisevich, M. (1979). Guanylate cyclase of isolated bovine retinal rod axonemes. *Biochemistry* **18,** 5060–5066.

Fleishman, D., Denisevich, M., Raveed; D., and Pannbacker, R. (1980). Association of guanylate cyclase with the axoneme of the retinal rods. *Biochim. Biophys. Acta* **630,** 176–186.

Fletcher, R. T., and Chader, G. J. (1976). Cyclic GMP: Control of concentration by light in retinal photoreceptors. *Biochem. Biophys. Res. Commun.* **70,** 1297–1302.

Frank, R. N., and Bensinger, R. E. (1974). Rhodopsin and light-sensitive kinase activity of retinal outer segments. *Exp. Eye Res.* **18,** 271–280.

Frank, R. N., and Buzney, S. M. (1975). Mechanism and specificity of rhodopsin phosphorylation. *Biochemistry* **14,** 5110–5117.

Fukada, Y., and Yoshizawa, T. (1981). Activation of phosphodiesterase in frog rod outer segment by an intermediate of rhodopsin photolysis II. *Biochim. Biophys. Acta* **675,** 195–200.

Fukada, Y., Kawamura, S., Yoshizawa, T., and Miki, N. (1981). Activation of phosphodiesterase in frog and rod outer segment by an intermediate of rhodopsin photolysis. *Biochim. Biophys. Acta* **675,** 188–194.

Fung, B.K.-K., and Stryer, L. (1980). Photolyzed rhodopsin catalyzes the exchange of GTP for bound GDP in retinal rod outer segments. *Proc. Natl. Acad. Sci. U.S.A.* **77,** 2500–2504.

Fung, B.K.-K., Hurley, J. B., and Stryer, L. (1981). Flow of information in the light-triggered cyclic nucleotide cascade of vision. *Proc. Natl. Acad. Sci. U.S.A.* **78,** 152–156.

Ghalayini, A., and Anderson, R. E. (1984). Phosphatidylinositol 4,5-bisphosphate: Light-mediated breakdown in the vertebrate retina. *Biochem. Biophys. Res. Commun.* **124,** 503–506.

Godchaux, W., and Zimmerman, W. F. (1979). Soluble proteins of intact bovine rod cell outer segments. *Exp. Eye Res.* **28,** 483–500.

Gold, G. H., and Korenbrot, J. I. (1980). Light-induced calcium release by intact retinal rods. *Proc. Natl. Acad. Sci. U.S.A.* **77,** 5557–5561.

Goldberg, N. D., Ames, A., Gander, J. E., and Walseth, T. F. (1983). Magnitude of increase in retinal cGMP metabolic flux determined by ^{18}O incorporation into nucleotide alpha-phosphoryls corresponds with intensity of photic stimulation. *J. Biol. Chem.* **258,** 9213–9219.

Goridis, C., Virmaux, N., Urban, P. F., and Mandel, P. (1973). Guanyl cyclase in a mammalian photoreceptor. *FEBS Lett.* **30,** 163–166.

Goridis, C., Virmaux, N., Cailla, M. L., and Delaage, M. A. (1974). Rapid, light-induced changes of retinal cyclic GMP levels. *FEBS Lett.* **49,** 167–169.

Goridis, C., Virmaux, N., Weller, M., Coquil, J. F., and Mandel, P. (1975). Guanylate cyclase and cyclic GMP phosphodiesterase in vertebrate photoreceptor organelles. *Proc. Congr. Collegium Int. Neuropsychopharmacol. 9th, Paris 1974* pp. 920–931.

Goridis, C., Virmaux, N., Weller, M., and Urban, P. F. (1976). Role of cyclic nucleotides in photoreceptor function. *In* "Transmitters in the Visual Process" (S. L. Bonting, ed.), pp. 27–58. Pergamon, Oxford.

Goridis, C., Urban, P. F., and Mandel, P. (1977). The effect of flash illumination on the endogenous cyclic GMP content in isolated frog retinae. *Exp. Eye Res.* **24,** 171–177.

Govardovskii, V. I., and Berman, A. L. (1981). Light-induced changes of cyclic GMP content in frog retinal rod outer segments measured with rapid freezing and microdissection. *Biophys. Struct. Mech.* **7,** 125–130.

Greengard, P. (1976). Possible role for cyclic nucleotides and phosphorylated membrane proteins in postsynaptic actions of neurotransmitters. *Nature (London)* **260,** 101–108.

Hagins, W. A. (1972). The visual process: Excitatory mechanisms in the primary photoreceptor cells. *Annu. Rev. Biophys. Bioeng.* **1,** 131–158.

Hargrave, P. A., Fong, S. L., McDowell, J. A., Mas, M. T., Curtis, P. R., Wang, J. K., Juszcazak, E., and Smith, D. P. (1980). The partial primary structure of bovine rhodopsin and its topography in the retinal rod cell disc membrane. *Neurochem. Int.* **1,** 231–244.

Haynes, L., and Yau, K. W. (1985). Cyclic GMP-sensitive conductance in outer segment membrane of catfish cones. *Nature (London)* **317,** 61–64.

Hemminki, K. (1975). Light-induced decrease in calcium binding to isolated bovine photoreceptors. *Vision Res.* **15,** 69–72.

Hermolin, J., Karell, M. A., Hamm, H. E., and Bownds, M. D. (1982). Calcium and cyclic GMP regulation of light-sensitive protein phosphorylation in frog photoreceptor membranes. *J. Gen. Physiol.* **79,** 633–655.

Herron, W. L., Riegel, B. W., Meyers, O. E., and Rubin, M. L. (1969). Retinal dystrophy in the rat: A pigment epithelial disease. *Invest. Ophthalmol.* **8,** 595–604.

Hesketh, J. E., Virmaux, N., and Mandel, P. (1978). Evidence for a cyclic nucleotide-dependent phosphorylation of retinal myosin. *FEBS Lett.* **94,** 357–360.

Hollyfield, J. G., Mottow, L. S., and Ward, A. (1975). Autoradiographic study of [^{3}H] glucosamide incorporation by the developing retina of the clawed toad, *Xenopus laevis*. *Exp. Eye Res.* **20,** 383–391.

Hollyfield, J. G., Raeburn, M. E., Farber, D. B., and Lolley, R. N. (1982). Selective photoreceptor degeneration during retinal development: The role of altered phosphodiesterase activity and increased levels of cyclic nucleotides. *In* "The Structure of the Eye" (J. G. Hollyfield, ed.), pp. 97–114. Elsevier, Amsterdam.

Hubbell, W. L., and Fung, B.K.-K. (1978). The structure and chemistry of rhodopsin: Relationship to models of function. *In* "Membrane Transduction Mechanisms" (R. A. Cone and J. Dowling, eds.), pp. 17–25. Raven, New York.

Hubbell, W. L., Fung, B.K.-K., Chen, Y., and Hong, K. (1977). Molecular anatomy and light-dependent processes in photoreceptor membranes. *In* "Vertebrate Photoreception" (H. B. Barlow and P. Fatt, eds.), pp. 41–59. Academic Press, New York.

Hughes, B. A., Miller, S. S., and Machen, T. E. (1984). Effects of cyclic AMP on fluid absorption and ion transport across frog retinal pigment epithelium. Measurements in the open-circuit state. *J. Gen. Physiol.* **83,** 875–899.

Hurley, J. B. (1980). Isolation and recombination of bovine rod outer segment cGMP phosphodiesterase and its regulators. *Biochem. Biophys. Res. Commun.* **92,** 505–510.

Hurley, J. B., and Ebrey, T. G. (1979). Regulation of rod outer segment phosphodiesterase. *Biophys. J.* **25,** 314a.

Hurley, J. B., and Stryer, L. (1982). Purification and characterization of the alpha regulatory subunit of the cyclic GMP phosphodiesterase from retinal rod outer segments. *J. Biol. Chem.* **257,** 11094–11099.

Hurwitz, R. L., Bunt-Milam, A. H., and Beavo, J. A. (1984). Immunologic characterization of the photoreceptor outer segment cyclic GMP phosphodiesterase. *J. Biol. Chem.* **259,** 8612–8618.

Iuvone, M., Rauch, A., Marshburn, P., Glass, D., and Neff, N. (1982). Activation of retinal tyrosine hydroxylase *in vitro* by cyclic AMP-dependent protein kinase: Characterization and comparison to activation *in vivo* by photic stimulation. *J. Neurochem.* **39,** 1632–1640.

Kaupp, V. B., and Junge, W. (1977). Rapid calcium release by passively loaded retinal discs on photoexcitation. *FEBS Lett.* **81,** 229–232.

Kaupp, V. B., and Schnetkamp, P. P. H. (1982). Calcium metabolism in vertebrate photoreceptors. *Cell Calcium* **3,** 83–112.

Kavipurapu, P. R., Farber, D. B., and Lolley, R. N. (1982). Degradation and resynthesis of cyclic 3′,5′-guanosine monophosphate in truncated rod photoreceptors from bovine retina. *Exp. Eye Res.* **34,** 181–189.

Keirns, J. J., Miki, N., Bitensky, M. W., and Keirns, M. (1975). A link between rhodopsin and disc membrane cyclic nucleotide phosphodiesterase. Action spectrum and sensitivity to illumination. *Biochemistry* **14,** 2760–2765.

Khoury, S. A., and Farber, D. B. (1981). Adenylate and guanylate cyclases of vertebrate retinas: Biochemical and histochemical properties. *Soc. Neurosci. Abstr.* **7,** 922.

Kilbride, P. (1980). Calcium effects on frog retinal cyclic guanosine 3′,5′-monophosphate levels and their light-initiated rate of decay. *J. Gen. Physiol.* **75,** 457–465.

Kilbride, P., and Ebrey, T. G. (1979). Light-initiated changes of cyclic guanosine monophosphate levels in the frog retina measured with quick-freezing techniques. *J. Gen. Physiol.* **74,** 415–426.

Koh, S.-W., and Chader, G. J. (1984). Agonist effects on the intracellular cyclic AMP concentration of retinal pigment epithelial cells in culture. *J. Neurochem.* **42,** 287–289.

Koh, S.-W.M., Kyritsis, A., and Chader, G. J. (1984). Interaction of neuropeptides and cultured glial (Müller) cells of the chick retina. Elevation of intracellular cyclic AMP by vasoactive intestinal peptide and glucagon. *J. Neurochem.* **43,** 199–203.

Kohnken, R. E., Eadie, D. M., Revzin, A., and McConnell, D. G. (1981a). The light-activated GTP-dependent cyclic GMP phosphodiesterase complex of bovine retinal rod outer segments. Dark resolution of the catalytic and regulatory proteins. *J. Biol. Chem.* **256,** 12502–12509.

Kohnken, R. E., Eadie, D. M., and McConnell, D. G. (1981b). The light-activated GTP dependent cyclic GMP phosphodiesterase complex of bovine retinal rod outer segments. Reconstitution from catalytic and regulatory proteins in the presence of membranes depleted of soluble proteins. *J. Biol. Chem.* **256,** 12510–12516.

Krishna, G., Krishnan, N., Fletcher, R. T., and Chader, G. (1976). Effects of light on cyclic GMP metabolism in retinal photoreceptors. *J. Neurochem.* **27,** 717–722.

Krishnan, N., Fletcher, R. T., Chader, G. J., and Krishna, G. (1978). Characterization of guanylate cyclase of rod outer segments of the bovine retina. *Biochim. Biophys. Acta* **523,** 506–515.

Kühn, H. (1978). Light-regulated binding of rhodopsin kinase and other proteins to cattle photoreceptor membranes. *Biochemistry* **17,** 4389–4395.

Kühn, H. (1980a). Light-induced reversible binding of proteins to bovine photoreceptor membranes.

Influence of nucleotides. *In* "Neurochemistry of the Retina" (N. Bazan and R. N. Lolley, eds.), pp. 269–286. Pergamon, Oxford.

Kühn, H. (1980b). Light- and GTP-regulated interaction of GTPase and other proteins with bovine photoreceptor membranes. *Nature (London)* **283,** 587–589.

Kühn, H. (1982). Light-regulated binding of proteins to photoreceptor membranes and its use for the purification of several rod cell proteins. *In* "Methods in Enzymology" (L. Packer, ed.), Vol. 81, pp. 556–564. Academic Press, New York.

Kühn, H. (1984). Interactions between photoexcited rhodopsin and light-activated enzymes in rods. *In* "Progress in Retinal Research" (N. Osborne and G. Chader, eds.), pp. 123–156. Pergamon, Oxford.

Kühn, H., and Bader, S. (1976). The rate of rhodopsin phosphorylation in isolated retinas of frog and cattle. *Biochim. Biophys. Acta* **428,** 13–18.

Kühn, H., and Dreyer, W. J. (1972). Light-dependent phosphorylation of rhodopsin by ATP. *FEBS Lett.* **20,** 1–6.

Kühn, H., McDowell, J. H., Leser, K.-H., and Bader, S. (1977). Phosphorylation of rhodopsin as a possible mechanism of adaptation. *Biophys. Struct. Mech.* **3,** 175–180.

Kühn, H., Bennett, N., Michel-Villaz, M., and Chabre, M. (1981). Interaction between photoexcited rhodopsin and GTP-binding protein. Kinetic and stoichiometric analyses from light-scattering changes. *Proc. Natl. Acad. Sci. U.S.A.* **78,** 6873–6877.

Kuwamaya, Y., Ishimoto, I., Fukuda, M., Shiosaka, Y. S. S., Inagaki, S., Senba, E., Sakanaka, M., and Tagaki, H. (1982). Overall distribution of glucagon-like immunoreactivity in the chicken retina: An immunohistochemical study with flat-mounts. *Invest. Ophthalmol. Visual Sci.* **22,** 681–686.

LaVail, M. M., and Sidman, R. L. (1974). Retinal degeneration in the mouse. *Arch. Ophthalmol.* **91,** 394–400.

LaVail, M. M., Sidman, R. L., and O'Neil, D. (1972). Photoreceptor-pigment epithelial cell relationships in rats with inherited retinal degeneration: Radio-autographic and electron microscope evidence for a dual source of extra lamellar material. *J. Cell Biol.* **53,** 185–209.

Lee, R. H., Brown, B. M., and Lolley, R. N. (1981). Protein kinases of retinal rod outer segments: Identification and partial characterization of cyclic nucleotide dependent protein kinase and rhodopsin kinase. *Biochemistry* **20,** 7532–7538.

Liebman, P. (1978). Rod disk calcium movement and transduction: A poorly illuminated story. *Ann. N.Y. Acad. Sci.* **307,** 642–644.

Liebman, P. A., and Pugh, E. N. (1979). The control of phosphodiesterase in rod disk membranes-kinetics, possible mechanisms and significance for vision. *Vision Res.* **19,** 375–380.

Liebman, P. A., and Pugh, E. N. (1982). Gain, speed and sensitivity of GTP binding vs PDE activation in visual excitation. *Vision Res.* **22,** 1475–1480.

Lipton, S. A., and Dowling, J. E. (1981). The relation between Ca^{2+} and cyclic GMP in rod photoreceptors. *Curr. Top. Membr. Trans.* **15,** 381–392.

Lipton, S. A., Ostroy, S. E., and Dowling, J. E. (1977a). Electrical and adaptive properties of rod photoreceptors in *Bufo marinus.* I. Effects of altered extracellular Ca^{2+} levels. *J. Gen. Physiol.* **70,** 747–770.

Lipton, S. A., Rasmussen, H., and Dowling, J. E. (1977b). Electrical and adaptive properties of rod photoreceptors in *Bufo marinus.* II. Effects of cyclic nucleotides and prostaglandins. *J. Gen. Physiol.* **70,** 771–791.

Lisman, J., Levine, E., Crain, E., and Robinson, P. (1985). Non-transducing rhodopsin. *Invest. Ophthalmol. Visual Sci. Suppl.* **26,** 43.

Liu, Y. P., and Wong, V. G. (1979). A heat-stable inhibitor of cyclic 3′,5′-nucleotide phosphodiesterase. *Biochim. Biophys. Acta* **583,** 273–278.

Liu, Y. P., Krishna, G., Aguirre, G., and Chader, G. J. (1979). Involvement of cyclic GMP phosphodiesterase activator in an hereditary retinal degeneration. *Nature (London)* **280,** 62–64.

Lolley, R. N., and Farber, D. B. (1975). Cyclic nucleotide phosphodiesterase in dystrophic rat retinas: Guanosine 3′,5′-cyclic monophosphate anomalies during photoreceptor cell degeneration. *Exp. Eye Res.* **20,** 585–597.

Lolley, R. N., and Farber, D. B. (1976). A proposed link between debris accumulation, guanosine 3′,5′-cyclic monophosphate changes and photoreceptor cell degeneration in retina of RCS rats. *Exp. Eye Res.* **22,** 477–486.

Lolley, R. N., and Farber, D. B. (1980). Cyclic GMP metabolic defects in inherited disorders of *rd* mice and RCS rats. *In* "Neurochemistry of the Retina" (N. G. Bazan and R. N. Lolley, eds.), pp. 427–441. Pergamon, Oxford.

Lolley, R. N., and Racz, E. (1982). Calcium modulation of cyclic GMP synthesis in rat visual cells. *Vision Res.* **22,** 1481–1486.

Lolley, R. N., Schmidt, S. Y., and Farber, D. B. (1974). Alterations in cyclic AMP metabolism associated with photoreceptor cell degeneration in the C3H mouse. *J. Neurochem.* **22,** 701–707.

Lolley, R. N., Brown, B. M., and Farber, D. B. (1977a). Protein phosphorylation in rod outer segments from bovine retina: Cyclic nucleotide-activated protein kinase and its endogenous substrate. *Biochem. Biophys. Res. Commun.* **78,** 572–578.

Lolley, R. N., Farber, D. B., Rayborn, M. E., and Hollyfield, J. G. (1977b). Cyclic GMP accumulation causes degeneration of photoreceptor cells: Simulation of an inherited disease. *Science* **196,** 664–666.

McDowell, J. H., and Kuhn, H. (1977). Light-induced phosphorylation of rhodopsin in cattle photoreceptor membrane: Substrate activation and inactivation. *Biochemistry* **16,** 4054–4060.

Makman, M. H., Brown, J. H., and Mishra, R. K. (1975). Cyclic AMP in retina and caudate nucleus: Influence of dopamine and other agents. *Adv. Cyclic Nucleotide Res.* **5,** 661–679.

Manthorpe, M., and McConnell, D. G. (1975). Cyclic nucleotide phosphodiesterases associated with bovine retinal outer-segment fragments. *Biochim. Biophys. Acta* **403,** 438–445.

Matthews, H. R., Torre, V., and Lamb, T. D. (1985). Effects on the photoresponse of calcium buffers and cyclic GMP incorporated into the cytoplasm of retinal rods. *Nature (London)* **313,** 582–585.

Merin, E., and Auerbach, E. (1976). Retinitis pigmentosa. *Survey Ophthalmol.* **20,** 303–346.

Meyertholen, I., Wilson, M., and Ostroy, S. E. (1980). Removing bicarbonate/CO_2 reduces the cGMP concentration of the vertebrate receptor to the levels normally observed on illumination. *Biochem. Biophys. Res. Commun.* **96,** 785–792.

Miki, N., Baraban, J. M., Keirns, J. J., Boyce, J. J., and Bitensky, M. W. (1975). Purification and properties of the light-activated cyclic nucleotide phosphodiesterase of rod outer segments. *J. Biol. Chem.* **250,** 6320–6327.

Miller, J. A., Paulsen, R., and Bownds, M. D. (1977). Control of light-activated phosphorylation in frog photoreceptor membranes. *Biochemistry* **16,** 2633–2639.

Miller, J. L., and Dratz, E. A. (1984). Phosphorylation at sites near rhodopsin's carboxyl-terminus regulates light initiated cGMP hydrolysis. *Vision Res.* **24,** 1509–1521.

Miller, S., and Farber, D. (1984).Cyclic AMP modulation of ion transport across frog retinal pigment epithelium. Measurements in the short-circuit state. *J. Gen. Physiol.* **83,** 853–874.

Miller, W. H., and Nicol, G. D. (1979). Evidence that cyclic GMP regulates membrane potential in rod photoreceptors. *Nature (London)* **280,** 64–66.

Mitzel, D. L., Hall, A. I., De Vries, G. W., Cohen, A. I., and Ferrendelli, J. A. (1978). Comparison of cyclic nucleotide and energy metabolism of intact mouse retina *in situ* and *in vitro*. *Exp. Eye Res.* **27,** 27–37.

Nathanson, J. A. (1977). Cyclic nucleotides and nervous system function. *Physiol. Rev.* **57,** 157–256.

Newsome, D. A., Fletcher, R. T., Robison, W. G., Kenyon, K. R., and Chader, G. J. (1974). Effects of cyclic AMP and Sephadex fractions of chick embryo extract on cloned retinal pigmented epithelium in tissue culture. *J. Cell Biol.* **61,** 369–382.

Newsome, D. A., Fletcher, R. T., and Chader, G. J. (1980). Cyclic nucleotides vary by area in the human retina and pigmented epithelium of the human and monkey. *Invest. Ophthalmol. Visual Sci.* **19,** 864–869.

Nicol, G. D., and Miller, W. H. (1978). Cyclic GMP injected into retinal rod outer segments increases latency and amplitude of response to illumination. *Proc. Natl. Acad. Sci. U.S.A.* **75,** 5217–5220.

Noell, W. K. (1952). The impairment of visual cell structure by iodoacetate. *J. Cell. Comp. Physiol.* **40,** 25–55.

Noell, W. K. (1965). Aspects of experimental and hereditary retinal degeneration. *In* "Biochemistry of the Retina" (C. N. Graymore, ed.), pp. 51–72. Academic Press, New York.

Orr, H. T., Lowry, O. H., Cohen, A. I., and Ferrendelli, J. A. (1976). Distribution of 3′,5′-cyclic AMP and 3′,5′-cyclic GMP in rabbit retina *in vivo:* Selective effects of dark and light adaptation and ischemia. *Proc. Natl. Acad. Sci. U.S.A.* **73,** 4442–4445.

Pannbacker, R. G. (1973). Control of guanylate cyclase activity in the rod outer segment. *Science* **182,** 1138–1140.

Pannbacker, R. G. (1974). Cyclic nucleotide metabolism in human photoreceptors. *Invest. Ophthalmol.* **13,** 535–538.

Pannbacker, R. G., and Lovett, K. (1977). Localization of cyclic nucleotide phosphodiesterase activity within the bovine photoreceptor cell. *Invest. Ophthalmol. Visual Sci.* **16,** 166–168.

Pannbacker, R. G., Fleischman, D. E., and Reed, D. W. (1972). Cyclic nucleotide phosphodiesterase: High activity in a mammalian photoreceptor. *Science* **175,** 757–758.

Paulsen, R., and Bentrop, J. (1984). Reversible phosphorylation of opsin induced by irradiation of blowfly retinae. *J. Comp. Physiol.* **155,** 39–45.

Pfister, C., Kuhn, H., and Chabre, M. (1983). Interaction between photoexcited rhodopsin and peripheral enzymes in frog retinal rods. Influence on the post metarhodopsin II decay and phosphorylation rate of rhodopsin. *Eur. J. Biochem.* **136,** 489–499.

Polans, A. S., Hermolin, J., and Bownds, M. D. (1979). Light-induced dephosphorylation of two proteins in frog rod outer segments. Influence of cyclic nucleotides and calcium. *J. Gen. Physiol.* **74,** 595–613.

Polans, A. S., Kawamura, S., and Bownds, M. D. (1981). Influence of calcium on guanosine 3′,5′-cyclic monophosphate levels in frog rod outer segments. *J. Gen. Physiol.* **77,** 41–48.

Poo, M., and Cone, R. A. (1974). Lateral diffusion of rhodopsin in the photoreceptor membrane. *Nature (London)* **247,** 438–441.

Redfern, N., Israel, P., Bergsma, D., Robison, W. G., Whikehart, D., and Chader, G. (1976). Neural retinal and pigment epithelial cells in culture: Patterns of differentiation and effects of prostaglandins and cyclic AMP on pigmentation. *Exp. Eye Res.* **22,** 559–568.

Robb, R. M. (1974). Histochemical evidence of cyclic nucleotide phosphodiesterase in photoreceptor outer segments. *Invest. Ophthalmol.* **13,** 740–747.

Robb, R. M. (1978). Histochemical demonstration of cyclic guanosine 3′,5′-monophosphate phosphodiesterase activity in retinal photoreceptor outer segments. *Invest. Ophthalmol. Visual Sci.* **17,** 476–480.

Robinson, P. R., Kawamura, S., Abramson, B., and Bownds, M. D. (1980). Control of the cyclic GMP phosphodiesterase of frog photoreceptor membranes. *J. Gen. Physiol.* **76,** 631–645.

Robinson, W. E., and Hagins, W. A. (1979a). A light-activated GTPase in retinal rod outer segments. *Photochem. Photobiol.* **29,** 693.

Robinson, W. E., and Hagins, W. A. (1979b). GTP hydrolysis in intact rod outer segments and the transmitter cycle in visual excitation. *Nature (London)* **280,** 398–400.

Roof, D. J., and Heuser, J. E. (1982). Surfaces of rod photoreceptor disk membranes: Integral membrane components. *J. Cell Biol.* **95,** 487–500.

Salceda, R., VanRoosmalen, G. R. E. M., Jansen, P. A. A., Bonting, S. L., and Daemen, F. J. M. (1982). Nucleotide content of isolated bovine rod outer segments. *Vision Res.* **22,** 1469–1474.

Sanyal, S., and Jansen, H. (1981). Absence of receptor outer segments in the retina of *rds* mutant mice. *Neurosci. Lett.* **21,** 23–26.

Sanyal, S., Fletcher, R., Liu, Y., Aguirre, G., and Chader, G. (1984). Cyclic nucleotide content and phosphodiesterase activity in the *rds* mouse (020/A) retina. *Exp. Eye Res.* **38,** 247–256.

Schmidt, S. Y. (1983). Light enhances the turnover of phosphatidylinositol in rat retinas. *J. Neurochem.* **40,** 1630–1638.

Schmidt, S. Y., and Lolley, R. N. (1973). Cyclic-nucleotide phosphodiesterase: An early defect in inherited retinal degeneration of C3H mice. *J. Cell Biol.* **57,** 117–123.

Schorderet, M., and Magistretti, P. J. (1980). The isolated retina of mammals: A useful preparation for enzymatic-(adenylyl cyclase) and/or binding studies of dopamine receptors. *In* "Neurochemistry of the Retina" (N. G. Bazan and R. N. Lolley, eds.), pp. 337–354. Pergamon, Oxford.

Schroeder, W. H., and Fain, G. L. (1984). Light-dependent calcium release from photoreceptors measured by laser micro-mass analysis. *Nature (London)* **309,** 268–270.

Schubert, G., and Bornschein, H. (1951). Specific damage to retinal elements by iodine acetate. *Experientia* **7,** 461–462.

Schwarcz, R., and Coyle, J. T. (1976). Adenylate cyclase activity in chick retina. *Gen. Pharmacol.* **7,** 349–354.

Shichi, H., and Somers, R. L. (1978). Light-dependent phosphorylation of rhodopsin: Purification and properties of rhodopsin kinase. *J. Biol. Chem.* **253,** 7040–7046.

Shinozawa, T., Sen, I., Wheeler, G., and Bitensky, M. W. (1979). Predictive value of the analogy between hormone-sensitivity adenylate cyclase and light-sensitive cyclic GMP phosphodiesterase: A specific role for a light-sensitive GTPase as a component in the activation sequence. *J. Supramol. Struct.* **10,** 185–190.

Shuster, T. A., and Farber, D. B. (1984). Phosphorylation in sealed rod outer segments: Effects of cyclic nucleotides. *Biochemistry* **23,** 515–521.

Shuster, T. A., and Farber, D. B. (1986). Rhodopsin phosphorylation in developing normal and degenerative mouse retinas. *Invest. Ophthalmol. Visual Sci.,* in press.

Sidman, R. L., and Green, M. C. (1965). Retinal degeneration in the mouse; location of the *rd* locus in linkage group XVII. *J. Hered.* **56,** 23–29.

Sitaramayya, A., and Liebman, P. A. (1983a). Mechanism of ATP quench of phosphodiesterase activation in rod disc membranes. *J. Biol. Chem.* **258,** 1205–1209.

Sitaramayya, A., and Liebman, P. A. (1983b). Phosphorylation of rhodopsin and quenching of cyclic GMP phosphodiesterase activation by ATP at weak bleaches. *J. Biol. Chem.* **258,** 12106–12109.

Sitaramayya, A., Virmaux, N., and Mandel, P. (1977). On the mechanism of light activation of retinal rod outer segments cyclic GMP phosphodiesterase (light activation-influence of bleached rhodopsin and KF-deinhibition). *Exp. Eye Res.* **25,** 163–169.

Smith, H. G., Fager, R. S., and Litman, B. J. (1977). Light-activated segment disks. *Biochemistry* **16,** 1399–1405.

Sonohara, O., and Shiose, Y. (1968). Electron microscopic study of the visual cell of inherited retinal dystrophic mice. *Folia Ophthalmol. Jpn.* **19,** 77–86.

Sorbi, R. T., and Cavaggioni, A. (1975). Effects of strong illumination on the ion efflux from the isolated discs of frog photoreceptors. *Biochim. Biophys. Acta* **394,** 577–585.

Spano, P. F., Kumakura, K., and Trabucchi, M. (1976). Dopamine-sensitive adenylate cyclase in the retina: A point of action for D-LSD. *Adv. Biochem. Psychopharmacol.* **15,** 357–365.

Szuts, E. Z., and Cone, R. A. (1977). Calcium content of frog rod outer segments and discs. *Biochim. Biophys. Acta* **468,** 194–208.

Takemoto, L. J., Hansen, J., Farber, D. B., Souza, D. W., and Takemoto, D. J. (1984). Cyclic GMP phosphodiesterase from bovine retina: Evidence for interspecies conservation. *Biochem. J.* **217,** 129–133.

Takeuchi, Y. K., and Kajishima, T. (1976). Inhibitory effects of dibutyryl cyclic AMP and theophylline on the melanosome transformation in the embryonic chick pigmented retina cultured *in vitro*. *Dev. Biol.* **53,** 178–189.

Toibana, M., Tsukahara, I., and Ogawa, K. (1982). Cytochemical demonstration of guanylate cyclase activity in retinal photoreceptors with special reference to changes under light and dark adaptation. *Acta Histochem. Cytochem.* **15,** 5–20.

Trabucchi, M., Govoni, S., Tonon, G. C., and Spano, P. F. (1976). Dopamine receptor supersensitivity in rat retina after light deprivation. *In* "Catecholamines and Stress" (E. Usdin, ed.), pp. 225–234. Pergamon, Oxford.

Troyer, E. W., Hall, I. A., and Ferrendelli, J. A. (1978). Guanylate cyclases in CNS: Enzymatic characteristics of soluble and particulate enzymes from mouse cerebellum and retina. *J. Neurochem.* **31,** 825–833.

Tunnicliff, G. (1977). Increase in the tyrosine hydroxylase activity of the chick retina by an apparent phosphorylation. *Union Med. Can.* **106,** 472–474.

Uchida, S., Wheeler, G. L., Yamazaki, A., and Bitensky, M. W. (1981). A GTP-protein activator of phosphodiesterase which forms in response to bleached rhodopsin. *J. Cyclic Nucleotide Res.* **7,** 95–104.

Ueno, S., Bambauer, H. J., Umar, H., and Ueck, M. (1984). A new histo- and cytochemical method for demonstration of cyclic 3′,5′-nucleotide phosphodiesterase activity in retinal rod photoreceptor cells in the rat. *Histochemistry* **81,** 445–451.

Uhl, R., and Abrahamson, E. W. (1981). Dynamic processes in visual transduction. *Chem. Rev.* **81,** 291–312.

Ulshafer, R. H., Garcia, C. A., and Hollyfield, J. C. (1980). Sensitivity of photoreceptors to elevated levels of cGMP in the human retina. *Invest. Ophthalmol. Visual Sci.* **19,** 1236–1241.

van Nie, R., Ivanyi, D., and Demant, P. (1978). A new H-2 linked mutation, *rds,* causing retinal degeneration in the mouse. *Tissue Antigens* **12,** 106–108.

Virmaux, N., Nullans, G., and Goridis, C. (1976). Guanylate cyclase in vertebrate retina: Evidence for specific association with rod outer segments. *J. Neurochem.* **26,** 233–235.

Waloga, G., and Anderson, R. E. (1985). Effects of inositol-1,4,5-triphosphate injections into salamander rods. *Biochem. Biophys. Res. Commun.* **126,** 59–62.

Waloga, G., and Brown, J. E. (1979). Effects of cyclic nucleotides and calcium ions on *Bufo* rods *Invest. Ophthalmol. Visual Sci. (Suppl.)* **18,** 5.

Walter, A. E., Shuster, T. A., and Farber, D. B. (1985). Light-induced phosphorylation of cone visual pigments in lizard retina. *Invest. Ophthalmol. Visual Sci. (Suppl.)* **26,** 291.

Watling, K. J., and Dowling, J. E. (1983). Effects of vasoactive intestinal peptide and other peptides on cyclic AMP accumulation in intact pieces and isolated horizontal cells of the teleost retina. *J. Neurochem.* **41,** 1205–1213.

Watling, K. J., Dowling, J. E., and Iversen, L. L. (1980). Dopaminergic mechanisms in the carp retina: Effects of dopamine, K^+ and light on cyclic AMP synthesis. *In* "Neurochemistry of the Retina" (N. G. Bazan and R. N. Lolley, eds.), pp. 519–537. Pergamon, Oxford.

Weller, M., Goridis, C., Virmaux, N., and Mandel, P. (1975a). A hypothetical model for the possible involvement of rhodopsin phosphorylation in light and dark adaptation in the retina. *Exp. Eye Res.* **21,** 405–408.

Weller, M., Virmaux, N., and Mandel, P. (1975b). Light-stimulated phosphorylation of rhodopsin in the retina: The presence of a protein kinase that is specific for photobleached rhodopsin. *Proc. Natl. Acad. Sci. U.S.A.* **72,** 381–385.

Werblin, F. S. (1974). Organization of the vertebrate retina: Receptive fields and sensitivity control. *In* "The Eye" (H. Davson and L. T. Graham, eds.), pp. 257–281. Academic Press, New York.

Wheeler, G. L., and Bitensky, M. W. (1977). A light-activated GTPase in vertebrate photoreceptors: Regulation of light-activated cyclic GMP phosphodiesterase. *Proc. Natl. Acad. Sci. U.S.A.* **74,** 4238–4242.

Wheeler, G. L., Matuo, Y., and Bitensky, M. W. (1977). Light-activated GTPase in vertebrate photoreceptors. *Nature (London)* **269,** 822–824.

Wiglusz, Z. (1973). Investigations on the role of cyclic 3′,5′-AMP and active sodium transport in generation of action potential *in vivo.* (in Polish) *Acta Biol. Med. Soc. Sci. Gedan.* **17,** 7–63.

Wiglusz, Z. (1975a). Investigations of the role of cyclic AMP in generation of action potential of frog isolated retina. (in Polish) *Klin. Oczna.* **45,** 765–771.

Wiglusz, Z. (1975b). The influence of pharmacological agents acting by lowering the membrane potential difference on the amplitude of b wave in ERG of frog isolated eye. (in Polish) *Klin. Oczna.* **45,** 1297–1304.

Wilden, U., and Kuhn, H. (1982). Light-dependent phosphorylation of rhodopsin: Number of phosphorylation sites. *Biochemistry* **21,** 3014–3022.

Williams, D. S., Colley, N. J., Anderson, D. H., Farber, D. B., and Fisher, S. K. (1986a). *In vitro* maintenance of a pure-cone retina. *Invest. Ophthalmol. Visual Sci.,* in press.

Williams, D. S., Colley, N. J., and Farber, D. B. (1986b). Photoreceptor degeneration of a pure cone retina: Effects of cyclic nucleotides and inhibitors of phosphodiesterase and protein synthesis. *Invest. Ophthalmol. Visual Sci.,* submitted.

Winkler, B., Fletcher, R., and Chader, G. (1984). Effects of diamide on cyclic nucleotide levels in rat retina. *Invest. Ophthalmol. Visual Sci.* **25,** 461–463.

Woodford, B., Liu, Y., Fletcher, R., Chader, G., Farber, D., Santos-Anderson, R., and T'so, M. (1982). Cyclic nucleotide metabolism in inherited retinopathy in Collies: A biochemical and histochemical study. *Exp. Eye Res.* **34,** 703–714.

Woodruff, M. L., and Bownds, M. D. (1979). Amplitude, kinetics and reversibility of a light-induced decrease in guanosine 3′,5′-cyclic monophosphate in frog photoreceptor membranes. *J. Gen. Physiol.* **73,** 629–653.

Woodruff, M. L., and Fain, G. L. (1982). Ca^{2+}-dependent changes in cyclic GMP levels are not correlated with opening and closing of the light-independent permeability of toad photoreceptors. *J. Gen. Physiol.* **80,** 537–555.

Woodruff, M. L., Bownds, M. D., Green, S. H., Morrisey, J. L., and Shedlovsky, A. (1977). Guanosine 3′,5′-cyclic monophosphate and the *in vitro* physiology of frog photoreceptor membranes. *J. Gen. Physiol.* **69,** 667–679.

Yamamoto, K., and Shichi, H. (1983). Rhodopsin phosphorylation occurs at metarhodopsin II level. *Biophys. Struct. Mech.* **9,** 259–267.

Yamazaki, A., Sen, I., Bitensky, M. W., Casnellie, J., and Greengard, P. (1980). Cyclic GMP-specific, high affinity, non-catalytic binding sites on light-activated phosphodiesterase. *J. Biol. Chem.* **255,** 11619–11624.

Yamazaki, A., Bartucca, F., Ting, A., and Bitensky, M. W. (1982a). Reciprocal effects of an inhibitory factor on catalytic activity and non-catalytic cGMP binding sites in rod phosphodiesterase. *Proc. Natl. Acad. Sci. U.S.A.* **79,** 3702–3706.

Yamazaki, A., Miki, N., and Bitensky, M. W. (1982b). Purification and characterization of a light-activated rod outer segment PDE. *In* "Methods in Enzymology" (L. Packer, ed.), Vol. 81, pp. 526–532. Academic Press, New York.

Yamazaki, A., Stein, P. J., Chernoff, N., and Bitensky, M. W. (1983). The regulatory mechanism of rod cGMP phosphodiesterase. *Invest. Ophthalmol. Visual Sci.* (*Suppl.*) **24,** 245.

Yau, K.-W., and Nakatani, K. (1985). Light-induced reduction of cytoplasmic free calcium in retinal rod outer segment. *Nature (London)* **313,** 579–582.

Yee, R., and Liebman, P. A. (1978). Light-activated phosphodiesterase of the rod outer segment. Kinetics and parameters of activation and deactivation. *J. Biol. Chem.* **253,** 8902–8909.

Yeh, H. H., Batelle, B.-A., and Puro, D. G. (1983). Maturation of neurotransmission at cholinergic synapses formed in culture by rat retinal neurons: Regulation by cyclic AMP. *Dev. Brain Res.* **10,** 63–72.

Yoshikami, S., and Hagins, W. A. (1971). Ionic basis of dark current and photocurrent of retinal rods. *Biophys. J.* **10,** 60a.

Yoshikami, S., George, J. S., and Hagins, W. A. (1980). Light-induced calcium fluxes from outer segment layer of vertebrate retinas. *Nature (London)* **286,** 395–398.

Zimmerman, W. F., and Daemen, F. J. (1976). Distribution of enzyme activities in subcellular fractions of bovine retina. *J. Biol. Chem.* **25,** 4700–4705.

PHOTOSENSITIVE MEMBRANE TURNOVER: DIFFERENTIATED MEMBRANE DOMAINS AND CELL–CELL INTERACTION

JOSEPH C. BESHARSE

Department of Anatomy and Cell Biology
Emory University School of Medicine
Atlanta, Georgia

I. Introduction

Recently, cell biologists have focused on the fact that cell function often depends on the generation of membrane domains with distinctive structural prop-

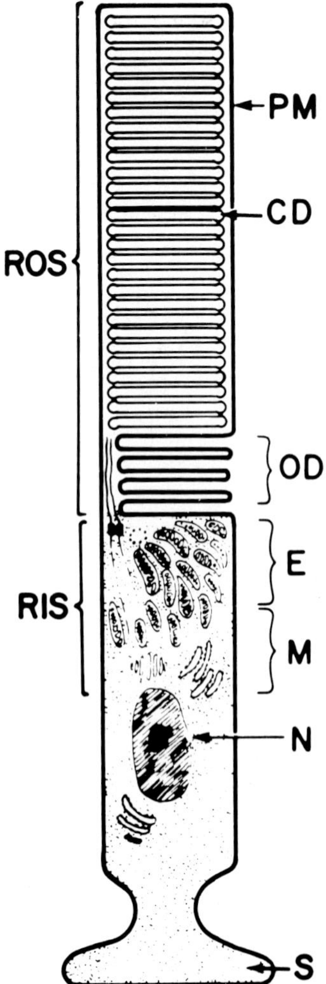

FIG. 1. Diagram illustrating the highly polarized structure of a rod photoreceptor. Note that cone photoreceptors differ principally in that all or most disks maintain continuity between intradiscal space and extracellular space, like that seen in the region of open disks in this diagram. Abbreviations: CD, closed disk; E, ellipsoid; M, myoid; N, nucleus; OD, open disks; PM, plasma membrane; RIS, rod inner segment; ROS, rod outer segment; S, synaptic terminal. (Reproduced from Besharse *et al.*, 1977a, *The Journal of Cell Biology*, **75,** 507–527 by copyright permission of The Rockefeller University Press.)

erties. Central to this concept are the mechanisms by which cells generate and maintain their distinctive regional properties while providing mechanisms that ensure turnover. In this regard photosensitive membrane turnover, as it occurs in vertebrate photoreceptors, has unique and analytically useful properties. First, photoreceptors are highly polarized so that the processes of synthesis, transport, and assembly of membrane components are vectorially organized along the axis of the cell. Second, the photosensitive outer segment is relatively simple structurally and is dominated by a single intrinsic membrane protein, the apoprotein of visual pigment. Third, the photosensitive membrane is organized into discrete lamellae or disks which in rod cells become separated from the plasma membrane. Fourth, degradation of opsin, and to some extent other membrane components as well, occurs within a separate cell type, the retinal pigment epithelial cell.

Vertebrate photoreceptors consist essentially of a cell body, a light-sensitive outer segment and a synaptic terminal (Fig. 1). The cell body is divided into nuclear, myoid, and ellipsoid regions and contains much of the cells metabolic machinery. It is confluent at opposite ends with two major functional domains, the synaptic terminal and outer segment. The outer segment is a membranous organelle derived from plasma membrane and designed for photon capture and transduction of the photochemical event of vision into a voltage change in the cell's plasma membrane. It is highly enriched in a single intrinsic protein, the visual pigment apoprotein, that comprises as much as 90% of the disk membrane protein (Hall *et al.*, 1969; Papermaster and Dryer, 1974; Robinson *et al.*, 1972; Daeman, 1973). This protein, opsin,[1] is a transmembrane glycoprotein that reversibly combines with retinaldehyde to form the photosensitive visual pigment (Hargrave, 1982, and this volume). In addition to opsin, photosensitive membranes also contain a high molecular weight intrinsic glycoprotein, an array of membrane-associated proteins which are probably involved in visual excitation (see articles by Hargrave and Farber and Shuster, this volume), and an array of highly unsaturated phospholipids (Anderson and Maude, 1970; Fliesler and Anderson, 1983). The latter feature is responsible for the high degree of fluidity of disk membranes, compared to that of most other plasma membranes (Poo and Cone, 1974; Liebman and Entine, 1974).

Structurally, the photosensitive membrane system takes on a flattened, lamellar organization with each lamella or disk exhibiting two domains: a flat surface and an edge. The high molecular weight glycoprotein appears to be confined to the latter (Papermaster *et al.*, 1978b). In rods the disks are separated from and surrounded by the plasma membrane, whereas in cones many or all of the disks remain confluent with the plasma membrane (Cohen, 1968; Anderson *et al.*, 1978). The plasma membrane of the outer segment is confluent with the cell body via the membrane of a connecting cilium. The latter is a nonmotile, sensory cilium with structural features comparable to those of the proximal or

[1]Throughout this review "opsin" refers to the visual pigment apoprotein regardless of the cell type (rod or cone) or absorption properties of the pigment. Opsins may combine with 11-*cis*-retinal to form rhodopsin or with 11-*cis*-dehydroretinal to form porphyropsin (Bridges, 1972). The different absorption properties of cone pigments are thought to result from variations in opsin structure (Wald, 1968; Nathans and Hogness, 1983).

transitional zone of motile cilia throughout the animal kingdom (Röhlich, 1975). This structure represents the zone of demarcation between the outer segment and cell body. The latter contains relatively little opsin but does contain a Na^+,K^+-ATPase (Stirling and Lee, 1980) which is repsonsible for maintenance of membrane potential (Hagins, 1972). It has been suggested that the cilium may be involved in both generation and maintenance of the distinctive properties of inner and outer segment membranes.

Distally, outer segments are intimately associated with interdigitating apical processes of the pigment epithelium. This monolayer epithelium intervenes between the photoreceptors and the choriocapillaris, and constitutes one aspect of the blood–retinal barrier (Peyman and Bok, 1972; Raviola, 1977). This intimate association is required for both photoreceptor development and function. Outer segments fail to form embryonically in the absence of pigment epithelium (Hollyfield and Witkovsky, 1974), and they eventually degenerate when separated from the pigment epithelium in retinal detachment (Kroll and Machemer, 1968). Among the major functions of pigment epithelium are homeostasis of the photoreceptor microenvironment (Steinberg and Miller, 1979), recycling and metabolism of vitamins A in support of the visual pigment cycle of photoreceptors (Bridges, 1976), and the disposal of effete photoreceptor disks in support of outer segment turnover (Young and Bok, 1969).

The purpose of this article is to review our current understanding of the cellular and molecular aspects of photosensitive membrane turnover. The analysis can be conveniently divided into two separate areas. First is photosensitive membrane assembly, which involves those events localized in the inner segment and proximal outer segment (Section II). Second is photoreceptor disk shedding, which includes those phenomena occurring at the distal end of the outer segment that collectively result in detachment and disposal of disk membranes (Section III). In addition to reviewing cellular mechanisms, it is important to realize that membrane turnover is temporally organized (LaVail, 1976). In recent years this has led to a variety of studies which have attempted to understand the regulation, particularly of disk shedding. Recent advances in this area will be reviewed in Section IV.

II. Synthesis and Assembly of Photosensitive Membrane

The early work of Young (1967) and later of Hall and colleagues (1969) and Young and Bok (1969) elegantly established that opsin in disk membranes of rod outer segments (ROS) is renewed by a process involving incorporation into basal disks proximally, displacement of disk membranes distally, and eventual detachment and phagocytosis by adjacent pigment epithelial cells (Fig. 2). The correct

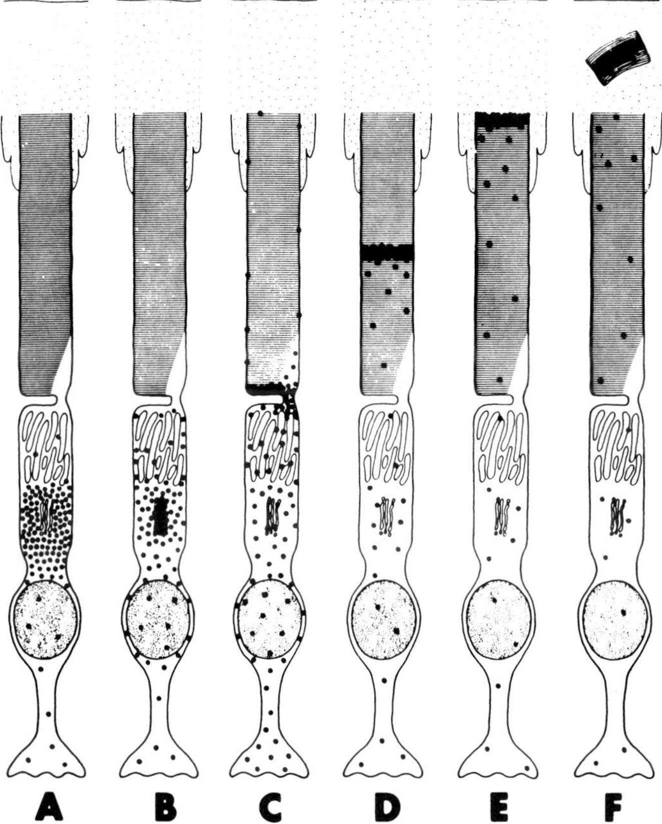

FIG. 2. Diagram illustrating the renewal of rod outer segment membrane as revealed in the classic autoradiographic studies of Young (1967) and Young and Bok (1969). (A–C) Radioactive protein synthesized in the inner segment and represented as black dots is transferred to the base of the outer segment where it is incorporated into forming disks. (D and E) Continued incorporation of unlabeled protein into forming disks displaces the labeled disks distally. (F) Disks at the distal tip are discarded (shed) and phagocytosed by adjacent pigment eithelium. (From Young, 1976, with permission of the C. V. Mosby Company, St. Louis, Missouri.)

interpretation of those studies was directly related to the discrete, isolated nature of rod disk membranes which trapped radiolabeled opsin for the duration of its sojourn through the outer segment. Although we now understand that cone photoreceptors exhibit an analogous process of turnover involving disk formation and shedding (Anderson and Fisher, 1975, 1976), this was difficult to establish in the early work because distinct radioactive bands did not form in cone outer segments (Young and Droz, 1968; Young, 1971a,b; Bok and Young, 1972). This may have been due to the continuity of disk and plasma membranes (Cohen,

1968) and to mobility of newly inserted cone opsin throughout the cone outer segment (Poo and Cone, 1974).

Photosensitive membrane turnover exhibits another level of complexity with regard to its lipid component. In rod outer segments (ROS), fatty acids and glycerol are incorporated into phospholipids that are distributed throughout the ROS rather than only in newly formed disks (Bibb and Young, 1974a,b). This has led to the concept that ROS disk membranes, during their transit from base to apex, exhibit a superimposed process of lipid turnover (see Anderson *et al.*, 1980b). Thus, in considering turnover of photosensitive membrane we must consider fundamental differences between rod and cone cells related to differences in their structure, and within each cell type we must consider fundamental differences in the mechanisms of intrinsic protein and lipid turnover. Very little is known about turnover of the pool of soluble and peripheral membrane proteins of the outer segment (Bok and Young, 1972).

A. *Synthesis of Photoreceptor Opsin*

In the early studies of membrane turnover (Young and Droz, 1968; Young, 1969; Hall *et al.*, 1969), it was apparent that the bulk of the protein destined for the ROS was synthesized in association with the rough endoplasmic reticulum of the myoid region. Electron microscope autoradiography revealed initial incorporation of radioactive amino acids in the myoid (Fig. 2), followed by transfer through the Golgi complex and perimitochondrial cytoplasm in transit to the disk-forming region of the connecting cilium (Young and Droz, 1968). Although it was initially proposed that opsin passed in a soluble form through the connecting cilium (Young, 1969) it was subsequently found that opsin, like many other membrane proteins, exhibits a transbilayer structure (reviewed by Hargrave, 1982), and in the post-Golgi transfer to the ROS remained associated with easily sedimentable particles (Papermaster *et al.*, 1975). These observations led directly to a model in which opsin remains membrane associated throughout its life in the cell (Fig. 3).

The synthesis of opsin appears to involve cotranslational insertion into the lipid bilayer. It had been proposed that synthesis of transmembrane proteins would involve the elaboration of a transient signal peptide at the amino-terminus, like that seen for many secretory proteins (Blobel *et al.*, 1979). However, the earliest cell free synthesis of opsin, using a wheat germ extract and partially purified messenger RNA, showed that the amino-terminus of the immunoprecipitable translation product exhibited sequence homology to mature ROS opsin for all radiolabeled amino acids utilized in the analysis (Schecter *et al.*, 1979; Papermaster *et al.*, 1980). Despite the lack of an amino-terminal signal, it has been shown that the cell-free translation product is transferred into dog pancreas

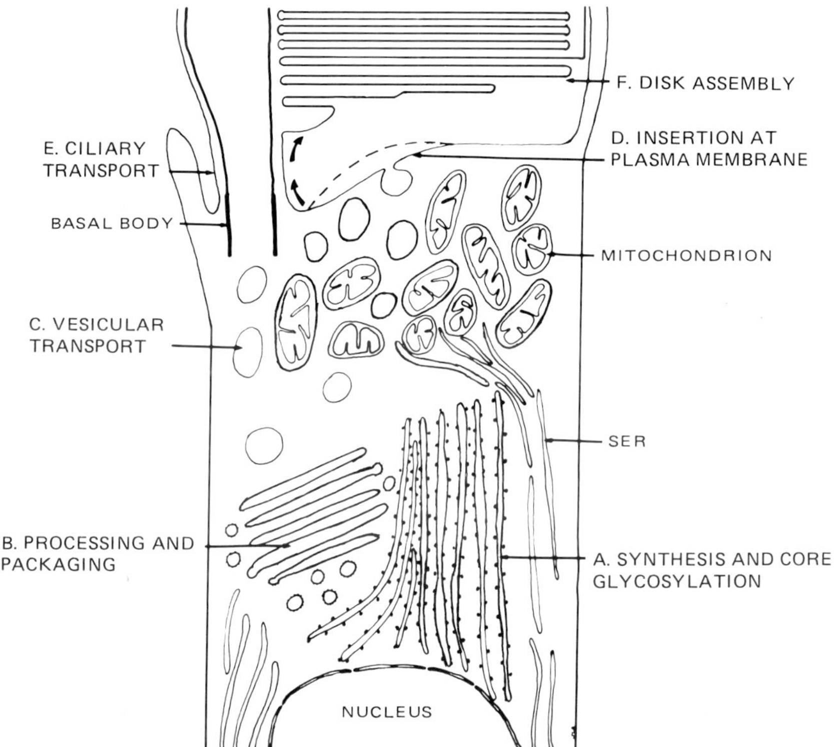

FIG. 3. Diagram illustrating the a hypothetical pathway for opsin synthesis and incorporation into disks. (A) Synthesis, membrane insertion, and core glycosylation occurs in the rough endoplasmic reticulum. (B) Processing of the oligosaccharide chain and packaging into vesicles occurs in the Golgi. (C) Post-Golgi transfer to the outer segment appears to involve vesicles which accumulate in the periciliary cytoplasm. (D) Vesicular membrane is incorporated into plasma membrane near the cilium. (E) Opsin is transferred to forming disks, possibly via the ciliary plasma membrane. (F) Disk morphogenesis occurs at the distal end of the connecting cilium. Evidence for each step is discussed in the text.

microsomal membranes and glycosylated when membranes are present during translation (Goldman and Blobel, 1981). This has led to the suggestion that opsin contains an internal signal sequence which facilitates cotranslational insertion into membrane (Goldman and Blobel, 1981). The latter point is particularly significant, given the fact that a number of membrane proteins are posttranslationally transferred into membrane (Wickner, 1979). The lack of oligosaccharides on the cell-free translation product, its cotranslational insertion into membrane *in vitro,* and the lack of posttranslational transfer *in vitro* all suggest that membrane insertion in intact cells occurs cotranslationally (Goldman and Blobel, 1981).

Opsin is a glycoprotein (Heller, 1968; Plantner and Kean, 1976; Liang *et al.*, 1979; Fukuda *et al.*, 1979) that binds the plant lectins, concanavalin A, and wheat germ agglutinin (Steinemann and Stryer, 1973; Molday and Molday, 1979). The oligosaccharides are linked to asparagine residues 2 and 15 near the amino-terminus (Hargrave, 1982). Sequence analysis indicates that the dominant oligosaccharide is of a simple form and contains only mannose and *N*-acetylglucosamine (Fukuda *et al.*, 1979; Liang *et al.*, 1979). Although opsin oligosaccharides have not been reported to contain sialic acid, galactose, or fucose, it has been shown that in intact retinas and isolated outer segments they act as an acceptor for galactosyl and fucosyl transferases (O'Brien, 1976, 1978).

A large body of evidence indicates that the formation of asparagine-linked oligosaccharides involves the transfer of a mannose-rich, core oligosaccharide from a dolichol phosphate precursor onto nascent polypeptides in the endoplasmic reticulum (Waechter and Lennarz, 1976; Struck and Lennarz, 1980). Oligosaccharides initially formed in this manner are then processed in endoplasmic reticulum and Golgi complex (Tabas *et al.*, 1978; Struck and Lennarz, 1980; Kornfeld and Kornfeld, 1980). Processing to form complex oligosaccarides involves enzymatic removal of glucose and mannose residues and the sequential addition in the Golgi complex of *N*-acetylglucosamine, galactose, and sialic acid. Synthesis of the simple oligosaccharide of opsin can be interpreted according to this scheme if it is assumed that terminal glycosylation in the Golgi of most of the molecules is limited to the addition of a single *N*-acetylglucosamine (see Figs. 3 and 4 in Hargrave, 1982). This view is supported by observations demonstrating involvement of a lipid-linked intermediate in opsin glycosylation (Kean, 1977; Plantner *et al.*, 1980; Fliesler *et al.*, 1984; see below), by the demonstration of core glycosylation of the cell-free translation product in the presence of microsomes (Goldman and Blobel, 1981), by autoradiography showing tritiated mannose incorporation only in endoplasmic reticulum, with *N*-acetylglucosamine incorporation in both endoplasmic reticulum and Golgi complex (Bok *et al.*, 1977) and by cytochemistry of intracellular lectin-binding sites (Wood and Napier-Marshall, 1985a).

Recently, tunicamycin, which blocks core glycosylation (Tkacz and Lampen, 1975), has been used to analyze the potential role of opsin oligosaccharides in transport to the outer segment (Plantner *et al.*, 1980; Fliesler *et al.*, 1983). Tunicamycin treatment blocks opsin glycosylation and provides conditions that permit synthesis of unglycosylated opsin (Plantner *et al.*, 1980; Fliesler *et al.*, 1984; Fliesler and Basinger, 1985). While cell fractionation studies of bovine retinas initially suggested transfer of radiolabeled, unglycosylated opsin to the outer segment and insertion into disks (Plantner *et al.*, 1980), recent autoradiographic studies on *Xenopus laevis* show the transfer of radiolabeled protein to the ciliary region with subsequent failure of the disk-forming mechanism (Fliesler *et al.*, 1985). Instead, membranes of the ciliary region exhibit an abnormal, vesiculated appearance. This observation has led to the conclusion that unglycosylated opsin is transported to the ciliary region but that oligosaccharides are essential for normal disk assembly. Whether the lack of oligosaccharides on opsin or on some other less abundant glycoprotein directly affects a key event in disk morphogenesis remains to be determined.

In addition to the posttranslational modification of its oligosaccharide, opsin has been shown to have an acetylated amino-terminus (Tsunasawa *et al.*, 1980; reviewed in Hargrave, 1982), and recent evidence indicates that some of the opsin may be acylated as well (O'Brien and Zatz, 1984). Acetylation blocks the amino-terminus of mature ROS opsin (Tsunasawa *et al.*, 1980) but not that of the cell-free translation product (Schecter *et al.*, 1979), suggesting that the process occurs during or after insertion into the endoplasmic reticulum (Papermaster *et al.*, 1980). Acylation is turning out to be a relatively common modification of membrane proteins. Recent experiments show the covalent incorporation of radiolabeled palmitate into opsin (O'Brien and Zatz, 1984), and dual label studies using both palmitate and leucine labels suggest that acylation occurs as an early posttranslational modification (P. O'Brien, personal communication). The role of these modifications in the function of opsin remains to be determined. Opsin also combines with 11-*cis*-retinal to form the functional photopigment (Wald, 1968) and can be phosphorylated at several sties at its cytoplasmic exposure (Wilden and Kuhn, 1982). These are both reversible posttranslational modifications directly related to opsin's role as a photopigment. Although the question may be raised as to whether opsin in transit to the ROS in the vicinity of the cilium either binds chromophore or is phosphorylated, both processes seem to occur primarily in the outer segment (Defoe and Bok, 1983; Besharse *et al.*, 1982a).

B. Opsin Transfer to the Periciliary Region

Opsin has been found in cytomembranes of the inner segment using cell fractionation (Papermaster *et al.*, 1975, 1976), autoradiographic (Young and Droz, 1968; Papermaster *et al.*, 1979, 1985), and immunocytochemical (Papermaster *et al.*, 1979, 1985; Defoe and Besharse, 1983, 1985) procedures. The earliest cell fractionation analysis (Papermaster *et al.*, 1975) showed that newly synthesized opsin remains associated with a particulate fraction during its transport through the Golgi complex and perimitochondrial cytoplasm. In a subsequent study, this was extended to include the high molecular weight intrinsic membrane protein of ROS disks, and the idea that a post-Golgi membrane fraction transports components to the ROS was further developed (Papermaster *et al.*, 1976). Subsequent immunocytochemistry using tissue embedded in cross-linked bovine serum albumin provided direct confirmation of autoradiographic

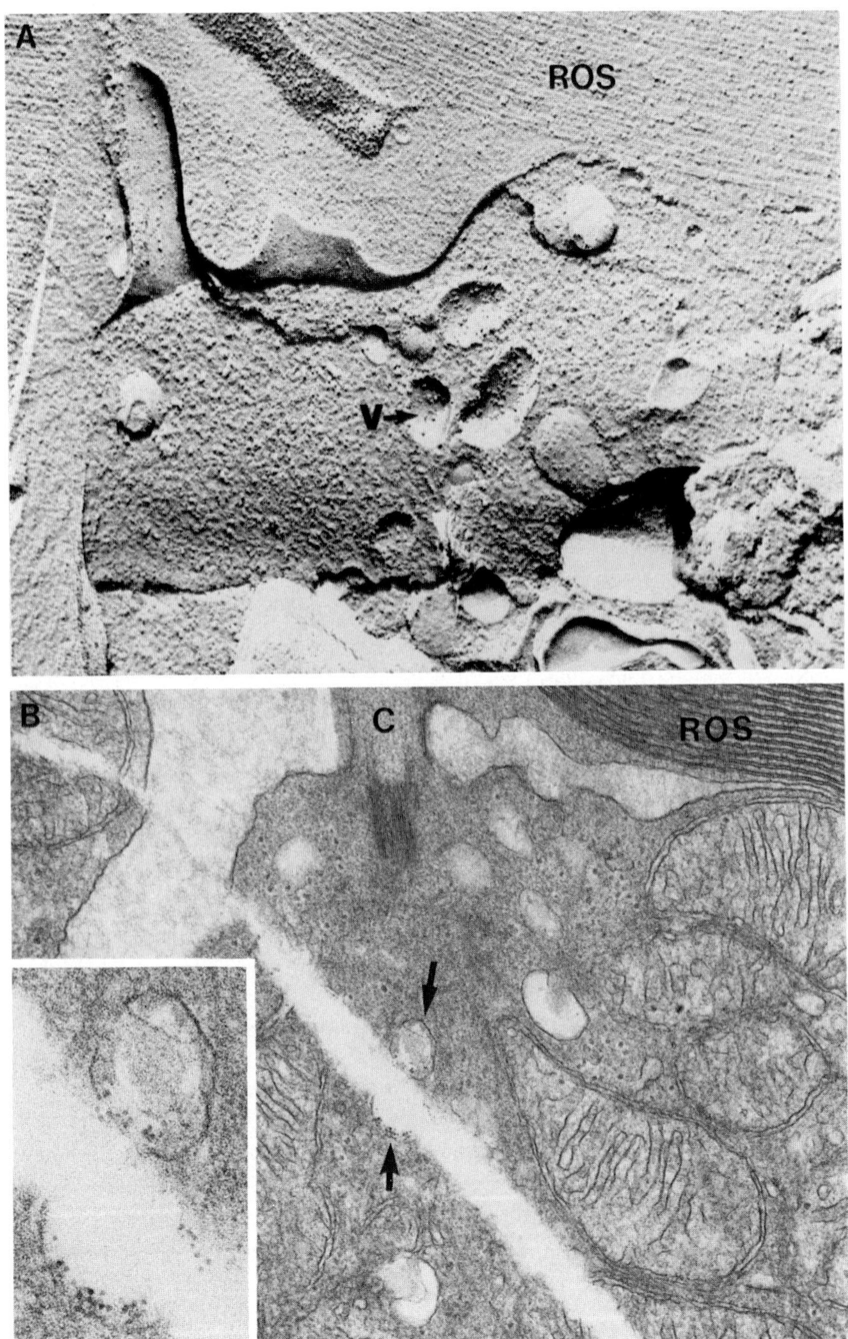

FIG. 4. Electron micrographs illustrating the periciliary region of freeze-fractured photoreceptors from *Xenopus laevis*. (A) Conventional freeze-fracture replica showing similarity of IMP of vesicles, V, and outer segment membrane as described by Besharse and Pfenninger (1980). ×62,000.

and cell-fractionation evidence for passage through the Golgi complex (Papermaster *et al.*, 1978a). The recent demonstration that monensin, a Na^+/H^+ ionophore which disrupts Golgi transport of membrane proteins (Griffiths *et al.*, 1983), blocks opsin transport to the ROS with little effect on opsin synthesis (Matheke *et al.*, 1984; Matheke and Holtzman, 1984) is consistent with the trans-Golgi pathway.

Several morphological studies (Holtzman *et al.*, 1977; Kinney and Fisher, 1978b; Besharse and Pfenninger, 1980) of anuran photoreceptors have identified a perimitochondrial and periciliary population of vesicles thought to be the morphological counterpart of the carrier vesicle suggested by the cell fractionation studies (Papermaster *et al.*, 1975, 1976). These membranes are distinct from a subellipsoidal population of smooth endoplasmic recticulum identified in recent structural studies of the inner segment (Holtzman *et al.*, 1977; Mercurio and Holtzman, 1982a,b). Furthermore, quantitative analysis of intramembrane particle (IMP) size and distribution in freeze-fractured rod photoreceptors (Besharse and Pfenninger, 1980) has provided direct evidence for a precursor–product relationship between inner segment cytomembranes and ROS disk membranes (Fig. 4A). Evidence for the vehicular role of vesicles has also come from the autoradiographic demonstration that newly synthesized protein destined for the ROS is closely associated with vesicles and from immunocytochemical localization of opsin over such vesicles using retinal tissue embedded in bovine serum albumin or Lowicryl (Papermaster *et al.*, 1979, 1985). Recently, those observations were extended using immunocytochemical detection of opsin in freeze-fractured membranes (Defoe and Besharse, 1985). In the latter study retinal tissue was freeze-fractured in liquid nitrogen and then thawed for incubation in antibody prior to embedment in epoxy resins. The major advantage was that antibody gained direct access to cytomembrane compartments along the fissures formed during fracturing, and positive contrast images typical of those in standard thin-section electron microscopy were obtained (see Fig. 4B). Opsin was localized in the membrane of trans-Golgi, perimitochondrial, and periciliary vesicles, providing confirmation of predictions about opsin's distribution based on the conventional freeze-fracture technique (Besharse and Pfenninger, 1980).

Although opsin-containing vesicles accumulate in the periciliary cytoplasm, the subsequent steps in opsin transfer to the ROS are poorly defined. It has been suggested that opsin-containing vesicles fuse with the inner segment plasmalemma adjacent to the cilium and that opsin is transferred via the ciliary plasma membrane (Kinney and Fisher, 1978b; Besharse and Pfenninger, 1980; Peters *et al.*, 1983). Related to this pathway is the recent description of a periciliary ridge

(B) Fracture-label picture showing a sectioned photoreceptor which had been fractured and labeled with antiopsin antibody. ×44,500. Note that the inset shows specific labeling of a periciliary vesicle with antiopsin. ×74,900. The same fractured vesicle is seen in the lower power field (arrows). C shows a cilium. (A is modified from Besharse, 1980, and B was prepared by Dennis M. Defoe for this review.)

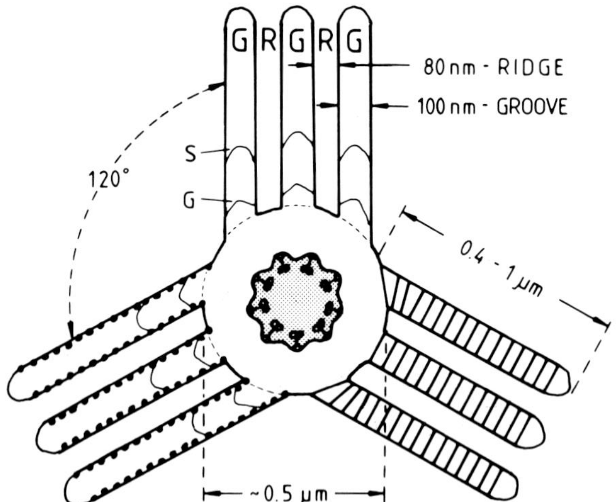

FIG. 5. Diagram of the periciliary ridge complex as seen on the apical plasma membrane of rod photoreceptors of the frog. The cilium (center) is surrounded by a groove that is continuous with sets of grooves (G) extending away from the cilium for a distance of 0.4–1 μm. Grooves are separated by ridges and generally occur as three sets of three. In some replicas edges of ridges are lined with particles (illustrated on left side) and in high-resolution images grooves are spanned by filaments (illustrated on right side). (Reproduced from Peters *et al.*, 1983, *The Journal of Cell Biology*, **96,** 265–276 by copyright permission of The Rockefeller University Press.)

complex (PRC) with ninefold radial symmetry as a distinct membrane domain adjacent to the cilium (Peters *et al.*, 1983). This structure, although adumbrated in previous thin-section (Besharse and Pfenninger, 1980) and freeze-fracture (Andrews, 1982) images, was revealed in its entirety in three-dimensional images obtained using high-resolution scanning electron microscopy (Peters *et al.*, 1983). Its essential feature is a series of nine radially arranged grooves continuous with a circumferential groove around the ciliary base (Fig. 5). It has been suggested that these grooves are the sites of vesicle fusion (Peters *et al.*, 1983). Consistent with this, Andrews (1982) has noted the similarities of these structures to the active zones of the presynaptic component of neuromuscular junctions (Heuser *et al.*, 1979) and has provided evidence of vesicle attachment sites in replicas.

In addition to their putative role in opsin transport, cytoplasmic vesicles in the inner segment may also reflect secretory and endocytic activity. For example, horseradish peroxidase (HRP) is taken up by the distal inner segment of isolated retinas (Holtzman *et al.*, 1977; Holtzman and Mercurio, 1980) where it rapidly achieves a steady-state level within the vesicle population (Hollyfield *et al.*, 1985a). HRP uptake also occurs *in vitro* during periods when net disk assembly also occurs (Besharse and Forestner, 1981). These findings imply that membrane vesicles of the ellipsoid are heterogeneous and that membrane retrieval via endocytosis normally accompanies disk assembly in photoreceptors. Whether endocytic vesicles also contain opsin has not been determined. However, it should be pointed out that vesicles containing HRP are often smaller than those implicated in opsin delivery (Holtzman and Mercurio, 1980; Besharse and Forestner, 1981) and that the IMP density in the putative periciliary carrier vesicles is only about half that found in ROS disks (Besharse and Pfenninger, 1980). Furthermore, rod photoreceptors appear to secrete interstitial retinol-binding protein (IRBP), a constituent of interphotoreceptor matrix (Hollyfield *et al.*, 1985b). They also specifically accumulate colloidal gold particles coated with IRBP by endocytosis (Hollyfield *et al.*, 1985c), suggesting that one role of endocytic vesicles may be in the retrieval and turnover of components of the interphotoreceptor matrix. It appears likely that opsin delivery to the PRC represents but one component of a complex network of vesicles involved in membrane recycling, turnover of extracellular components, and secretion.

C. Opsin Transfer to the Outer Segment

The connecting cilium of vertebrate photoreceptors delineates the distinctive structural and functional domains of inner and outer segment. Both plasma membrane and disks of outer segments are highly enriched in opsin, yet opsin is found in only small quantities and in restricted distribution in the inner segment plasma membrane (Jan and Revel, 1974; Papermaster *et al.*, 1978a; Basinger *et al.*, 1976). In contrast, the inner segment has a distinctive ultrastructure (Besharse and Pfenninger, 1980) consistent with its unique content of a Na^+,K^+-ATPase (Stirling and Lee, 1980). The continued function of photoreceptors depends on the generation and maintenance of these distinctive regions. Given the fluid nature of biological membranes (see Cherry, 1979) and the high degree of fluidity characteristic of disk membranes (Poo and Cone, 1974; Liebman and Entine, 1974; Wey *et al.*, 1981), a mechanism must exist to prevent the random insertion and redistribution of the membrane components of each region. The strategic location of the cilium and the unique structure and cytoskeletal support of its membrane suggest that it may play a role in both the generation and the maintenance of inner and outer segment membrane domains.

The mechanism of transfer of membrane components to forming disks must ultimately account for the extraordinary magnitude of disk assembly. For example, in rats, turnover data (Young, 1967) indicate that disk assembly requires the addition of about 0.1 μm^2 of membrane containing some 2000 opsin molecules, per minute. In the much larger photoreceptors of *X. laevis* which forms disks at about the same daily rate as the rat (Besharse *et al.*, 1977a), the demand is far greater. Here membrane is incorporated into disks at an average rate of about 3.2

μm^2 containing some 6×10^4 opsin molecules per minute.[2] Since all components are delivered from the inner segment, photoreceptors clearly require a highly efficient system for transfer of membrane precursors to the disk-forming region. It appears that the only direct and persistent membrane connection between inner and outer segment is that of the connecting cilium.

Although recent studies have emphasized the ciliary plasma membrane as the conduit through which opsin is transferred to forming disks (Matsusaka, 1974; Röhlich, 1975; Kinney and Fisher, 1978; Besharse and Pfenninger, 1980; Peters *et al.*, 1983), direct evidence for this route is lacking. Furthermore, its highly ordered structure (see below) and restricted dimensions, at least in frogs, may be incompatible with bulk transfer of membrane precursors via the ciliary membrane in sufficient quantities to support assembly. For example, in *X. laevis* the membrane of the short cilium (total surface area ~0.33 μm^2) would have to be exchanged on average about 10 times per minute.

Other possible routes for opsin delivery include (1) transfer through the ciliary cytoplasm (Young, 1969), (2) the formation of a cytoplasmic bridge between inner and outer segment (Richardson, 1969), or (3) the formation of extracellular membrane vesicles as vehicles between inner and outer segment. The first possibility appears to be ruled out by observations showing the transmembrane nature of opsin (see above) and a lack of immunoreactive opsin or vesicles in the ciliary cytosol (Kinney and Fisher, 1978b; Papermaster and Schneider, 1982). The second and third possibilities are difficult to confirm or eliminate. Extensive electron microscopic observations by a variety of scientists would seem to rule out persistent interconnections between inner and outer segment as claimed by Richardson (1969). Nevertheless, we have occasionally seen fusion of inner and outer segments in *X. laevis* rods (unpublished data) similar to those reported by Richardson (1969) and cannot presently distinguish between the possibilities that they are artifactual or represent transient, normal interconnections. Likewise, membranous structures are frequently found in the extracellular space around photoreceptors including the space between inner and outer segment (unpublished data). Although such structures could be taken as evidence for the third possibility, it is also possible that they form through artifactual membrane blebbing during fixation. It should be emphasized that each of these alternative pathways have in common a mechanism by which opsin could bypass the ciliary plasmalemma in transit to the ROS and that only the first of the three possibilities can be eliminated by current data.

As pointed out by Röhlich (1975), the connecting cilium of rat photoreceptors

[2]Estimates are based on published turnover data (Young, 1967; Besharse *et al.*, 1977a; Kinney and Fisher, 1978a), average dimensions of ROS disks, and an average of 20,000 opsins/μm^2 disk membrane (Corless *et al.*, 1976). The estimates represent an average over a 24-hr period. The large differences between rat and *Xenopus* is related to the much larger size of *Xenopus* disks. The estimates for *Rana*, where disk assembly is slower, would be about half those for *Xenopus*.

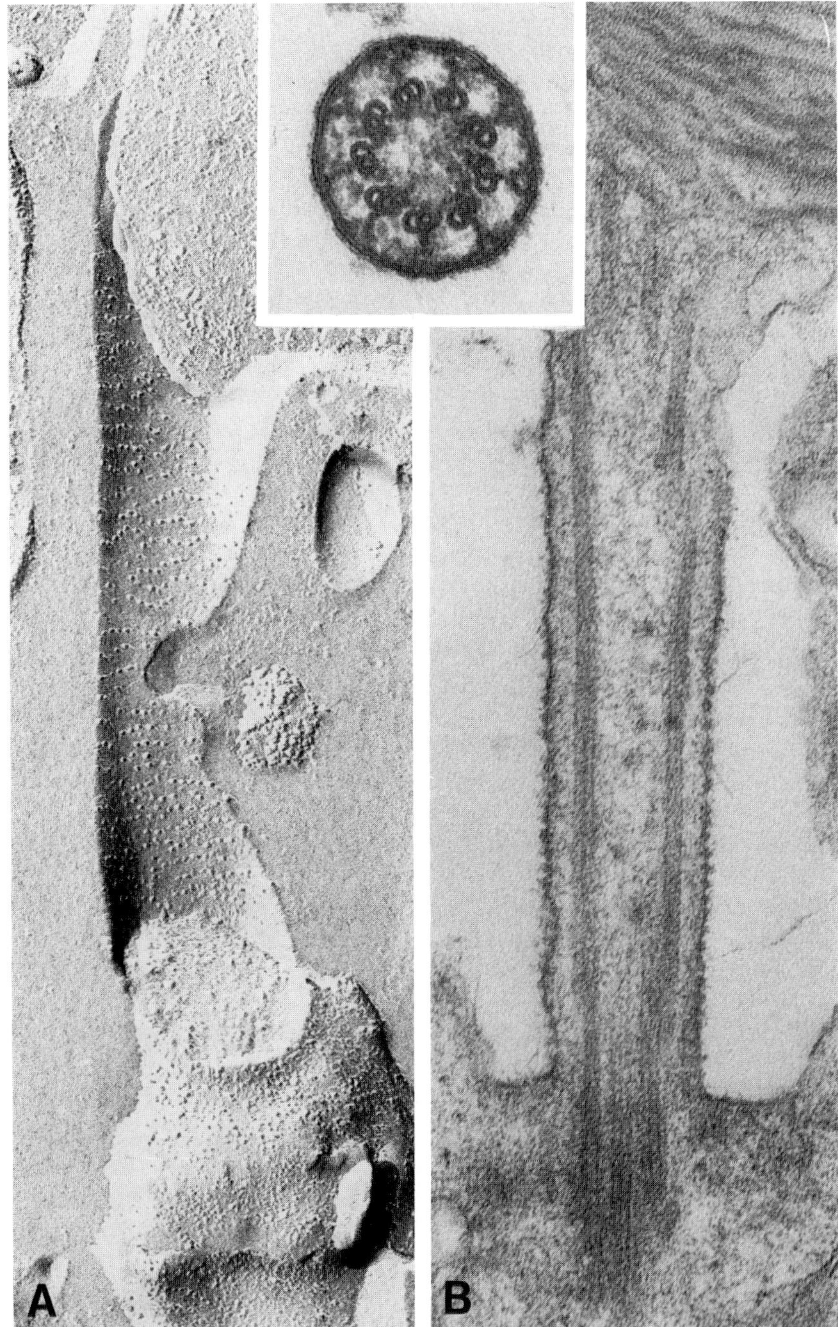

FIG. 6. Electron micrographs illustrating ciliary ultrastructure of rat photoreceptors. (A) Freeze-fracture replica illustrating ciliary necklaces on the EF leaflet. ×75,750. Note the horizontal rows of

is structurally similar to an extended transitional zone characteristic of motile cilia (Fig. 6). It lacks the central pair of microtubules seen in motile cilia (but see Matsusaka, 1976) and exhibits an array of membrane surface globules, axonemal–membrane cross-links, and ciliary necklaces characteristic of motile cilia (Gilula and Satir, 1972). Although these features are best defined in the elongated cilium of rat photoreceptors (Matsusaka, 1974; Röhlich, 1975), evidence of similar structures in the short cilia of frog photoreceptors has been presented as well (Peters *et al.*, 1983). Immunocytochemical localization of opsin in both rat and frog photoreceptors has revealed either no immunoreactive opsin or a very low concentration in the most proximal ciliary region, which contrasts sharply with that in the outer segment plasma membrane and the distal cilium from which disks form (Papermaster and Schneider, 1982; Nir and Papermaster, 1983; Papermaster *et al.*, 1985). Although the low level of antiopsin labeling in proximal ciliary membrane may result from rapid opsin transport through preferred tracks (Nir and Papermaster, 1983; Nir *et al.*, 1984), it should be emphasized that a similar distribution would be expected if opsin were delivered to the ROS by a nonciliary route (Besharse *et al.*, 1985; see above).

Recent studies have shown that a similar distribution of immunoreactive opsin characterizes photoreceptors of developing rats during the period of initial outer segment formation (Nir *et al.*, 1984; Besharse *et al.*, 1985). Both studies found opsin mainly in distal ciliary membrane, with lower concentrations detected in the proximal cilium and inner segment plasma membrane. Clearly, delivery of opsin via the inner segment plasma membrane is expected based on studies of mature photoreceptors (Peters *et al.*, 1983). However, Nir and colleagues (1984) have emphasized more extensive labeling of inner segment membrane at earlier developmental stages. In contrast, our study demonstrates that from developmental stages prior to the initial formation of disks, the distribution of opsin exhibits a pattern remarkably like that of adult photoreceptors. For example, in 3-day-old rats large numbers of photoreceptors exhibit a high density of opsin in distal ciliary membrane even though disk morphogenesis has not begun (see Fig. 7). This observation demonstrates that prior to the period of disk morphogenesis, photoreceptors are capable of concentrating opsin in the distal cilium from which disks form.

Assuming a ciliary membrane pathway, it is of some importance to consider possible mechanisms for the delivery of opsin against a large concentration gradient. Peters *et al.* (1983) have suggested that simple diffusion could quan-

---

IMP in the EF leaflet of ciliary membrane between cross-fractured ROS (above) and PF leaflet of inner segment (below). (B) Thin section illustrating surface globules with a repeat period similar to that of ciliary necklaces. ×75,200. Inset: Cross-section of cilium fixed in the presence of tannic acid to illustrate the axoneme membrane cross-links. ×111,700. These micrographs illustrate the features common to both mature (Matsusaka, 1974; Rohlich, 1975) and developing (Besharse *et al.*, 1985) photoreceptors. (Micrographs are modified from Besharse *et al.*, 1985.)

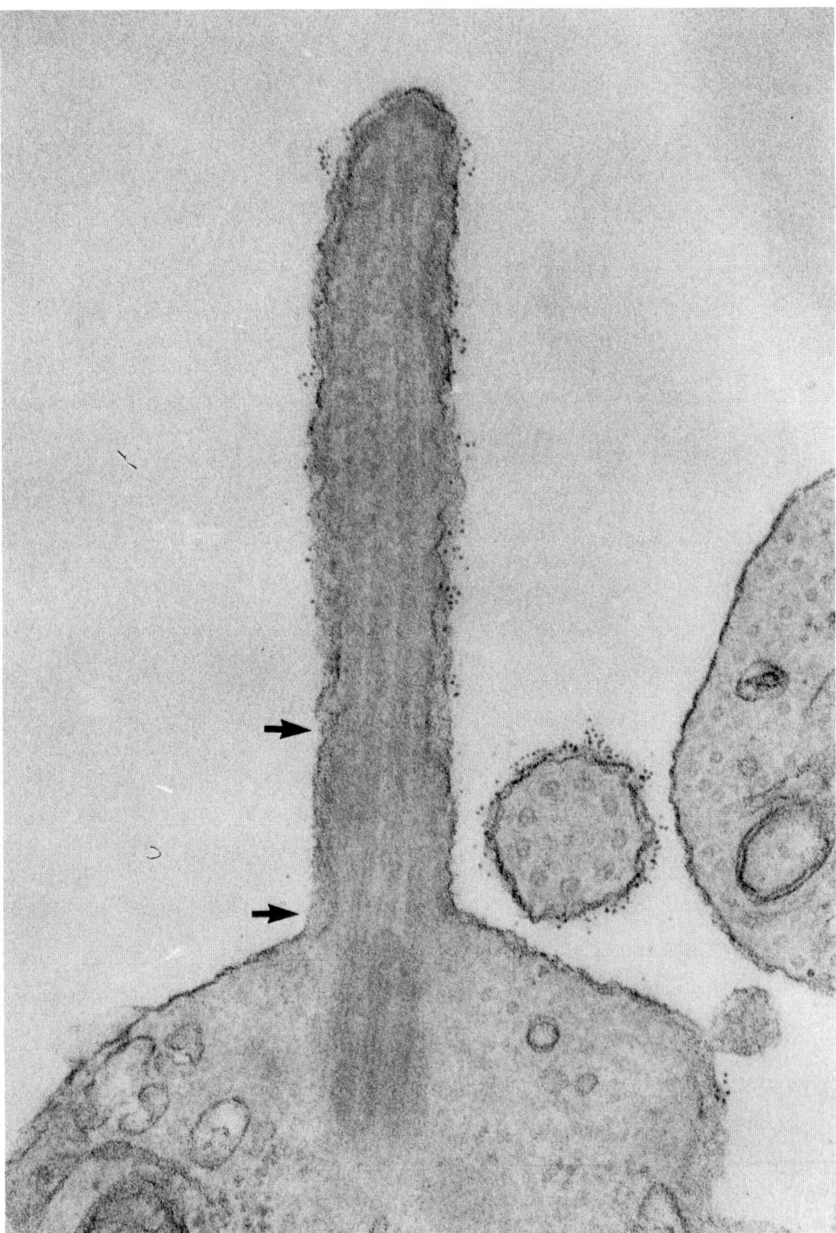

FIG. 7. Electron micrograph illustrating the distribution of immunoreactive opsin on the ciliary membrane of a 3-day-old rat. ×85,000. Retina was incubated in affinity-purified sheep anti-bovine opsin IgG which had been biotinylated. Its distribution was detected by avidin–ferritin. Note that opsin is found concentrated in the distal cilium even though disk morphogenesis has not yet begun. Arrows indicate proximal ciliary region. This micrograph is similar to those presented in Besharse *et al.* (1985).

titatively account for the transfer if a sink for opsin were located in the forming disks or distal cilium. With such a mechanism the elaborate structures of the proximal cilium could act as a barrier to back-diffusion of opsin into the inner segment. Related to this, Andrews and Cohen (1983) have suggested that a higher content of cholesterol detected in inner segment plasma membranes outside the pariciliary region would favor segregation of opsin into the periciliary domain. Another possibility is that the proximal ciliary region actively transfers opsin to the distal zone. This idea is consistent with several observations on motile cilia which demonstrate surface membrane mobility (Bloodgood, 1977). In this regard it is of some interest that axoneme membrane cross-links, similar structurally to those observed in the proximal zone of photoreceptor cilia, have been shown to connect the axoneme to a membrane-bound dynein-like ATPase (Dentler *et al.*, 1980). It has been suggested that mobility of surface membrane components in ciliary membrane could result from the ATP driven action of the cross-links (Dentler, 1981).

D. Disk Morphogenesis

The detection of opsin immunoreactivity in photoreceptor cilia well in advance of the period of disk morphogenesis has enabled us to distinguish between two fundamentally distinct processes: opsin delivery to the distal ciliary region and the formation of disks from distal ciliary membrane (Besharse *et al.*, 1985). The latter process had long been considered to occur through membrane invagination (Nilsson, 1964). Recently, however, the observation that forming disks increase in diameter in a proximal to distal direction has led directly to a model (Steinberg *et al.*, 1980) in which disks form by evagination of distal ciliary membrane (Fig. 8A). In considering this model, it is useful to realize that disk morphogenesis initially results in the formation of disks whose lumens are confluent with the extracellular space (see Fig. 1). In cones this continuity is maintained, while in rod cells an additional process results in formation of a disk edge between adjacent evaginations (Fig. 8B). This model is consistent with morphological data and provides a basis for understanding the mechanisms that are involved. For example, it is now known that disk assembly is stimulated by light in *X. laevis* (Besharse *et al.*, 1977a,b; Hollyfield, 1979; Hollyfield *et al.*, 1982). Our observation that open disks accumulate at the base of the ROS during stimulation was originally interpreted in terms of a model in which open disk formation (i.e., evagination) and maturation (i.e., edge formation) occurred at different rates (Besharse *et al.*, 1977a). The model is also particularly useful in that it provides a basis for understanding the significance of the distal ciliary lip region in frog photoreceptors (see Peters *et al.*, 1983; Papermaster *et al.*, 1979, 1985; Nir and Papermaster, 1983).

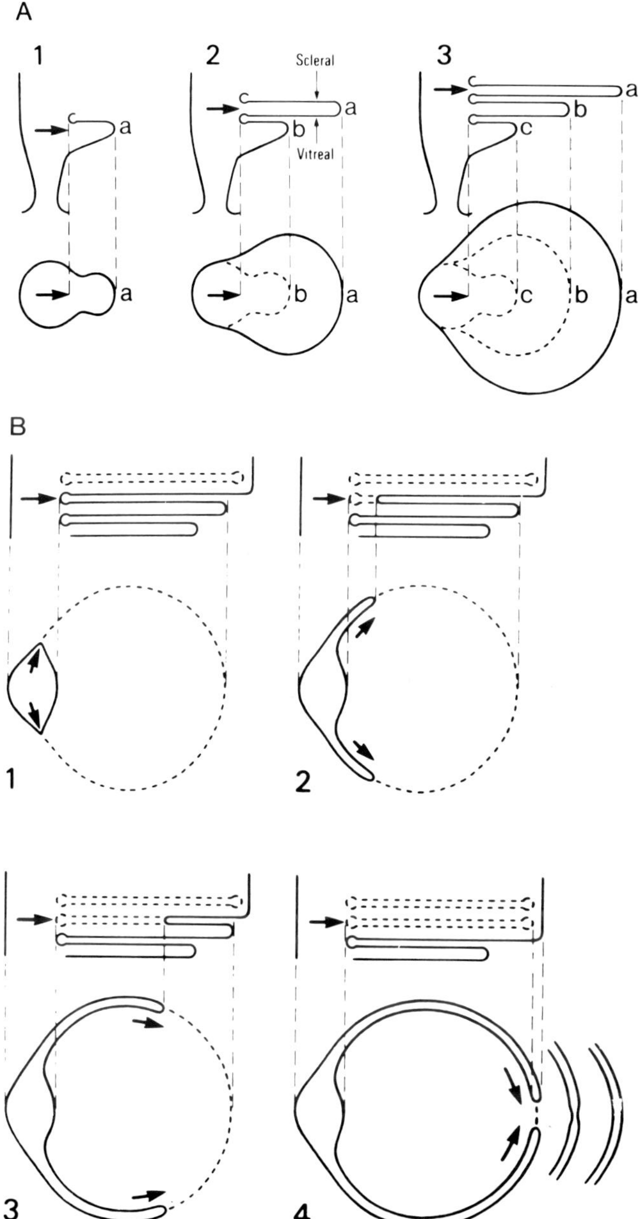

FIG. 8. Diagrammatic illustration of disk morphogenesis as described by Steinberg *et al.* (1980). (A) Disk surface formation of three successive disks (a through c) as seen in longitudinal and cross-sectional views. (B) Four consecutive stages of disk rim (edge) formation in rods. Note that the rim forms between two adjacent disks and is thought to begin at the cilium and to extend peripherally around the evaginations. (From Figs. 5 and 9 in Steinberg *et al.*, 1980, with permission of Alan R. Liss, Inc., New York).

The preceding discussion is based on the assumption that disks form initially from distal ciliary or outer segment plasma membranes (Steinberg *et al.*, 1980). However, several observations suggest that disk morphogenesis may be more complicated. Although the outer segment plasma membrane clearly contains opsin (Jan and Revel, 1974; Basinger *et al.*, 1976; Papermaster *et al.*, 1978a; Clark and Hall, 1982; Nir and Papermaster, 1983), it may also contain other intrinsic components different from those in disks (Andrews and Cohen, 1979; Clark and Hall, 1982). For example, radioiodination of intact ROS has revealed two high molecular weight proteins in addition to opsin (Clark and Hall, 1982). Furthermore, freeze-fracture analysis coupled with the use of filipin to identify regions of high cholesterol indicate that in several species the ROS plasma membrane and most proximal disks contain IMP-free patches with demonstrable cholesterol (Andrews and Cohen, 1979; 1983). Although the overall distribution of IMP-free patches in rod photoreceptors differs among the species examined to date (Andrews and Cohen, 1981; Besharse and Pfenninger, 1980), observations on frog and toad photoreceptors are most interesting because they have revealed the presence of IMP-free patches restricted to the most proximal disks (Andrews and Cohen, 1981). Such observations[3] suggest that newly formed disks are not identical in structure and composition to the ROS plasma membrane or to mature disks. They imply that mechanisms exist for sorting membrane components during disk morphogenesis and maturation.

Recently, it has been found that the connecting cilium, ciliary lip, and basalmost disks contain immunoreactive actin (Chaitin *et al.*, 1984). This finding once again establishes a degree of analogy to the proximal region of motile cilia, where actin has also been localized (Sandoz *et al.*, 1982), and raises the possibility that actin may play a role in the final step of opsin delivery or in disk morphogenesis (Chaitin *et al.*, 1984). Actin polymerization could provide the force necessary for extension of the evaginating membrane sheet (Fig. 8A). Such a model is similar to that proposed for extension of the acrosomal process of echinoderm spermatozoa, where actin polymerization is thought to provide the motive force for extension (Tilney *et al.*, 1973; Tilney and Kallenbach, 1979). Although it is too early to suggest a mechanism for edge formation, it is of some interest that the high molecular weight glycoprotein characteristic of disk membranes (Papermaster *et al.*, 1976) has been localized to the disk edge of both rod and cone photoreceptors, and that fine filaments extend between disks at their edge (Roof and Heuser, 1982). This raises the possibility that such components play a morphogenetic role in extension and stabilization of the disk. However, the fact that the large intrinsic membrane protein is found in cone photoreceptors

[3]We have emphasized the similarity of basal disks and ROS plasma membrane in *X. laevis* (Besharse and Pfenninger, 1980). However, IMP-free patches lacking in ROS plasma membrane are occasionally seen in basal disks of *Xenopus* rods (unpublished data).

(Papermaster *et al.*, 1982) where disk closure does not frequently occur (Cohen, 1968) suggests that its presence is not sufficient for disk closure.

E. Lipid Turnover in Outer Segments

Once incorporated into a disk, opsin remains associated with that disk and exhibits no apparent turnover until it is lost at the ROS distal tip (Hall *et al.*, 1969; Young and Bok, 1969). In contrast, phospholipid turnover appears to be an exponential process directly related to its randomizaion throughout the entire ROS (Bibb and Young, 1974a,b; Anderson *et al.*, 1980a). Furthermore, recent studies suggest that turnover of ROS phospholipid is more rapid than would be predicted from a process that involved disk shedding as the sole means of removal (Anderson *et al.*, 1980a–d). These observations have led to the idea that as disks are displaced in a proximal to distal direction within the ROS, phospholipids exchange among disks and that molecular turnover accounts for the shorter half-lives of phospholipids. These mechanisms remain poorly defined and deserve greater emphasis.

Glycerol, because it is metabolized quickly, is useful for studies of phospholipid turnover since a true ''pulse'' of radioactivity can be incorporated into phospholipids after injection (Anderson *et al.*, 1980a–d). Bibb and Young (1974b) showed that glycerol was initially incorporated into photoreceptors in the myoid region prior to its appearance in the outer segment. Results of subsequent EM autoradiography (Mercurio and Holtzman, 1982b; Matheke and Holtzman, 1984) and cell fractionation analysis (Anderson *et al.*, 1980a,b; Anderson and Kelleher, 1981) are consistent with the view that the endoplasmic reticulum (predominantly rough endoplasmic reticulum of the myoid) is a major site of phospholipid synthesis by both a *de novo* pathway and by base exchange. Using tritiated choline as precursor in an isolated *in vitro* preparation, it has been shown that phosphatidylcholine is synthesized and transferred to the ROS with kinetics similar to that for opsin (Hall *et al.*, 1973; Basinger and Hoffman, 1976). Similar results have been obtained using glycerol as precursor (Mercurio and Holtzman, 1982b). Although such observations are consistent with a mechanism involving delivery of phospholipid in vesicular form along with intrinsic membrane protein (see Anderson *et al.*, 1980a; Fliesler and Anderson, 1983), several recent studies indicate that phospholipid and protein delivery to ROS can be dissociated. For example, inhibition of opsin synthesis and transfer to the ROS with puromycin does not inhibit phospholipid synthesis or delivery (Basinger and Hoffman, 1976; Matheke and Holtzman, 1984). Furthermore, the inhibition of opsin transport to the ROS by monensin does not inhibit phospholipid delivery to the ROS (Matheke *et al.*, 1984; Matheke and Holtzman, 1984). Monensin is thought to have its effect by disrupting Golgi function

(Griffiths *et al.*, 1983). This raises the possibility that a major pathway of phospholipid delivery to the ROS bypasses the Golgi complex. The nature of this pathway remains undefined.

After incorporation into ROS, phospholipids are apparently free to exchange among disks so that the radioactivity becomes randomized. This would account for the generalized labeling pattern detected in all autoradiographic studies to date (Bibb and Young, 1974b; Anderson *et al.*, 1980b; Masland and Mills, 1979). The mechanism of randomization has not been identified, but data showing that a retinal phospholipid transfer enzyme catalyzes the transfer of phosphatidylcholine into the ROS suggest that such enzymes may play a role (Dudley and Anderson, 1978). The conclusion from autoradiographic studies is also consistent with the observation that after peak labeling of phospholipid with tritiated glycerol, the radioactivity in phospholipid declines exponentially. This is the result that would be expected if phospholipids were randomly distributed throughout the ROS and were disposed of by intermittent disk shedding. However, the reported half-lives of the major phospholipids (see Anderson *et al.*, 1980b–d) are shorter by a substantial factor than would be predicted from the assumption that phospholipid is removed only in the course of disk shedding. For example, in *Rana pipiens*, such a model predicts that 35% of the labeled phospholipid would remain after one entire turnover period of outer segment opsin (Anderson *et al.*, 1980b). The actual measured values, which probably overestimate the true turnover time, averaged 23% or less (Anderson *et al.*, 1980b–d). Such data suggest that in addition to disk shedding, ROS phospholipid may be degraded by mechanisms operating within the photoreceptor. The nature of such mechanisms remain completely undefined, but could involve transfer of phospholipid to inner segments for degradation or the action of local lipases. The observation that ROS contain high levels of diacylglycerols that appear to form during local phosphatidylinositol turnover (Anderson *et al.*, 1980d) is consistent with the recent report of phospholipases associated with ROS membranes (Zimmerman, 1984). A specific phospholipase C has been implicated in the light-stimulated breakdown of phosphatidylinositol 4,5-bisphosphate in frog ROS (Ghalayini and Anderson, 1984; Hayashi and Amakawa, 1985).

The high rate of net membrane accumulation in ROS means that net rates of synthesis of phospholipid would be high. To support such synthesis photoreceptors appear to share with cholinergic neurons an ability to accumulate choline provided at low extracellular concentrations (Masland and Mills, 1979, 1980). Unlike cholinergic neurons, however, much of the choline is converted to phosphorylcholine (Masland and Mills, 1980). Furthermore, recent studies using the choline analog, hemicholinium-3, indicate that photoreceptor metabolism is easily disrupted (see Masland, 1982). From work on cholinergic neurons, hemicholinium is thought to function as a choline uptake blocker. Although it has little if any immediate effect on phosphatidylcholine synthesis by base exchange or from radiolabeled glycerol as precursor, it does reduce both accumulation and formation of phosphorylcholine in photoreceptors (Pu and Anderson, 1983). Disruption of this pathway seems sufficient to greatly impair photoreceptor outer segment structure and results in severe degeneration (Pu and Masland, 1984; Masland, 1982). Although it is not presently clear whether outer segment degeneration results solely from the reduced uptake and phosphorylation of choline, its rapid and predictable occurrence after hemicholinium treatment is consistent with the view that photoreceptor phospholipid metabolism is in a precarious balance (Masland, 1982).

The foregoing does not adequately account for the unique and dynamic features of retinal phosphatidylinositol metabolism. This phospholipid is particularly interesting because of its putative role in intracellular signal processing (Michell, 1975). Its synthesis using glycerol as a precursor occurs more rapidly than other phospholipids (Hall *et al.*, 1973; Anderson *et al.*, 1980d; Schmidt, 1983b) and is stimulated by light (de Bazan and Bazan, 1977; Schmidt, 1983a,b; Dudley *et al.*, 1984; Anderson *et al.*, 1983). Its half-life within ROS is much shorter than that of other ROS phospholipids (Anderson *et al.*, 1980d), and its turnover in the ROS appears to be accompanied by increased levels of 1,2-diacylglycerol (Anderson *et al.*, 1980d), a fusogenic lipid (Allan and Michell, 1975). Furthermore, the breakdown of phosphatidyl inositol 4,5-bisphosphate (Ghalayini and Anderson, 1984) presumably results in release of inositol 1,4,5-trisphosphate (IP_3). Diacylglycerol could promote the several membrane fusion events thought to be essential in the processes of disk assembly and shedding (Anderson *et al.*, 1980d) or the activation of protein kinase C (see Kapoor and Chader, 1984). IP_3 release could result in mobilization of intracellular calcium (Streb *et al.*, 1983).

An understanding of the significance of phosphatidylinositol metabolism in the retina will have to take into account the observation that it is stimulated in light (de Bazan and Bazan, 1977; Anderson and Hollyfield, 1981; Anderson *et al.*, 1983; Schmidt, 1983a,b; Dudley *et al.*, 1984). In *X. laevis* light-stimulated synthesis of phosphatidyl inositol using inositol or phosphate as precursor has the characteristics of a classic ''phosphatidylinositol cycle'' (Anderson *et al.*, 1983). The lack of detectible stimulation when glycerol was used as precursor suggests that *de novo* synthesis was not involved. In this case the light-stimulated response was autoradiographically localized to horizontal cells, a neuronal population postsynaptic to photoreceptors (Anderson *et al.*, 1983).

Light enhances the incorporation of glycerol into phoshoinositides and other glycerolipids in rat retinas (Schmidt, 1983a,b), leading to the suggestion that, in addition to the initiation of phosphatidylinositol turnover, *de novo* synthesis of glycerolipids was stimulated by light. Furthermore, fractionation and autoradiography experiments suggested that the pathway involving *de novo* synthesis occurred predominantly in photoreceptors (Schmidt, 1983b), as a cytidine dependent pathway (Schmidt, 1983c,d). The implication was that the light stimulated lipid metabolism in photoreceptors was dominated by a pathway involving cytidine diphosphate-diacylglyceride. Recently, Anderson and colleagues (1985) have confirmed the finding that light stimulates radiolabeled glycerol incorporation into phospholipids. However, it was also reported that light had no effect on palmitic acid incorporation into phosphatidic acid or other glycerolipids with the exception of phosphatidylinositol. This observation has led to the suggestion that the light effect on glycerol incorporation might be explained by a light dependent alteration of specific activity of the intracellular pool of glycerol rather than to a true stimulation of phospholipid synthesis (Anderson *et al.*, 1985). Although there is some question regarding the nature of light's effect on *de novo* synthesis of phospholipids in photoreceptors, recent evidence that light stimulates the breakdown of phosphatidylinositol 4,5-bisphosphate (Ghalayini and Anderson, 1984; Hayashi and Amakawa, 1985) raises the possibility that inositol 1,4,5-trisphosphate or the 1,2-diacylglycerol formed in this reaction may play an as yet undefined role in some aspect of light dependent photoreceptor physiology.

F. Turnover of Cone Outer Segments

Cone photoreceptors differ from rods in that disks generally maintain continuity with the plasma membrane (Cohen, 1968, 1970). This structural difference is correlated with failure of radiolabeled proteins to form a proximal band in cone outer segments (COS). Instead, COS become diffusely and uniformly labeled (Young and Droz, 1968; Young, 1971b; Bok and Young, 1972; Ditto, 1975). In the earliest studies of outer segment membrane turnover, the lack of a radioactive band at the base of COS in both developing and mature cone photoreceptors (Young and Droz, 1968; Young, 1971a; Ditto, 1975) led to the suggestion (Young, 1971b) that cone photoreceptors lacked a mechanism for membrane renewal comparable to that of rods. Although we now know less about the general properties of COS turnover, several observations suggest that membrane assembly comparable to that in rods must occur. For example, disk shedding is a constant feature of cone photoreceptors in a variety of species (Hogan *et al.*, 1974; Anderson and Fisher, 1975, 1976; Steinberg *et al.*, 1977). Furthermore, analysis of cytomembrane compartments of cone inner segments suggests a precursor–product relationship between inner segment vesicular membranes and COS comparable to that seen for rods (Besharse and Pfenninger, 1980).

The uniform labeling pattern routinely seen in cones in autoradiographic studies of protein in COS is generally interpreted as reflecting randomization of newly inserted protein. Based on the high degree of lateral mobility of opsin within rod disk membranes, it was suggested that continuity of cone disks with each other via the plasma membrane would permit free diffusion of cone opsin among disks (see Poo and Cone, 1974). Available photobleach-recovery data for axial diffusion in cones, however, suggests that redistribution of cone pigment occurs quite slowly (Liebman and Entine, 1974; Liebman *et al.*, 1982). This conclusion is consistent with a recent autoradiographic study (Anderson *et al.*, 1986), using the unique fucosylated protein recently identified as an intrinsic component of COS (Bunt and Saari, 1982). Anderson and colleagues (1986) found that a fucosylated component was initially localized over the proximal region of COS and became completely randomized in its distribution only after several hours. The time course of randomization indicates that free axial diffusion within the COS membrane cannot account for the distribution of fucosylated protein and suggests that the edge region of cone disks might present a significant barrier to redistribution in the COS.

It has been found that COS are uniformly labeled in a variety of species when tritiated fucose is used as a tracer (Bunt, 1978; Saari and Bunt, 1980; Bunt and Klock, 1980a,b; Bunt and Saari, 1982). Although it has not been unequivocally identified as a cone visual pigment, the fucosylated glycoprotein has properties consistent with that role. It appears to be an intrinsic membrane component and is similar in molecular weight in SDS–PAGE to rod opsin seen in the same gels (Saari and Bunt, 1980; Bunt and Saari, 1982). In goldfish, however, its slightly smaller size makes it clearly resolvable from rod opsin in the same preparation (Bunt and Saari, 1982). Variations in labeling pattern of cones with different spectral properties raise the possibility that different cone opsins may differ from each other and from rod opsin in the structure of their oligosaccharide chains. Further study of the properties of COS proteins may provide insight into the unique features of COS turnover as well as the predicted structural differences in cone opsins related to their function in photopic and color vision (Wald *et al.*, 1955; Wald, 1968).

III. Disk Shedding and Phagocytosis

Disk assembly is balanced by the periodic detachment of groups of disks from the photoreceptor distal tip and phagocytosis of those disks by cells of the pigment epithelium (Young and Bok, 1969). Although little is known about underlying cellular–molecular mechanisms, it is convenient to consider three separate but interrelated processes: disk detachment, phagocytosis, and phagosome degradation. Phagocytosis in particular has been studied *in situ* (Hollyfield, 1976; Hollyfield and Ward, 1974a,b) and in organ and cell culture (Feeney and Mixon, 1976; Rosenstock *et al.*, 1980; Philp and Bernstein, 1981; Edwards and Szamier, 1977; Hall, 1978). It is thought to involve mechanisms similar to those currently under investigation in other phacogytic cells (Chaitin and Hall, 1983b; Silverstein and Loike, 1980). The unique features of the retinal pigment epithelium (RPE) are that its cells remain attached to their basement membrane and to each other at junctional complexes and that phagocytosis occurs at the distinctive apical membrane domain (Bok, 1982). The emphasis of this section is on the mechanism of disk detachment and phagocytosis of ROS disks. The general phagocytic properties of pigment epithelial cells and the mechanism of phagosome transport and degradation have been covered in recent reviews (Bok and Young, 1979; Besharse, 1982), and are described in Part II by Clark.

A. *Mechanism of Disk Detachment*

Disk detachment occurs through a process that maintains the continuity of the outer segment plasma membrane during and after detachment (Young and Bok, 1969; Bok and Young, 1979). Electron microscopic studies of mammalian rod and cone shedding have led to the view that disk detachment occurs first and is followed by uptake of the shed fragment (see Fig. 9; Young, 1971a; Steinberg *et al.*, 1977; Bok and Young, 1979). This is supported by the frequent observation of intermediate stages of detachment showing that distal disks curl toward the RPE or are separated from more proximal disks by apparent invagination of outer segment plasma membrane (Young, 1967, 1971a; Bok and Young, 1979). However, a level of ambiguity is introduced by the observation that RPE processes often intrude into the outer segment prior to complete detachment of the fragment (Spitznas and Hogan, 1970; Anderson and Fisher, 1976; Steinberg *et al.*, 1977; Anderson *et al.*, 1978). Furthermore, EM images of incipient phagosomes (i.e., completely detached fragments) in the space between photoreceptors and RPE have generally not been reported. Although the former observation led to the suggestion that RPE cells actively remove disks from the ROS distal tip (Spitznas and Hogan, 1970), Steinberg and colleagues (1977) have pointed out that similar images would be expected if RPE intrusion represented a quick response to a detachment process inherent in the photoreceptor. The same could be said for the lack of images showing incipient phagosomes in the subretinal space.

We have evaluated the relative roles of the rod photoreceptor and RPE by analysis of inhibitors of disk shedding using eye cups from *X. laevis* which sustain light-evoked rod disk shedding *in vitro* (Besharse *et al.*, 1980). The rationale was that an inhibitor of phagocytosis should cause an accumulation of incipient phagosomes in the space between photoreceptors and RPE if (1) disk detachment occurs through a photoreceptor active process and (2) the inhibitor affects only phagocytosis. The data show (Table I) that disk shedding is blocked by disruption of an actin filament system (Besharse and Dunis, 1982) and of

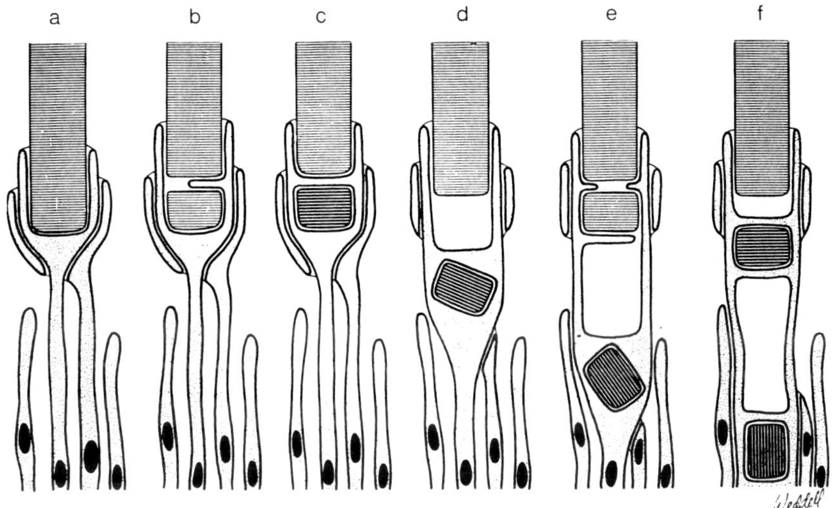

FIG. 9. Diagram illustrating the arrangement of RPE sheaths surrounding COS and their role in disk shedding and phagocytosis in the human retina. (a) COS tips permanently invested in processes that extend from RPE to COS tips. Ensheathing processes are distinct from villous processes that contain melanin granules. (b–e) COS disk shedding and phagocytosis occurs within the innermost ensheathment. Pseudopods extend into the COS to engulf the fragment. Regions of intrusion are often characterized by disk separation and curling. After detachment the phagosome is transported into the RPE cell body. (e–f) Evidence for a second round of disk shedding and phagocytosis is often seen. (From Steinberg *et al.*, 1977, with permission of The Royal Society of London.)

oxidative phosphorylation as well as by conditions which raise cAMP levels (Besharse *et al.*, 1982b) or reduce extracellular Ca^{2+} (Greenberger and Besharse, 1983). Although it can be argued that these inhibitors block an active process related to shedding in photoreceptors, the data emphasize our consistent failure to establish a treatment that leads to the formation of completely detached, incipient phagosomes.[4] Most interesting among these were the experiments involving cytochalasins (Besharse and Dunis, 1982). Although at low concentration occasional examples of complete or near-complete detachment without phagocytosis were detected, such examples were never found at higher concentrations where disk shedding was virtually eliminated (see Table I). Cytochalasin D is widely used because it disrupts actin microfilament systems (Tanenbaum *et al.*, 1977), and our data are consistent with a role for an actin-based

[4]A demonstration of disk shedding in the absence of pigment epithelium has not been published. Although I reported in a discussion (Besharse and Dunis, 1982) and a review (Besharse, 1982) that colchicine in light caused shortening of ROS of isolated *Xenopus* retinas, we have subsequently been unable to reproduce the observation.

TABLE I
INHIBITORS OF DISK SHEDDING IN *Xenopus* EYE CUPS GENERALLY BLOCK DETACHMENT FROM THE ROS

Compound	Concentration (μ*M*)	Phagosomes in RPE (% of control)	Incipient phagosomes[a]
Control	—	100	–
Cytochalasin D[b]	5	25	– (+)[c]
	25	10	–
Dinitrophenol[d]	100	25	–
IBMX[e]	100	30	–
db cAMP[e]	2000	30	–
(Low Ca^{2+})[f]	(10)	20	–

[a]Identifiable phagosome-sized fragments, completely detached from the ROS and contained in the subretinal space.

[b]Data from Besharse and Dunis (1982).

[c]Rarely seen, only at low drug concentration.

[d]Unpublished data of Besharse.

[e]Data from Besharse *et al.* (1982).

[f]Data from Greenberger and Besharse (1983). Omission of Ca^{2+} from culture medium without use of chelators generally results in a final concentration of 10 μ*M*.

system in disk detachment. Immunocytochemical observations, however, have failed to detect immunoreactive actin in the distal outer segment (Chaitin *et al.*, 1984). Because a well-developed actin based system develops at sites of phagocytosis in both macrophages (Hartwig *et al.*, 1980) and RPE (Chaitin and Hall, 1983b; see below), it must be suggested that the RPE normally plays an active role in detachment.

We have also used *Xenopus* eye cups to obtain electron microscopic images of intermediate stages of disk detachment. Most useful in this regard were eye cups treated with aspartate or glutamate. Greenberger (1984) has shown that these photoreceptor neurotransmitter candidates induce a massive and near-synchronous shedding response (see Section IV for further discussion). Using such preparations we have found numerous examples of intrusion of RPE processes into distal regions of the ROS (Fig. 10). Such processes contain few mem-

FIG. 10. Electron micrographs illustrating the pseudopod-like processes that form around ROS distal tips during disk shedding in *X. laevis*. (A) Partially detached distal tip of an ROS surrounded by a pseudopod containing a rich array of cross-linked 7- to 8-nm filaments. Note that the pseudopod membrane is closely apposed to the ROS plasma membrane. (B) Newly formed phagosome contained within a pseudopod from RPE. Note that the processes in both A and B lack most of the membranous organelles characteristic of the RPE. Micrographs are from an experiment in which disk shedding was activated by treating the eye cup with 10 m*M* aspartate for 1 hr (Greenberger, 1984). Bars, 1 μm. (Micrographs prepared by Donna M. Forestner for this review.)

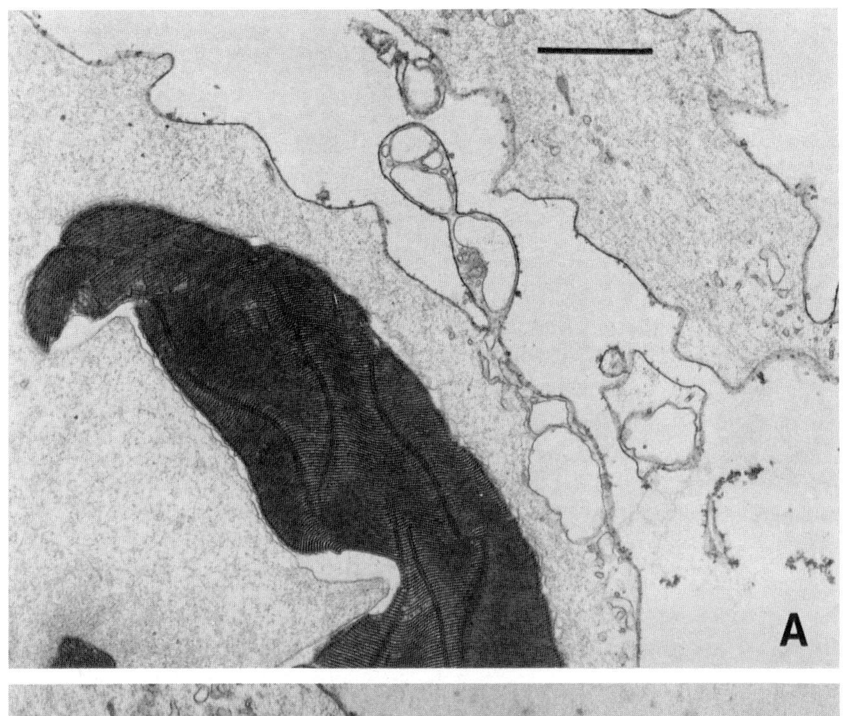
A

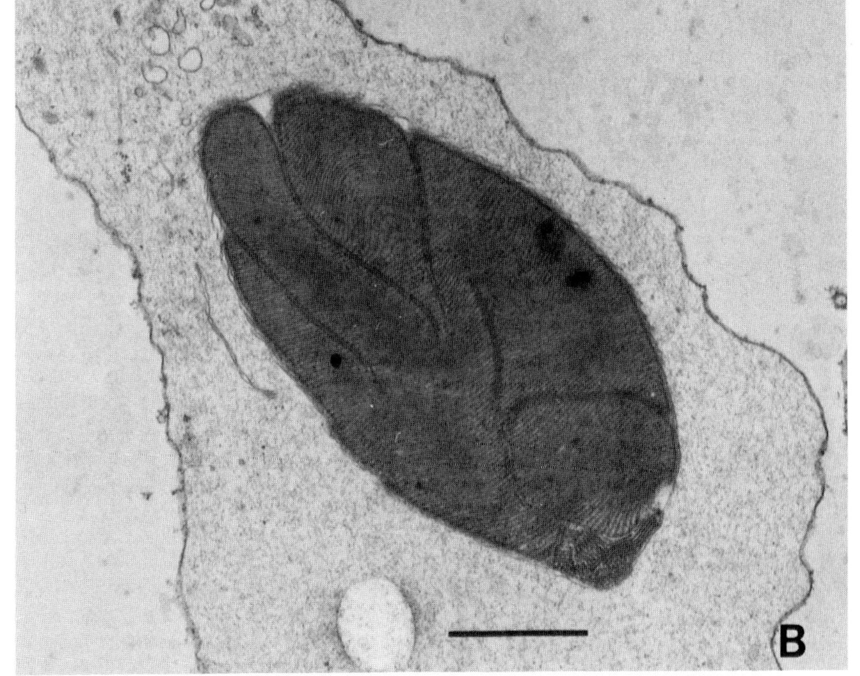
B

branous organelles but do exhibit abundant 7- to 8-nm microfilaments in a diffuse, cross-linked array. Consistent with their identification as actin microfilaments, the processes do not form in the presence of cytochalasin D, and are stained intensively with rhodamine phalloidin (Matsumoto *et al.*, 1986), an actin-binding toxin used widely to study the distribution of polymerized actin (Barak *et al.*, 1980). The intruding processes involved in phagocytosis are distinct from villous processes that contain longitudinal bundles of cytochalasin-insensitive actin filaments (Murray and Dubin, 1975; Burnside and Laties, 1976). They are similar, however, to the pseudopods formed by macrophages during phagocytosis (Hartwig *et al.*, 1980).

The inhibition of disk detachment by cytochalasins and intrusion of pseudopods into ROS suggest an active role for the RPE in disk detachment. However, the same cytochalasin experiments provided evidence of an active process in the ROS. Structural changes (vesiculation and tubulation of disks) were observed in many ROS in regions where disk detachment would have been expected (see Besharse, 1982). These areas were similar to the areas of disk vesiculation in distal ROS reported in other studies (Currie *et al.*, 1978; Tsukamoto and Yamada, 1982), and their detection led me to suggest that an active process in the ROS tip at least determines the site where detachment will occur. Although a highly localized artifact related to chemical fixation cannot be ruled out, the structural data suggest that disk detachment may occur as disks fuse with the plasma membrane and with each other. Such a process could be promoted through the generation of fusogenic diacylglycerols within the ROS (see Section II,D; Anderson *et al.*, 1980d).

Perhaps related to the structural changes noted above, we recently found that the fluorescent dye, Lucifer Yellow, labels the distal tip of many of the ROS in isolated retinas (Matsumoto and Besharse, 1985). The staining pattern is restricted to a zone averaging about 4 μm long at the ROS distal tip (Fig. 11), and, within this zone, the dye penetrates the ROS in a banded pattern. Three observations are important. First, in obviously damaged ROS, Lucifer Yellow penetrates throughout the ROS cytosol, labeling the entire structure. Thus, the restricted labeling of ROS distal tips indicates that the fluorescent dye is sequestered in a compartment and does not gain free access to the cytosol. Second, the labeling is greatly enhanced in retinas isolated in light. Third, the labeling is greatly reduced in retinas isolated at room temperature and then incubated in the dye at 3°C. The latter observations suggest that the uptake of Lucifer Yellow at the ROS distal tip is an active process. Although the necessity of retinal isolation for detecting Lucifer Yellow staining raises the possibility that ROS distal tips are damaged, the restricted distribution and size of the stained region, the light dependence, and active nature of the process suggest that the dye detects a feature of the ROS distal tip-related to shedding.

The above experimental observations lead to a model for rod disk shedding in amphibians in which disk detachment results from processes occurring in both

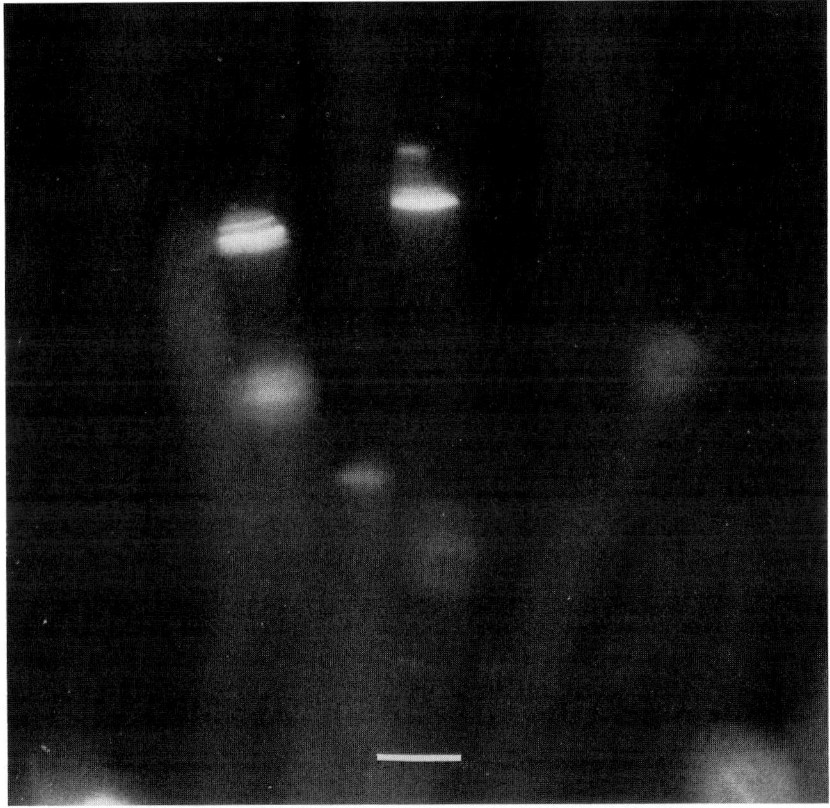

FIG. 11. Fluorescence photomicrograph illustrating the labeling of distal tips of ROS with Lucifer Yellow CH. This is a retinal piece in which ROS tips are oriented toward the top. Unlabeled tips and more proximal regions of the ROS are barely visible in the micrograph. The retina was isolated and then incubated in light with Lucifer Yellow for 5 min prior to fixation with aldehydes. Bar, 10 μm. (Micrograph prepared by Brian Matsumoto for this review.)

ROS and RPE. In the quiescent state (Fig. 12A) ROS tips interdigitate with villous processes of the RPE. In this state polymerized actin is largely restricted to the longitudinal arrays in villous processes (Burnside and Laties, 1976), to a band at the level of junctional complexes (Owaribe *et al.*, 1981), and to a thin subplasmalemmal region (Owaribe *et al.*, 1981). When disk shedding is activated (Fig. 12B) actin-containing RPE processes with a structure distinct from that in villous processes form in close apposition to the ROS plasma membrane (Fig. 12B). The model also shows that an active process in the ROS tip begins in the zone where disk detachment will occur. This process may involve fusion of disk membranes with plasma membrane and may precede development of the

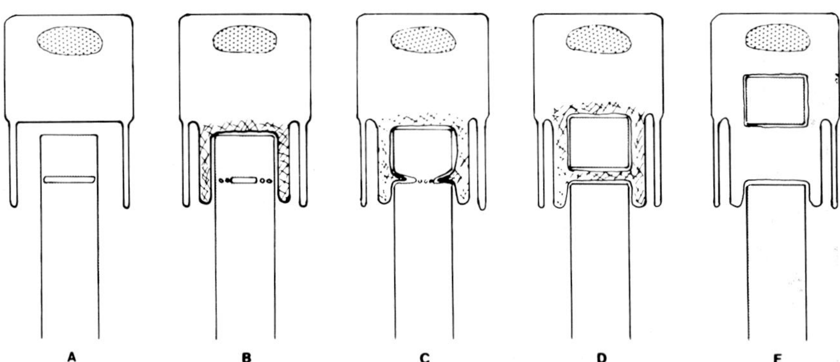

FIG. 12. Diagram illustrating basic features of ROS disk shedding as seen in eye cups of *X. laevis*. For convenience, only one ROS disk in the zone of detachment is illustrated. (A) Prior to shedding, ROS interdigitate with RPE villous processes that contain longitudinal bundles of actin filaments. (B) Actin polymerization results in formation of processes containing an abundance of actin in a cross-linked array. Sites where disk detachment will occur are seen as sites of disk vesiculation. (C) RPE processes are closely apposed to the ROS surface and intrude into the ROS at the site of detachment. (D) The detached fragment is contained within the RPE cytoplasm and is surrounded by an abundance of cross-linked actin. (E) The actin array surrounding the phagosome disappears as the phagosome is transported basally. (Based on analysis from the authors laboratory.)

ensheathing processes. The pseudopod-like ensheathing processes intrude into the ROS tip maintaining close apposition to the ROS plasmalemma (Fig. 12C and D). Whether the pseudopods provide the motive force for detachment or simply respond to a detachment process inherent in the rod remains to be determined. After complete engulfment the phagosome is contained in apical cytoplasm largely free of membranous organelles, but as the phagosome is displaced basally the cross-linked actin array breaks down (Fig. 12E).

The above model based on amphibian material (Fig. 12) emphasizes the transient formation of processes that ensheath the ROS. Their close apposition to the ROS is reminiscent of the permanent ensheathing processes of mammalian cone photoreceptors (Steinberg *et al.*, 1977; Fisher and Steinberg, 1982). Cone ensheathment results from the elaboration of multiple, leaf-like processes from the RPE which are closely applied to the COS in concentric layers (Fisher and Steinberg, 1982). Ensheathing processes are structurally distinct from villous processes which also contain melanin granules (Fig. 9). In extreme cases (i.e., cat photoreceptors; Pfeffer and Fisher, 1981) ensheathing processes enclose the entire outer segment. The regular arrangement of filamentous structures in ensheathing processes differs from that in the transient pseudopods involved in phagocytosis described above. Nonetheless, COS disk shedding and phagocytosis are confined to the innermost ensheathment and exhibit features similar to those described for *Xenopus* eye cups (see Fig. 9). It involves the intrusion of RPE processes from the ensheathment into zones of disk detachment. Zones of detachment frequently exhibit regions of disk "curling" and occasional vesiculation (see Steinberg *et al.*, 1977), suggesting that an active process in photoreceptors is also involved.

B. Retinal Dystrophy and Phagocytosis

In theory, disruption of either disk shedding or disk assembly would be expected to alter rod outer segment length by altering the normal balance between the two processes (Young, 1976). This is elegantly demonstrated in the Royal College of Surgeons (RCS) rat, where a form of inherited photoreceptor degeneration (Dowling and Sidman, 1962) results from a failure of the phagocytic mechanism in RPE cells (Bok and Hall, 1971; LaVail, 1981). In RCS rats photoreceptors differentiate normally, but soon thereafter, excess ROS membranes accumulate at their distal regions (Dowling and Sidman, 1962; Herron *et al.*, 1969; Bok and Hall, 1971; LaVail *et al.*, 1973), apparently impairing the normal interchange between RPE and photoreceptors. The result is that photoreceptors gradually degenerate. The accumulation of debris adjacent to the apical RPE results from the failure of the RPE to phagocytose ROS membranes (Bok and Hall, 1971) even though particles other than outer segment membranes are readily phagocytosed (Custer and Bok, 1975). In the rhythmic shedding cycle the defect is expressed as greatly reduced levels of phagosomes in RPE at the time of the expected peak of disk shedding (Goldman and O'Brien, 1978).

The mutant allele which impairs phagocytosis in the dystrophic rat is expressed specifically in RPE cells. This was demonstrated in two chimeric rats whose eyes contained a mosaic of mutant and wild-type cells (Mullen and LaVail, 1976). In those animals, photoreceptor degeneration occurred in a mosaic pattern only in association with mutant RPE (Fig. 13). This conclusion was confirmed in a series of recombination experiments using retinas and RPE from wild-type and RCS rats (Tamai and O'Brien, 1979). Wild-type RPE in organ culture was able to phagocytose photoreceptor fragments from both wild-type and mutant retinas equally well, whereas mutant RPE in similar cultures was unable to phagocytose either. Although the mutant RPE is unable to phagocytose ROS membranes, it is capable of phagocytosing a variety of test particles other than ROS membranes (Custer and Bok, 1975; Reich-D'Almeida and Hockley, 1975; Edwards and Szamier, 1977). These observations suggest that the mutant RPE is defective in a part of the phagocytic mechanism that is specific for the uptake of photoreceptor membranes. The implication is that phagocytosis of photoreceptor membranes is initiated by a mechanism distinct from that for other test particles.

Recent studies using cultures of mutant and wild-type pigment epithelium suggest that the phagocytic defect in the mutant is in the mechanism that initiates

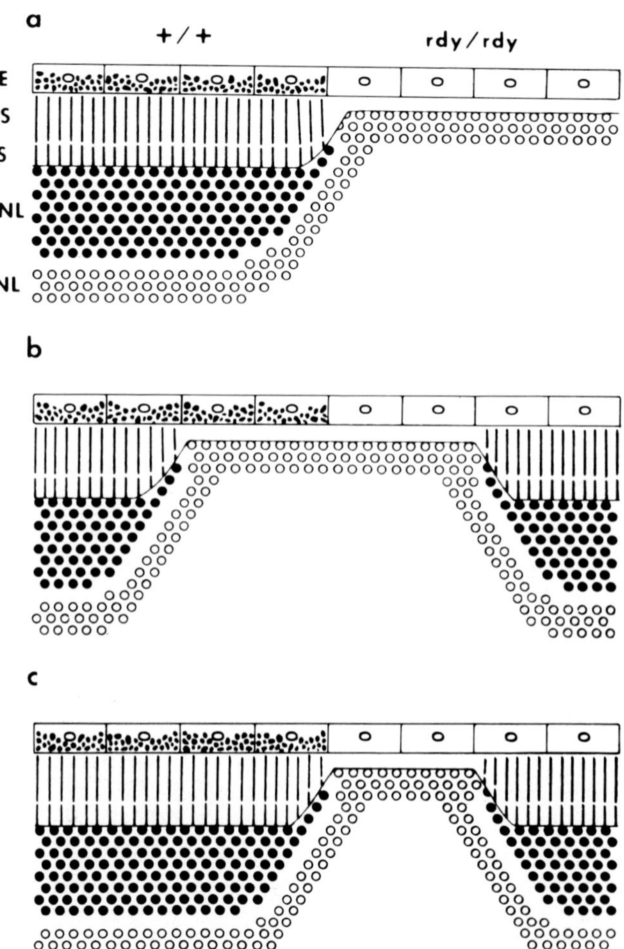

FIG. 13. Diagram illustrating the possible patterns of photoreceptor degeneration relative to mutant RPE cells in chimeras, based on different assumptions about the locus of the effect of the mutation. +/+ is wild-type phenotype marked by the presence of melanin pigment in the pigment epithelium. rdy/rdy is the mutant phenotype marked by the absence of melanin pigment. (a) Photoreceptors degenerate exclusively in association with mutant RPE if the gene defect effects pigment epithelial function only. This is the result obtained in two different chimeras. (b) Photoreceptor degeneration occurs as a mosaic pattern in association with either mutant or wild-type pigment epithelium if the locus of the defect is in the photoreceptor or retina. (c) Photoreceptor degeneration is observed only but not always in association with mutant RPE if it results from effects on both photoreceptor and RPE. [Diagram from Mullen and LaVail, Copyright 1976, by the American Association for the Advancement of Science (AAAS).]

engulfment. Cultures of wild-type RPE retain their ability to phagocytose substantial amounts of ROS membranes whereas mutant cultures are defective in this regard (Edwards and Szamier, 1977). Chaitin and Hall (1983a,b) have shown that the failure of uptake is not due to a failure of ROS membrane binding to the pigment epithelium. ROS membranes bind in numbers at least equal to those in wild-type cultures. Furthermore, immunofluorescence observations indicate that in both wild-type and mutant cultures an actin feltwork forms at sites of attachment (Chaitin and Hall, 1983b). The difference is that in mutant cultures membrane fragments remain attached to the surface, whereas in wild type the actin feltwork is extended into pseudopods which surround and engulf the particle. Such observations suggest that a receptor or transmembrane signaling mechanism may be defective in the mutant.

A difficulty with the foregoing analysis is that currently we know little about the putative recognition system for the uptake of ROS membranes (reviewed by Bok and Young, 1979; Besharse, 1982). Thus, the possibility that ROS membranes bind to the surfaces of mutant cells by a mechanism different from that of wild type cannot be ruled out. Phagocytosis by macrophages is generally studied as a receptor-mediated process that can be divided into three events: recognition–attachment, internalization, and degradation (Stossel, 1974; Silverstein and Loike, 1980). The best studied receptor–ligand systems are for phagocytes containing receptors for the Fc component of immunoglobulins and the third component of complement. Particles coated them with IgG or complement bind to the surfaces of macrophages inducing the local formation of an actin-containing feltwork. The particles are then engulfed by receptor-guided expansion of the zone of contact through a process called "zippering" (Griffin *et al.*, 1976). Interestingly, the only example of receptor-mediated uptake of particles by RPE cells is for red blood cells coated with IgG or complement (Elner *et al.*, 1981). The possibility that these receptor–ligand systems normally mediate the phagocytosis of disk membranes has not been evaluated nor has a plausible mechanism for the introduction of either ligand onto photoreceptor surfaces been suggested. Although several recent studies (O'Brien, 1976; Hall, 1978; Philp and Bernstein, 1982; McLaughlin and Wood, 1980; LaVail *et al.*, 1981; Wood and Napier-Marshall, 1985) raise the possibility of a specific receptor system mediating phagocytosis of disk membranes, detailed verification of such a system is needed (see Besharse, 1982).

C. Conservation of Outer Segment Length

Normally, the processes of disk assembly and disk shedding are maintained in relative balance, consistent with an average ROS length (see Young, 1976). A basic question is, what features of the system determine the size of the outer segment and of the fragment to be detached and internalized? In addition to identification of putative recognition markers (see previous section), such a mechanism must account for the observations that disk assembly and disk shedding can be dissociated experimentally, resulting in altered ROS length (Besharse *et al.*, 1977a; Currie *et al.*, 1978; Basinger and Hoffman, 1982). For example, disk shedding is blocked in constant light while disk assembly is maximized, resulting in elongation of the ROS (Besharse *et al.*, 1977a; Currie *et al.*, 1978). A shortened light period, however, results in reduced disk assembly while maintaining normal levels of disk shedding and reduced ROS length. Similarly, prolonged treatment at low temperature results in reduced outer segment length (O'Day and Young, 1979). Such observations led us to suggest (Besharse *et al.*, 1977a) that despite the relative balance which normally conserves ROS length, the processes of disk assembly and disk shedding are not directly coupled since they can be modulated independently.

Consistent with this are experiments showing conservation of both ROS length and phagosome size in frogs over the physiological temperature range (Hollyfield *et al.*, 1977). Because the distal ROS disks in animals kept at lower temperatures can be shown to be older than those at higher temperatures, Kaplan (1984) has emphasized that axial position within the ROS rather than disk age may be a primary determinant of shedding. Presumably a mechanism for sensing ROS length would play a role in determining when and where disks are detached. Although it is useful to think in these terms, it should be emphasized that ROS length is conserved only in the stochastic sense and that ROS exhibit a Gaussian distribution of lengths both before and after a disk-shedding event (Donnelly and Wyse, 1982). It has also been found that in frogs, where only a fraction of the rods shed their tips each day (Basinger *et al.*, 1976; Hollyfield *et al.*, 1976), the effect of increasing temperature is to increase the frequency of rod shedding rather than the size of the shed fragments (Hollyfield *et al.*, 1977). Recently, this finding was extended by the demonstration that average phagosome size is conserved even when temperature acclimation periods were sufficient for a complete turnover of the ROS (Basinger and Hoffman, 1983; personal communication). Since at the different temperatures, phagosomes of a given size contain disks synthesized over widely different time periods (see Hollyfield *et al.*, 1977), the data suggest that shedding signals that later determine phagosome size are probably not incorporated into disks during disk assembly. Although the nature of the process that determines the site of disk detachment remains to be determined, available data make plausible the possibility that it occurs through a process involving photoreceptor–RPE interaction.

IV. Regulation of Membrane Turnover

Beginning with the work of LaVail (1976), it has become evident that photoreceptor membrane turnover exhibits remarkable temporal regulation. Rod photoreceptor disk shedding occurs immediately after light onset in a variety of species (LaVail, 1976; Hollyfield *et al.*, 1976; Basinger *et al.*, 1976), while a similar process involving cone photoreceptors occurs after light offset in most species examined to date (Young, 1977, 1978). Furthermore, the assembly of disks appears to be stimulated by light at least in frogs (Besharse *et al.*, 1977a,b; Hollyfield, 1979; Hollyfield *et al.*, 1982) and to be influenced by a circadian process in rats (Dudley *et al.*, 1984). The observation of temporal organization in the regulation of photoreceptor membrane turnover has greatly stimulated efforts to understand the control of disk turnover. The general view has developed that the daily light–dark cycle influences photoreceptors both directly (Basinger *et al.*, 1982) and indirectly (LaVail, 1976, 1980; Besharse *et al.*, 1977a) through entrainment of the organisms circadian oscillators. A general review of the relationship of both disk assembly and disk shedding to light–dark cycle has recently appeared (Besharse, 1982), and for the general properties of the regulation by light the reader is referred to that review. The objective of this section is to review recent observations relevant to the mechanism of the effect of light. It should be emphasized that while recent advances have been made in this area, the mechanism of temporal regulation is complex and cannot yet be extended in a rigorous way to the cellular–molecular level.

A. *The Role of Light*

The role of light in disk shedding varies among the species analyzed to date (see Besharse, 1982, for a complete review). It acts as an entrainment signal in those species exhibiting circadian rhythmicity (LaVail, 1976; Besharse *et al.*, 1977a; Tabor *et al.*, 1982), but also acts as a direct stimulus for shedding, most notably in *R. pipiens* (Basinger *et al.*, 1976; Hollyfield *et al.*, 1976; Basinger and Hoffman, 1982). Furthermore, constant light of sufficient intensity blocks rhythmic disk shedding altogether (Besharse *et al.*, 1977a; Currie *et al.*, 1978; Goldman *et al.*, 1980; Goldman, 1982). In the rat this effect seems to be independent of the circadian clock. Recently, the characteristics of light as an entraining signal and as an inhibitor of disk shedding were analyzed in the rat. The sensitivity of the two processes differs by at least two orders of magnitude (Goldman, 1982). The light intensity necessary for an entraining signal is comparable to that necessary for entrainment of circadian behavioral rhythms in mammals (Takahashi *et al.*, 1984) whereas that necessary for inhibition of disk shedding is greater.

The characteristics of light as a direct stimulus for disk shedding in *R. pipiens* have not been well characterized. However, preliminary data suggest an action spectrum matching the absorption spectrum for a rhodopsin-like photopigment (S. Basinger, personal communication). Interestingly, the light stimulus necessary to induce shedding through whole retinal illumination, must be of sufficient intensity and duration to bleach 5–10% of the visual pigment (Basinger and Hollyfield, 1980). It is evident, therefore, that the signal for light-evoked shedding results in sustained saturation of the rod photoresponse. Detailed experimental analysis of the nature of this effect is needed.

Another peculiar characteristic of the effect of light in the frog is that a spot of light focused on a small region of the retina causes shedding throughout the retina (Basinger and Gordon, 1982). This unexpected result contrasts with previous observations on cone retinomotor movement in fish, in which a local contractile response occurred in the area of the local stimulus only (Easter and Macy, 1978). Pan-retinal shedding has major implications for the mechanism of shedding control. It suggests that control signals for disk shedding may involve lateral spread, either through the retina or pigment epithelium, or the local release of a diffusible substance that spreads throughout the retina (Basinger and Gordon, 1982). As pointed out by Basinger and Gordon (1982) spread mediated by intercellular contacts in retina would not necessarily be expected to result in pan-retinal shedding. Although there is currently no direct evidence for any pathway, a mechanism involving a diffusible substance is consistent with other observations (see below) on the control of shedding.

B. Possible Role of Indoleamines

In the earliest study of rhythmic disk shedding, attention was drawn to its similarities to pineal indoleamine metabolism (LaVail, 1976). Subsequently, the circadian nature of disk shedding was well documented in the rat (LaVail, 1980; Goldman *et al.*, 1980; Tierstein *et al.*, 1980), and evidence for a circadian component in the control of disk shedding was obtained for both *X. laevis* (Besharse *et al.*, 1977a) and the Eastern gray squirrel (Tabor *et al.*, 1982). The role of the pineal gland, however, was brought into question by the demonstration that pinealectomy failed to alter light-evoked shedding (Currie *et al.*, 1978) or circadian shedding (LaVail and Ward, 1978; Tamai *et al.*, 1978), and by several lines of evidence indicating that both processes were controlled locally within the eye (Hollyfield and Basinger, 1978; Tierstein *et al.*, 1980; Besharse *et al.*, 1980). The idea that indoleamine metabolism might be involved in the control of the circadian component of shedding remained tenable, however, because of the substantial evidence for a local indoleamine generating system in retina similar to that in pineal gland (Besharse *et al.*, 1984).

Recent efforts in this area have focused on attempts to analyze the effects of melatonin and related compounds on retinal physiology (Besharse and Dunis, 1983; Pierce *et al.*, 1984; Pierce and Besharse, 1985; Dubocovich, 1983) and to analyze the regulation of melatonin synthesis in the retina (Iuvone and Besharse, 1983). A possible role for melatonin is best documented in *X. laevis* where disk shedding can be activated *in vitro* by melatonin and related compounds (Besharse and Dunis, 1983), and where the rate-limiting enzyme for melatonin biosynthesis varies in its activity in a circadian pattern (Binkley *et al.*, 1980; Hamm and Menaker, 1980; Iuvone and Besharse, 1983; see article by Iuvone, Part II). In the same *in vitro* preparation, melatonin has also been shown to induce dark-adaptive elongation of cone photoreceptors (Pierce *et al.*, 1984). Studies of the mechanism of melatonin's effects in retina are just beginning to emerge. Its effect on photoreceptors, while demonstrated in *X. laevis,* has not been extended to other species. Furthermore, the effect may not be direct. Dubocovich (1983) recently showed that melatonin is a potent modulator of dopamine release in retina. Melatonin's effect on *Xenopus* photoreceptors is blocked by dopamine, raising the possibility that the effect may be mediated by a dopaminergic neuron (Pierce *et al.*, 1984; Pierce and Besharse, 1985; Iuvone, Part II). As will be shown in a subsequent section (Section IV,D), other data also suggest a role for neurotransmitter release in the control of disk shedding in *X. laevis* (Greenberger, 1984).

The demonstration of circadian properties of the disk-shedding rhythm have led to attempts to localize the circadian clock that influences shedding. Circadian clocks are endogenous timing mechanisms that exhibit a persistent period of approximately 24 hr under constant conditions and whose phase of activity can be entrained to diurnal signals (i.e., light and dark) in the external environment (Besharse, 1982). In vertebrates circadian clocks have previously been documented in the pineal gland and the suprachiasmatic nucleus of the hypothalamus (see Besharse *et al.*, 1984, for a review). Tierstein and colleagues (1980) and Flannery and Fisher (1980, 1984) have provided evidence consistent with the idea that a local clock in the eye has direct influence on the disk-shedding rhythm. In the former study, it was proposed that entrainment of the local circadian clock involved interaction with a central circadian oscillator in the central nervous system. That the effects of the clock on disk shedding may be mediated by an indoleamine is suggested by the recent *in vitro* demonstration of a local circadian clock controlling the activity of serotonin *N*-acetyltransferase, the rate-limiting enzyme for melatonin biosynthesis, in the retina (Besharse and Iuvone, 1983). By demonstrating sustained circadian oscillation under constant conditions *in vitro,* the latter study provided unequivocal evidence for a local clock and also provided envidence for a local entrainment pathway within the eye.

C. Possible Role of Postreceptoral Elements

As indicated above, studies on regulation of cone movement raise the interesting possibility that dopamine, presumably from an inner retinal source, is involved in the control mechanism (Pierce *et al.*, 1984; Pierce and Besharse, 1985). The recent demonstration that disk shedding in the *Xenopus* eye cup is dependent on millimolar concentrations of calcium in the culture medium (Greenberger and Besharse, 1983) led Greenberger (1984) to consider the idea that Ca^{2+}-dependent neurotransmitter release was involved (Llinas, 1979). In order to evaluate the possible role of postreceptoral elements in the control of disk shedding, he tested the hypothesis that glutamate or aspartate, photoreceptor neurotransmitter candidates (Wu and Dowling, 1978; Lasater and Dowling, 1982; Slaughter and Miller, 1983; Miller and Schwartz, 1983), might influence disk shedding. The rationale for the experiment was based on the extensive use of these compounds to isolate the photoreceptor response from activity of postreceptoral neurons. Generally, these compounds eliminate the b wave of the electroretinogram (Dowling and Ripps, 1972; Shimazaki *et al.*, 1984), presumably by interacting with postsynaptic receptors for excitatory amino acids (Slaughter and Miller, 1983). They are also potent neurotoxins, with specific effects on the inner retina (Olney, 1982). Although direct effects on photoreceptors or RPE cannot be ruled out (see Brown and Pinto, 1974; Shimazaki *et al.*, 1984), electrophysiological and morphological studies have consistently demonstrated their effects on neurons postsynaptic to photoreceptors.

Surprisingly, both glutamate and asparate caused a dose-dependent stimulation of rod disk shedding over the same concentration range generally employed to block inner retinal physiological activity (Greenberger, 1984; Greenberger and Besharse, 1985). Disk shedding reached a peak in those experiments three- to fivefold higher than normally expected, and appeared to be synchronized so that the peak was achieved somewhat earlier than expected in the light evoked response. Furthermore, the increase in disk shedding was correlated with a dose-dependent increase in the extent of swelling in the inner plexiform and inner nuclear layers. There was no effect other than that related to disk shedding on rod photoreceptors or pigment epithelium. Further analysis revealed that kainic acid, a glutamate analog (Watkins, 1978), produced similar effects on disk shedding and inner retinal morphology but was effective at much lower concentrations.

Although the results with excitatory amino acids are consistent with a role for retinal neurons in the control of shedding, the D isomers of glutamate and aspartate as well as several other L-amino acids also stimulate shedding. This raises questions about specificity and mechanism of the effect. Interestingly, glutamine, taurine, cycloleucine, and α-aminobutyric acid all caused massive shedding with only moderate neurotoxic effects on the inner retina. Although we cannot rule out a direct effect on postreceptoral neurons, stimulation of disk shedding by amino acids may actually be mediated within the photoreceptor–pigment epithelial complex or through altered amino acid and ionic homeostasis through interuption of amino acid uptake mechanisms (Greenberger and Besharse, 1985). Regardless of the locus of the effect, however, the discovery of amino acid stimulated shedding has important implications for both the control of disk shedding and postreceptoral mechanisms in the vertebrate retina.

D. *Role of Ca^{2+} and cAMP*

The complicated interaction of photoreceptor and pigment epithelium culminating in disk shedding has made it difficult to resolve the relative roles of the two cell types. Previous studies using an *in vitro* eye cup preparation have shown that in addition to light, such factors as HCO_3^- concentration (Besharse *et al.*, 1980; Basinger and Hoffman, 1982), Ca^{2+} concentration (Greenberger and Besharse, 1983), cAMP (Besharse *et al.*, 1982b; Eckmiller and Burnside, 1983; Heath and Basinger, 1983), melatonin (Besharse and Dunis, 1983), ouabain (Williams *et al.*, 1984), and altered Na^+ concentration (Williams *et al.*, 1984) all have effects on disk shedding. The addition of a possible regulatory pathway involving activity in the neural retina amplifies the problem of localizing the site at which those effectors act. It must be realized that each perturbation of disk shedding could be mediated through postreceptoral neurons or glia, the photoreceptor, or the pigment epithelium.

That they may have an effect on postreceptoral neurons represents an interesting possibility for both melatonin and Ca^{2+}. In the former case the major established effect of the compound at the cellular level is in the modulation of dopamine release (Zisapel *et al.*, 1982) which was recently demonstrated in the retina as well (Dubocovich, 1983). In this regard, it is of some interest that the effect of melatonin on cone movement in eye cups is blocked by dopamine (Pierce *et al.*, 1984; Pierce and Besharse, 1985). The well-established requirement for Ca^{2+} in neurotransmitter release (Llinas, 1979), as pointed out above, was the immediate reason for our investigation of possible neuronal involvement in disk shedding and remains a plausible explanation for the extraordinary sensitivity of disk shedding to reduction of extracellular Ca^{2+} (Greenberger and Besharse, 1983). For example, disk shedding is also blocked by Co^{2+} and other Ca^{2+} antagonists (Besharse *et al.*, 1986). This is not to say that other Ca^{2+}-requiring events are not involved. Certainly, there is precedent for Ca^{2+} regulation of the complex reorganization of the pigment epithelial cytoskeleton that accompanies shedding. However, those intracellular events would not necessarily require high levels of extracellular Ca^{2+} in short-term experiments. Consistent with this expectation, reduction of Ca^{2+} to submicromolar levels has little or no effect on aspartate- or kainic acid-induced shedding (Greenberger, 1984; Besharse *et al.*, 1986). This shows that the entire process of disk detachment and internalization can occur when extracellular Ca^{2+} is reduced, and suggests that amino acids mimic the effects of an endogenous agent normally released in a Ca^{2+} dependent manner.

In contrast, a direct effect on the photoreceptor–pigment epithelial complex represents an attractive possible mechanism for the inhibitory effect of cAMP on disk shedding (Besharse *et al.*, 1982b; Eckmiller and Burnside, 1983). Conditions expected to alter cAMP levels have been shown to mimic darkness by promoting dark-adaptive retinomotor movements of rods, cones, and pigment epithelium (Burnside *et al.*, 1982; Besharse *et al.*, 1982; Burnside and Nagle, 1983). Parallel studies using detergent-lysed models indicate that the effects of cAMP on retinomotor movements are direct (see Burnside and Nagle, 1983; Burnside and Dearry, this volume). Furthermore, there is evidence (reviewed in Burnside and Nagle, 1983) that cAMP levels actually increase in the photoreceptor pigment epithelial complex at night. This has led to the view that dark-adaptive retinomotor movements (Burnside *et al.*, 1982) as well as the night-time inhibition of disk shedding (Besharse *et al.*, 1982b) are influenced by local levels of cAMP. For disk shedding, cAMP may be involved in the regulation of disk detachment or the phagocytic machinery itself (see Besharse *et al.*, 1982).

V. Summary

This review summarizes the unique features of the vertebrate photoreceptor as a model for cell biological studies of membrane turnover. The highly polarized structure and differentiated membrane domains of the photoreceptor have made it accessible to detailed analysis by both morphological and biochemical approaches. Such studies have led to an understanding of the phenomena involved in turnover and have begun to yield insight into the principal cellular–molecular processes. Three general areas of cell biological research are emphasized in considering these processes. First, the cotranslational incorporation of opsin into membrane and its vectorial delivery to the region of disk assembly represent a paradigm for membrane assembly that may be of general applicability in a wide range of polarized cells. Second, the extraordinary cytoskeleton–membrane interaction in the connecting cilium and periciliary region provides a model system for understanding how cytoskeleton and membranes interact to generate and maintain membrane microdomains. Third, the unique mechanism for turnover of disks, involving phagocytic activity of an adjacent epithelial cell, represents a unique case of highly regulated cell–cell interaction.

Our current understanding of the cellular and molecular events of turnover, make it possible to raise additional questions and to identify areas of further research. For example, the recent cloning of opsin cDNA (Nathans and Hogness, 1983) may make it possible in the immediate future to elucidate the molecular mechanisms of insertion and sorting of this membrane protein. Further studies on lipid metabolism in photoreceptors should lead to a better understanding of how membrane protein and lipid turnover are coordinated. A more complete understanding of the delivery of both lipid and protein precursors into outer segment disks will likely result from further research on the ciliary membrane as a barrier to diffusion and as a possible conduit for delivery of disk precursors to the outer segment. Finally, further research directed at the recognition signals used by RPE for disk phagocytosis and the mechanism of shedding activation should help in understanding how the extraordinary balance between assembly and degradation is achieved and maintained. Current active research in each of these areas suggests that immediate progress is likely.

The cell biological features of photosensitive membrane turnover occur with a high degree of temporal order. Light and darkness as well as an endogenous biological clock all influence turnover. The recent development of *in vitro* systems that sustain aspects of turnover characteristic of the intact eye have made it possible to address issues related to regulation at both the cellular and supracellular levels. Such studies have begun to elucidate the role of both intracellular messengers (i.e., Ca^{2+} and cAMP) as well as intercellular messengers (i.e., neurotransmitters and neurohormones) in the regulation of membrane turnover. This approach has led directly to the hypotheses that a retinal indoleamine may mediate the effects of a local biological clock, and that neurotransmitter release by postreceptoral neurons may influence turnover. Such studies may ultimately provide a basis for understanding the temporal control of specific cellular–molecular events of turnover.

Acknowledgments

Many people contributed to the preparation of this review. I would particularly like to thank the following people: Mary Pierce, Brian Matsumoto, and Dennis Defoe for their critical reading of an early version of the manuscript; Donna Forestner, Brian Matsumoto, and Gwen Spratt for preparation of figures; Holly Berry for checking on numerous references; and Sherry Wilson for her work at the word processor. In addition, I thank the following individuals for making available preprints of manuscripts, data from unpublished work, or sound advice: Gene Anderson, Don Anderson, Scott Basinger, Mike Chaitin, Steve Fliesler, Mike Iuvone, Mike Kaplan, Brian Matsumoto, and Win Sale. The recent work from my own laboratory was supported by National Institutes of Health Research Grants EY02414, EY03222 and Research Career Development Award EY00169.

References

Allan, D., and Michell, R. H. (1975). Accumulation of 1,2-diacylglycerol in the plasma membrane may lead to echinocyte transformation of erythrocytes. *Nature (London)* **258,** 348–349.

Anderson, D. H., and Fisher, S. K. (1975). Disc shedding in rodlike and conelike photoreceptors of tree squirrels. *Science* **187,** 953–955.

Anderson, D. H., and Fisher, S. K. (1976). The photoreceptors of diurnal squirrels: Outer segment structure disc shedding, and protein renewal. *J. Ultrastruct. Res.* **55,** 119–141.

Anderson, D. H., Fisher, S. K., and Steinberg, R. H. (1978). Mammalian cones: disc shedding, phagocytosis and renewal. *Invest. Ophthalmol. Visual Sci.* **17,** 117–133.

Anderson, D. H., Fisher, S. K., and Breding, D. J. (1986). A concentration of fucosylated glycoconjugates at the base of cone outer segments; quantitative electron microscope autoradiography. *Exp. Eye Res.* **42,** in press.

Anderson, R. E., and Hollyfield, J. G. (1981). Light stimulates the incorporation of inositol into phosphatidylinositol in the retina. *Biochim. Biophys. Acta* **665,** 619–622.

Anderson, R. E., and Kelleher, P. A. (1981). Biosynthesis of retinal phospholipids by base exchange reactions. *Exp. Eye Res.* 32, 729–736.

Anderson, R. E., and Maude, M. B. (1970). Phospholipids of bovine rod outer segments. *Biochemistry* **9,** 3624–3628.

Anderson, R. E., Kelleher, P. A., Maude, M. B., and Maida, T. M. (1980a). Synthesis and turnover of lipid and protein components of frog retinal rod outer segments. *In* "Neurochemistry of the Retina" (N. G. Bazan and R. N. Lolley, eds.), pp. 29–42. Pergamon, Oxford.

Anderson, R. E., Maude, M. B., Kelleher, P. A., Maida, T. M., and Basinger, S. F. (1980b). Metabolism of phosphatidylcholine in the frog retina. *Biochim. Biophys. Acta* **620,** 212–226.

Anderson, R. E., Kelleher, P. A., and Maude, M. B. (1980c). Metabolism of phosphatidylethanolamine in the frog retina. *Biochim. Biophys. Acta* **620,** 227–235.

Anderson, R. E., Maude, M. B., and Kelleher, P. A. (1980d). Metabolism of phosphatidylinositol in the frog retina. *Biochim. Biophys. Acta* **620,** 236–246.

Anderson, R. E., Maude, M. B., Kelleher, P. A., Rayborn, M. E., and Hollyfield, J. G. (1983). Phosphoinositide metabolism in the retina: Localization to horizontal cells and regulation by light and divalent cations. *J. Neurochem.* **41,** 764–771.

Anderson, R. E., Maude, M. B., Pu, G. A.-W., and Hollyfield, J. G. (1985). Effect of light on the metabolism of lipids in the rat retina. *J. Neurochem.* **44,** 773–778.

Andrews, L. D. (1982). Freeze-fracture studies of vertebrate photoreceptor membranes. *In* "The Structure of the Eye" (J. G. Hollyfield, ed.), pp. 11–23. Elsevier, Amsterdam.

Andrews, L. D., and Cohen, A. I. (1979). Freeze-fracture evidence for the presence of cholesterol in particle-free patches of basal disks and the plasma membrane of retinal rod outer segments of mice and frogs. *J. Cell Biol.* **81,** 215–228.

Andrews, L. D., and Cohen, A. I. (1981). Freeze-fracture studies of the structure of rod outer segment membranes: New observations regarding the distribution of particle-free patches and the location of fracture planes in conventionally prepared retinas. *Exp. Eye Res.* **33,** 1–10.

Andrews, L. D., and Cohen, A. I. (1983). Freeze-fracture studies of photoreceptor membranes: New observations bearing upon the distribution of cholesterol. *J. Cell Biol.* **97,** 749–755.

Barak, L. S., Yocum, R. R., Nothnagel, E. A., and Webb, W. W. (1980). Fluorescence staining of the actin cytoskeleton in living cells with 7-nitrobenz-2-oxa-1,3-diazole-phallicidin. *Proc. Natl. Acad. Sci. U.S.A.* **77,** 980–984.

Basinger, S. F., and Gordon, W. C. (1982). Local stimulation induces shedding throughout the frog retina. *Vision Res.* **22,** 1533–1538.

Basinger, S., and Hoffman, R. (1976). Phosphatidyl choline metabolism in the frog rod photoreceptor. *Exp. Eye Res.* **23,** 117–126.

Basinger, S. F., and Hoffman, R. T. (1982). Regulation of rod shedding in the frog retina. *In* "The Structure of the Eye" (J. G. Hollyfield, ed.), pp. 75–83. Elsevier, Amsterdam.

Basinger, S. F., and Hoffman, R. T. (1983). Phagosome size is independent of renewal rate. *Invest. Ophthalmol. Visual Sci. Suppl.* **24,** 279 (Abstr.).

Basinger, S. F., and Hollyfield, J. G. (1980). Control of rod shedding in the frog retina. *In*

"Neurochemistry of the Retina" (N. G. Bazan and R. W. Lolley, eds.), pp. 81–92. Pergamon, Oxford.

Basinger, S., Bok, D., and Hall, M. (1976). Rhodopsin in the rod outer segment plasma membrane. *J. Cell Biol.* **69,** 29–42.

Basinger, S. F., Hoffman, R., and Matthes, M. (1976). Photoreceptor shedding is initiated by light in the frog retina. *Science* **194,** 1074–1076.

Besharse, J. C. (1982). The daily light–dark cycle and rhythmic metabolism in the photoreceptor–pigment epithelial complex. *Prog. Retinal Res.* **1,** 81–124.

Besharse, J. C., and Dunis, D. A. (1982). Rod photoreceptor disc shedding *in vitro:* Inhibition by cytochalasins and activation by colchicine. *In* "The Structure of the Eye" (J. G. Hollyfield, ed.), pp. 85–96. Elsevier, Amsterdam.

Besharse, J. C., and Dunis, D. A. (1983). Methoxyindoles and photoreceptor metabolism: Activation of rod shedding. *Science* **219,** 1341–1343.

Besharse, J. C., and Forestner, D. M. (1981). Horseradish peroxidase uptake by rod photoreceptor inner segments accompanies outer segment disc assembly. *Annu. Proc. Electron Microsc. Soc. Am. 39th* 486–487.

Besharse, J. C., and Forestner, D. M. (1983). Membrane domains of developing photoreceptor cilia revealed by lectins and antiopsin. *J. Cell Biol.* **97,** 411a.

Besharse, J. C., and Iuvone, P. M. (1983). Circadian clock in *Xenopus* eye controlling retinal serotonin *N*-acetyltransferase. *Nature (London)* **305,** 133–135.

Besharse, J. C., and Pfenninger, K. H. (1980). Membrane assembly in retinal photoreceptors. I. Freeze-fracture analysis of cytoplasmic vesicles in relationship to disc assembly. *J. Cell Biol.* **75,** 507–527.

Besharse, J. C., Hollyfield, J. G., and Rayborn, M. E. (1977a). Turnover of rod photoreceptor outer segments. II. Membrane addition and loss in relationship to light. *J. Cell Biol.* **75,** 507–527.

Besharse, J. C., Hollyfield, J. G., and Rayborn, M. E. (1977b). Photoreceptor outer segments: Accelerated membrane renewal in rods after exposure to light. *Science* 196, 536–538.

Besharse, J. C., Terrill, R. O., and Dunis, D. A. (1980). Light evoked shedding by rod photoreceptors *in vitro:* Relationship to medium, bicarbonate concentration. *Invest. Ophthalmol. Visual Sci.* **19,** 1512–1517.

Besharse, J. C., Defoe, D. M., and Forestner, D. M. (1982a). Opsin phosphorylation in rod photoreceptors: Autoradiographic analysis using ^{33}P. *J. Cell Biol.* **95,**264a.

Besharse, J. C., Dunis, D. A., and Burnside, B. (1982b). Effects of cyclic adenosine 3′,5′-monophosphate on disc shedding and retinomotor movement. Inhibition of rod shedding and stimulation of cone elongation. *J. Gen. Physiol.* **79,** 775–790.

Besharse, J. C., Forestner, D. M., and Defoe, D. M. (1985). Membrane assembly in retinal photoreceptors. III. Distinct membrane domains of the connecting cilium of developing rods. *J. Neurosci.* **5,** 1035–1048.

Besharse, J. C., Dunis, D. A., and Iuvone, P. M. (1984). Regulation and possible role of serotonin *N*-acetyltransferase in the retina. *Fed. Proc. Fed. Am. Soc. Exp. Biol.* **43,** 34–38.

Besharse, J. C., Spratt, G., and Forestner, D. M. (1986). Light-evoked and kainic acid induced disc shedding by rod photoreceptors: differential sensitivity to extracellular calcium. Submitted.

Bibb, C., and Young, R. W. (1974a). Renewal of fatty acids in the membranes of visual cell outer segments. *J. Cell Biol.* **61,** 327–343.

Bibb, C., and Young, R. W. (1974b). Renewal of glycerol in the visual cells and pigment epithelium of the frog retina. *J. Cell Biol.* **62,** 378–389.

Binkley, S., Reilly, K. B., and Hryshchyshyn, M. (1980). *N*-Acetyltransferase in the chick retina. I. Circadian rhythms controlled by environmental lighting are similar to those in the pineal gland. *J. Comp. Physiol.* **139,** 103–108.

Blobel, G., Walter, P., Chang, C. N., Goldman, B. M., Erickson, A. H., and Lingappa, V. R. (1979). Translocation of proteins across membranes: The signal hypothesis and beyond. *Symp. Soc. Exp. Biol.* **33,** 9–36.

Bloodgood, R. A. (1977). lotility occurring in association with the surface of the *Chlamydomonas* flagellum. *J. Cell Biol.* **75,** 983–989.

Bok, D. (1982). Autoradiographic studies on the polarity of plasma membrane receptors in retinal pigment epithelial cells. *In* "The Structure of the Eye" (J. G. Hollyfield, ed.), pp. 247–256. Elsevier, Amsterdam.

Bok, D., and Hall, M. O. (1971). The role of the pigment epithelium in the etiology of inherited retinal dystrophy in the rat. *J. Cell Biol.* **49,** 664–682.

Bok, D., and Young, R. W. (1972). The renewal of diffusely distributed protein in the outer segments and rods and cones. *Vision Res.* **12,** 161–168.

Bok, D., and Young, R. W. (1979). Phagocytic properties of the retinal pigment epithelium. *In* "The Retinal Pigment Epithelium" (K. M. Zinn and M. F. Marmor, eds.), pp. 148–174. Harvard Univ. Press, Cambridge, Massachusetts.

Bok, D., Hall, M. O., and O'Brien, P. (1977). The biosynthesis of rhodopsin as studied by membrane renewal in rod outer segments. *In* "International Cell Biology 1976–1977" (B. R. Brinkley and K. R. Porter, eds.), pp. 608–617. The Rockefeller Univ. Press, New York.

Bridges, C. D. B. (1972). The rhodopsin–porphyropsin visual system. *In* "Handbook of Sensory Physiology. Vol. 7. Photochemistry of Vision" (H. J. A. Dartnall, ed.), pp. 417–480. Springer-Verlag, Berlin and New York.

Bridges, C. D. B. (1976). Vitamin A and the role of the pigment epithelium during bleaching and regeneration of rhodopsin in the frog eye. *Exp. Eye Res.* **22,** 435–455.

Brown, J. E., and Pinto, L. H. (1974). Ionic mechanism for the photoreceptor potential of the retina of *Bufo marinus. J. Physiol. (London)* **236,** 575–591.

Bunt, A. H. (1978). Fine structure and radioautography of rabbit photoreceptor cells. *Invest. Ophthalmol. Visual Sci.* **17,** 90–104.

Bunt, A. H. and Klock, I. B. (1980a). Fine structure and radioautography of retinal cone outer segments in goldfish and carp. *Invest. Ophthalmol. Visual Sci.* **17,** 707–719.

Bunt, A. H., and Klock, I. B. (1980b). Comparative study of ^{3}H-fucose incorporation into vertebrate photoreceptor outer segments. *Vision Res.* **20,** 739–747.

Bunt, A. H., and Saari, J. C. (1982). Fucosylated protein of retinal cone photoreceptor outer segments: Morphological and biochemical analysis. *J. Cell Biol.* **92,** 269–276.

Burnside, B., and Laties, A. M. (1976). Actin filaments in apical projections of the primate pigmented epithelial cell. *Invest. Ophthalmol.* **15,** 570–575.

Burnside, B., and Nagle, B. (1983). Retinomotor movements of photoreceptors and retinal pigment epithelium: Mechanisms and regulation. *Prog. Retinal Res.* **2,** 67–109.

Burnside, B., Evans, M., Fletcher, R. T., and Chader, G. J. (1982). Induction of dark-adaptive retinomotor movement (cell elongation) in teleost retinal cones by cyclic adenosine 3′,5′-monophosphate. *J. Gen. Physiol.* **79,** 759–774.

Chaitin, M. H., and Hall, M. O. (1983a). Defective ingestion of rod outer segments by cultured dystrophic rat pigment epithelium. *Invest. Ophthalmol. Visual Sci.* **24,** 812–820.

Chaitin, M. H., and Hall, M. O. (1983b). The distribution of actin in cultured normal and dystrophic rat epithelial cells during the phagocytosis of rod outer segments. *Invest. Ophthalmol. Visual Sci.* **24,** 821–831.

Chaitin, M. H., Schneider, B. G., Hall, M. O., and Papermaster, D. S. (1984). Actin in the photoreceptor connecting cilium: Immunocytochemical localization to the site of outer segment disk formation. *J. Cell Biol.* **99,** 239–247.

Cherry, R. J. (1979). Rotational and lateral diffusion of membrane proteins. *Biochim. Biophys. Acta* **559,** 289–327.

Clark, V. M., and Hall, M. O. (1982). Labeling of bovine rod outer segment surface proteins with ^{125}I. *Exp. Eye Res.* **34,** 847–859.

Cohen, A. I. (1968). New evidence supporting the linkage to the extracellular space of outer segment saccules of frog cones but not rods. *J. Cell Biol.* **37,** 424–444.

Cohen, A. I. (1970). Further studies on the question of patency of saccules in outer segments of vertebrate photoreceptors. *Vision Res.* **10,** 445–453.

Currie, J. R., Hollyfield, J. G., and Rayborn, M. E. (1978). Rod outer segments elongate in constant light; Darkness is required for normal shedding. *Vision Res.* **18,** 995–1003.

Custer, N. V., and Bok, D. (1975). Pigment epithelium–photoreceptor interactions in the normal and dystrophic rat retina. *Exp. Eye Res.* **21,** 153–166.

Daeman, F. J. M. (1973). Vertebrate rod outer segment membranes. *Biochim. Biophys. Acta* **300,** 255–288.

de Bazan, H. E. P., and Bazan, N. G. (1977). Effects of temperature, ionic environment, and light flashes on the glycerolipid neosynthesis in the toad retina. *In* "Function and Biosynthesis of Lipids" (N. G. Bazan, R. R. Brenner, and N. M. Giusto, eds.), pp. 489–495. Plenum, New York.

Defoe, D. M., and Besharse, J. C. (1983). Membrane assembly in retinal photoreceptors: Immunocytochemical analysis of freeze-fractured rod photoreceptor membranes using anti-opsin antibodies. *J. Cell Biol.* **97,** 412a.

Defoe, D. M., and Besharse, J. C. (1985). Membrane assembly in retinal photoreceptors. II. Immunocytochemical analysis of freeze-fractured rod photoreceptor membranes using antiopsin antibodies. *J. Neurosci.* **5,** 1023–1034.

Defoe, D. M., and Bok, D. (1983). Rhodopsin chromophore exchanges among opsin molecules in the dark. *Invest. Ophthalmol. Visual Sci.* **24,** 1211–1226.

Dentler, W. L. (1981). Microtubule–membrane interactions in cilia and flagella. *Int. Rev. Cytol.* **72,** 1–47.

Dentler, W. L., Pratt, M. M., and Stephens, R. E. (1980). Microtubule–membrane interactions in cilia. II. Photochemical cross-linking of bridge structures and the identification of a membrane-associated dynein-like ATPase. *J. Cell Biol.* **84,** 381–403.

Ditto, M. (1975). A difference between developing rods and cones in the formation of outer segment membranes. *Vision Res.* **15,** 535–536.

Donnelly, J. K., and Wyse, J. P. H. (1982). An underlying order to outer segment (OS) length: A new approach for investigation of renewal. *Invest. Ophthalmol. Visual Sci. Suppl.* **22,** 284 (Abstr.).

Dowling, J. E., and Ripps, H. (1972). Adaptation in skate photoreceptors. *J. Gen. Physiol.* **60,** 698–719.

Dowling, J. E., and Sidman, R. L. (1962). Inherited retinal dystrophy in the rat. *J. Cell Biol.* **14,** 73–109.

Dubocovich, M. L. (1983). Melatonin is a potent modulator of dopamine release in the retina. *Nature (London)* **306,** 782–784.

Dudley, P. A., and Anderson, R. E. (1978). Phospholipid transfer from bovine retina with high activity towards retinal rod disc membranes. *FEBS Lett.* **95,** 57–60.

Dudley, P. A., Alligood, J. P., and O'Brien, P. (1984). Biochemical events related to circadian photoreceptor shedding. *In* "Molecular and Cellular Basis of Visual Acuity" (S. R. Hilfer and J. B. Sheffield, eds.), pp. 13–30. Springer-Verlag, Berlin and New York.

Easter, S. S., Jr., and Macy, A. (1978). Local control of retinomotor activity in the fish retina. *Vision Res.* **18,** 937–942.

Eckmiller, M. S., and Burnside, B. (1983). Light induced photoreceptor shedding in teleost retina blocked by dibutyryl cyclic AMP. *Invest. Ophthalmol. Visual Sci.* **24,** 1328–1332.

Edwards, R. B., and Szamier, R. B. (1977). Defective phagocytosis of isolated rod outer segments by RCS rat retinal pigment epithelium in culture. *Science* **197,** 1001–1003.

Elner, V. M., Schaffer, T., Taylor, K., and Glagov, S. (1981). Immunophagocytic properties of retinal pigment epithelial cells. *Science* **211,** 74–76.

Feeney, L., and Mixon, R. N. (1976). An *in vitro* model of phagocytosis in bovine and human retinal pigment epithelium. *Exp. Eye Res.* **22,** 533–548.

Fisher, S. K., and Steinberg, R. H. (1982). Origin and organization of pigment epithelial apical projections to cones in cat retina. *J. Comp. Neurol.* **206,** 131–145.

Flannery, J. G., and Fisher, S. K. (1980). Evidence for an intraoculaar circadian oscillator in *Xenopus* eye explants *in vitro. Proc. Int. Soc. Eye Res.* **1,** 85 (Abstr.).

Flannery, J. G., and Fisher, S. K. (1984). Circadian disc shedding in *Xenopus* retina *in vitro. Invest. Ophthalmol. Visual Sci.* **25,** 229–232.

Fliesler, S. J., and Anderson, R. E. (1983). Chemistry and metabolism of lipids in the vertebrate retina. *Prog. Lipid Res.* **22,** 79–131.

Fliesler, S. J., and Basinger, S. F. (1985). Tunicamycin blocks the incorporation of opsin into retinal rod outer segment membranes. *Proc. Natl. Acad. Sci. U.S.A.* **82,** 1116–1120.

Fliesler, S. J., Tabor, G. A., and Hollyfield, J. G. (1984). Glycoprotein synthesis in the human retina: Localization of the lipid intermediate pathway. *Exp. Eye Res.* **39,** 153–173.

Fliesler, S. J., Rayborn, M. E., and Hollyfield, J. G. (1985). Membrane morphogenesis in retinal rod outer segments: Inhibition by tunicamycin. *J. Cell Biol.* **100,** 574–587.

Fukuda, M. N., Papermaster, D. S., and Hargrave, P. A. (1979). Rhodopsin carbohydrate: Structure of small oligosaccarides attached at two sites near the NH_2 terminus. *J. Biol. Chem.* **254,** 8201–8207.

Ghalayini, A., and Anderson, R. E. (1984). Phosphatidylinositol 4,5-bisphosphate: Light mediated breakdown in the vertebrate retina. *Biochem. Biophys. Res. Commun.* **124,** 503–506.

Gilula, N. B., and Satir, P. (1972). The ciliary necklace: A ciliary membrane specialization. *J. Cell Biol.* **53,** 494–509.

Goldman, A. I. (1982). The sensitivity of rat rod outer segment shedding to light. *Invest. Ophthalmol. Visual Sci.* **22,** 695–700.

Goldman, A. I., and O'Brien, P. J. (1978). Phagocytosis in the retinal pigment epithelium of the RCS rat. *Science* **201,** 1023–1025.

Goldman, A. I., Tierstein, P. S., and O'Brien, P. J. (1980). The role of ambient lighting in circadian disc sheding in the rod outer segment of the rat retina. *Invest. Ophthalmol. Visual Sci.* **19,** 1257–1267.

Goldman, B. M., and Blobel, G. (1981). *In vitro* biosynthesis, core glycosylation, and membrane integration of opsin. *J. Cell Biol.* **90,** 236–242.

Greenberger, L. M. (1984). Photoreceptor disc shedding in eye cups: Inhibition by deletion of extracellular divalent cations and stimulation by aspartate and other amino acids. Ph.D. dissertation, Emory University School of Medicine, Atlanta, Georgia.

Greenberger, L. M., and Besharse, J. C. (1983). Photoreceptor disc shedding in eye cups: Inhibition by deletion of extracellular divalent cations. *Invest. Ophthalmol. Visual Sci.* **24,** 1456–1464.

Greenberger, L. M., and Besharse, J. C. (1985). Stimulation of photoreceptor disc shedding by glutamate, aspartate and other amino acids. *J. Comp. Neurol.* **239,** 361–372.

Griffin, F. M., Griffin, J. A., and Silverstein, S. C. (1976). Studies on the mechanism of phagocytosis. II. The interaction of macrophages with anti-immunoglobulin IgG-coated bone marrow-derived lymphocytes. *J. Exp. Med.* **144,** 788–809.

Griffiths, G., Quinn, P., and Warren, G. (1983). Dissection of the Golgi complex. I. Monensin inhibits the transport of viral membrane proteins from medial to trans-Golgi cisternae in baby hamster kidney cells infected with Semliki forest virus. *J. Cell Biol.* **96,** 835–850.

Hagins, W. A. (1972). The visual process: Excitatory mechanisms in the primary photoreceptor cells. *Annu. Rev. Biophys. Bioeng.* **1,** 131–158.

Hall, M. O. (1978). Phagocytosis of light- and dark-adapted rod outer segments by cultured pigment epithelium. *Science* **202,** 526–528.

Hall, M. O., Bok, D., and Bacharach, A. D. E. (1969). Biosynthesis and assembly of the rod outer segment membrane system. Formation and fate of visual pigment in the frog retina. *J. Mol. Biol.* **45,** 397–406.

Hall, M. O., Basinger, S. F., and Bok, D. (1973). Studies on the assembly of rod outer segment disc membranes. *In* "Biochemistry and Physiology of Visual Pigments" (H. Langer, ed.), pp. 319–326. Springer-Verlag, Berlin and New York.

Hamm, H. E., and Menaker, M. (1980). Retinal rhythms in chicks: Circadian variation in melatonin and serotonin *N*-acetyltransferase activity. *Proc. Natl. Acad. Sci. U.S.A.* **77,** 4998–5002.

Hargrave, P. A. (1982). Rhodopsin chemistry, structure and topography. *Prog. Retinal Res.* **1,** 1–51.

Hartwig, J. H., Yin, H. L., and Stossel, T. P. (1980). Contractile proteins and the mechanism of phagocytosis in macrophages. *In* "Mononuclear Phagocytes Functional Aspects Part II" (R. van Furth, ed.), pp. 971–996. Nijhoff, The Hague.

Hayashi, F., and Amakawa, T. (1985). Light-mediated breakdown of phosphatidylinositol-4,5-bisphosphate in isolated rod outer segments of frog photoreceptor. *Biochem. Biophys. Res. Commun.* **128,** 954–959.

Heath, A. R., and Basinger, S. F. (1983). Frog rod outer segment shedding *in vitro. Invest. Ophthalmol. Visual Sci.* **24,** 277–284.

Heller, J. (1968). Structure of visual pigments. I. Purification, molecular weight, and composition of bovine visual pigment-500. *Biochemistry* **7,** 2906–2913.

Herron, W. L., Reigel, B. W., Myers, O. E., and Rubin, M. L. (1969). Retinal dystrophy in the rat—pigment epithelial disease. *Invest. Ophthalmol.* **8,** 595–604.

Heuser, J. E., Reese, T. S., Dennis, M. J., Jan, Y., Jan, L., and Evans, L. (1979). Synaptic vesicle exocytosis captured by quick freezing and correlated with quantal transmitter release. *J. Cell Biol.* **81,** 275–300.

Hogan, M. J., Wood, I., and Steinberg, R. H. (1974). Phagocytosis by pigment epithelium of human retinal cones. *Nature (London)* **252,** 305–307.

Hollyfield, J. G. (1976). Phagocytic capabilities of the pigment epithelium. *Exp. Eye Res.* **22,** 457–468.

Hollyfield, J. G. (1979). Membrane addition to photoreceptor outer segments: Progressive reduction in the stimulatory effect of light with increased temperature. *Invest. Ophthalmol. Visual Sci.* **18,** 977–981.

Hollyfield, J. G., and Basinger, S. F. (1978). Photoreceptor shedding can be initiated within the eye. *Nature (London)* **274,** 794–796.

Hollyfield, J. G., and Ward, A. (1974a). Phagocytic activity in the retinal pigment epithelium of the frog *Rana pipiens.* I. Uptake of polystyrene spheres. *J. Ultrastruct. Res.* **46,** 327–338.

Hollyfield, J. G., and Ward, A. (1974b). Phagocytic activity in the retinal pigment epithelium of the frog *Rana pipiens.* II. Exclusion of *Sarcina subflava. J. Ultrastruct. Res.* **46,** 339–350.

Hollyfield, J. G., and Witkovsky, P. (1974). Pigmented retinal epithelium involvement in photoreceptor development and function. *J. Exp. Zool.* **189,** 357–378.

Hollyfield, J. G., Besharse, J. C., and Rayborn, M. E. (1976). The effect of light on the quantity of phagosomes in the pigment epithelium. *Exp. Eye Res.* **23,** 623–635.

Hollyfield, J. G., Besharse, J. C., and Rayborn, M. E. (1977). Turnover of rod photoreceptor outer segments. I. Membrane addition and loss in relationship to temperature. *J. Cell Biol.* **75,** 490–506.

Hollyfield, J. G., Rayborn, M. E., Verner, G. E., Maude, M. B., and Anderson, R. E. (1982). Membrane addition to rod photoreceptor outer segments: Light stimulates membrane assembly in the absence of increased membrane biosynthesis. *Invest. Ophthalmol. Visual Sci.* **32,** 417–427.

Hollyfield, J. G., Varner, H. H., Rayborn, M. E., and Bridges, C. D. (1985a). Participation of photoreceptor cells in retrieval and degradation of components in the interphotoreceptor matrix. *In* "The Interphotoreceptor Matrix in Health and Disease" (C. D. Bridges and A. J. Adler, eds.), pp. 171–175. Liss, New York.

Hollyfield, J. G., Fliesler, S. J., Rayborn, M. E., Fong, A.-L., Landers, R. A., and Bridges, D. D. (1985b). Synthesis and secretion of interstitial retinol-binding protein by the human retina. *Invest. Ophthalmol. Visual Sci.* **26,** 58–67.

Hollyfield, J. G., Varner, H. H., Rayborn, M. E., Liou, G. I., and Bridges, C. D. (1985c). Endocytosis and degradation of interstitial retinol-binding protein: Differential capabilities of cells that border the interphotoreceptor matrix. *J. Cell Biol.* **100,** 1676–1681.

Holtzman, E., and Mercurio, A. M. (1980). Membrane circulation in neurons and photoreceptors: Some unresolved issues. *Int. Rev. Cytol.* **67,** 1–67.

Holtzman, E., Schacher, S., Evans, J., and Teichberg, S. (1977). Origin and fate of the membranes of secretion granules and synaptic granules and synaptic vesicles: Membrane circulation in neurons, gland cells and retinal photoreceptors. *In* "The Synthesis, Assembly, and Turnover of Cell Surface Components" (G. Poste and G. L. Nicholson, eds.), pp. 165–246. Elsevier, Amsterdam.

Iuvone, P. M., and Besharse, J. C. (1983). Regulation of indoleamine *N*-acetyltransferase activity in the retina: Effects of light and dark, protein synthesis inhibitors and cyclic nucleotide analogs. *Brain Res.* **273,** 111–119.

Jan, L. Y., and Revel, J. P. (1974). Ultrastructural localization of rhodopsin in the vertebrate retina. *J. Cell Biol.* **62,** 257–273.

Kaplan, M. W. (1984). Shedding is correlated with disk membrane axial position rather than disk age in *Xenopus laevis* rod outer segments. *Vision Res.* **24,** 1163–1168.

Kapoor, C. L., and Chader, G. J. (1984). Endogenous phosphorylation of retinal photoreceptor outer segment proteins by calcium phospholipid-dependent protein kinase. *Biochem. Biophys. Res. Commun.* **122,** 1397–1403.

Kean, E. L. (1977). The biosynthesis of mannolipids and mannose-containing complex glycans by retina. *J. Supramol. Struct.* **7,** 381–395.

Kinney, M. S., and Fisher, S. K. (1978a). The photoreceptors and pigment epithelium of the adult *Xenopus* retina: Morphology and outer segment renewal. *Proc. R. Soc. London Ser. B* **201,** 131–147.

Kinney, M. S., and Fisher, S. K. (1978b). The photoreceptors and pigment epithelium of the larval *Xenopus* retina: Morphogenesis and outer segment renewal. *Proc. R. Soc. London Ser. B* **201,** 149–167.

Kornfeld, R., and Kornfeld, S. (1980). Structure of glycoproteins and their oligosaccharide units. *In* "The Biochemistry of Glycoproteins and Proteoglycans" (W. J. Lennarz, ed.), pp. 1–34. Plenum, New York.

Kroll, A. J., and Machemer, R. (1968). Experimental retinal detachment in the owl monkey. III. Electron microscopy of retina and pigment epithelium. *Am. J. Ophthalmol.* **66,** 410–427.

Lasater, E. M., and Dowling, J. E. (1982). Carp horizontal cells in culture respond selectively to L-glutamate and its agonists. *Proc. Natl. Acad. Sci. U.S.A.* **79,** 936–940.

LaVail, M. M. (1976). Rod outer segment disc shedding in rat retina: Relationship to cyclic lighting. *Science* **194,** 1071–1074.

LaVail, M. M. (1980). Circadian nature of rod outer segment disc shedding in the rat. *Invest. Ophthalmol. Visual Sci.* **19,** 407–411.

LaVail, M. M. (1981). Analysis of neurological mutants with inherited retinal degeneration. *Invest. Ophthalmol. Visual Sci.* **21,** 638–657.

LaVail, M. M., and Ward, P. A. (1978). Studies on the hormonal control of circadian outer segment disc shedding in the rat retina. *Invest. Ophthalmol. Visual Sci.* **17,** 1189–1193.

LaVail, M. M., Sidman, R. L., and O'Neil, D. (1973). Photoreceptor pigment epithlial relationships in rats with inherited retinal degeneration. Radioautographic and electron microscopic evidence for a dual source of extra lamellar material. *J. Cell Biol.* **53,** 185–209.

LaVail, M. M., Pinto, L. H., and Yasumura, D. (1981). The interphotoreceptor matrix in rats with inherited retinal dystrophy. *Invest. Phthalmol. Visual Sci.* **21,** 658–668.

Liang, D.-J., Yamashita, K., Muellenberg, C. G., Shichi, H., and Kobata, A. (1979). Structure of the carbohydrate moieties of bovine rhodopsin. *J. Biol. Chem.* **254,** 6414–6418.

Liebman, P. A., and Entine, G. (1974). Lateral diffusion of visual pigment in photoreceptor disk membranes. *Science* **185,** 457–459.

Liebman, P. A., Weiner, H. L., Drzymala, R. E. (1982). Lateral diffusion of visual pigment in rod disk membranes. *In* "Methods in Enzymology" (L. Packer, ed.), Vol. 81, pp. 660–668. Academic Press, New York.

Llinas, R. (1979). The role of calcium in neuronal function. *In* "The Neurosciences 4th Study Program" (F. O. Schmitt and F. G. Worden, eds.), pp. 555–571. MIT Press, Cambridge, Massachusetts.

McLaughlin, B. J., and Wood, J. G. (1980). The localization of lectin binding sites on photoreceptor outer segments and pigment epithelium of dystrophic retinas. *Invest. Opthalmol. Visual Sci.* **19,** 728–742.

Masland, R. H. (1982). Choline metabolism and the maintenance of photoreceptor cell structure. *Retina* **2,** 282–287.

Masland, R. H.,and Mills, J. W. (1979). Autoradiographic identification of acetylcholine in the rabbit retina. *J. Cell Biol.* **83,** 159–178.

Masland, R. H., and Mills, J. W. (1980). Choline accumulation by photoreceptor cells of the rabbit retina. *Proc. Natl. Acad. Sci. U.S.A.* **77,** 1671–1675.

Matheke, M. L., and Holtzman, E. (1984). The effects of monensin and of puromycin on transport of membrane components in the frog retinal photoreceptor. II. Electron microscopic autoradiography of proteins and glycerolipids. *J. Neurosci.* **4,** 1093–1103.

Matheke, M. L., Fliesler, S. J., Basinger, S. F., and Holtzman, E. (1984). The effects of monensin on transport of membrane components in the frog retinal photoreceptor. I. Light microscopic autoradiography and biochemical analysis. *J. Neurosci.* **4,** 1086–1092.

Matsumoto, B., and Besharse, J. C. (1985). The staining of the distal tips of rod outer segments with Lucifer Yellow: A fluorescence microscope study. *Invest. Ophthalmol. Visual Sci.* **26,** 628–635.

Matsumoto, B., Defoe, D. M., and Besharse, J. C. (1986). Rod photoreceptor disks shedding induced by aspartate: Formation of epithelial pseudopods and the effects of cytochalasin D. Submitted.

Mercurio, A. M., and Holtzman, E. (1982a). Smooth endoplasmic reticulum and other agranular reticulum in frog retinal photoreceptors. *J. Neurocytol.* **11,** 263–293.

Mercurio, A. M., and Holtzman, E. (1982b). Ultrastructural localization of glycerolipid synthesis in rod cells of the isolated frog retina. *J. Neurocytol.* **11,** 295–322.

Matsusaka, T. (1974). Membrane particles of the connecting epithelium. *J. Ultrastruct. Res.* **48,** 305–312.

Matsusaka, T. (1976). Cytoplasmic fibrils of the connecting cilium. *J. Ultrastruct. Res.* **54,** 318–324.

Michell, R. H. (1975). Inositol phospholipids and cell surface receptor function. *Biochim. Biophys. Acta* **415,** 82–147.

Miller, A. M., and Schwartz, E. A. (1983). Evidence for the identification of synaptic transmitters released by photoreceptors of the toad retina. *J. Physiol. (London)* **334,** 325–349.

Molday, R. S., and Molday, L. L. (1979). Identification and characterization of multiple forms of rhodopsin and minor proteins in frog and bovine rod outer segment disc membranes. *J. Biol. Chem.* **254,** 4653–4660.

Mullen, R. J., and LaVail, M. M. (1976). Inherited retinal dystrophy: Primary defect in pigment epithelium determined with experimental rat chimeras. *Science* **192,** 799–801.

Murray, R. L., and Dubin, M. W. (1975). The occurrence of actinlike filaments in association with migrating pigment granules in frog retinal pigment epithelium. *J. Cell Biol.* **64,** 705–710.

Nathans, J., and Hogness, D. S. (1983). Isolation, sequence analysis, and Intron–exon arrangement of the gene encoding bovine rodopsin. *Cell* **34,** 807–814.

Nilsson, S. E. G. (1964). Receptor outer segment development and ultrastructure of the disk membranes in the retina of the tadpole (*Rana pipiens*). *J. Ultrastruct. Res.* **11,** 581–620.

Nir, I., and Papermaster, D. S. (1983). Differential distribution of opsin in the plasma membrane of frog photoreceptors: An immunocytochemical study. *Invest. Ophthalmol. Visual Sci.* **24,** 868–878.

Nir, I., Cohen, D., and Papermaster, D. S. (1984). Immunocytochemical localization of opsin in the cell membrane of developing rat retinal photoreceptors. *J. Cell Biol.* **98,** 1788–1795.

O'Brien, P. (1976). Rhodopsin as a glycoprotein: A possible role for the oligosaccharide in phagocytosis. *Exp. Eye Res.* **23,** 127–137.

O'Brien, P. (1978). Characteristics of galactosyl and fucosyl transfer to bovine rhodopsin. *Exp. Eye Res.* **26,** 197–206.

O'Brien, P. J., and Zatz, M. (1984). Acylation of bovine rhodopsin by ^{3}H palmitic acid. *J. Biol. Chem.* **259,** 5054–5057.

O'Day, W. T., and Young, R. W. (1979). The effects of prolonged exposure to cold on visual cells of the goldfish. *Exp. Eye Res.* **28,** 167–187.

Olney, J. W. (1982). The toxic effects of glutamate and related compounds in the retina and brain. *Retina* **2,** 341–359.

Owaribe, K., Kodama, R., and Eguchi, G. (1981). Demonstration of contractility of circumferential actin bundles and its morphogenetic significance in pigmented epithelium *in vitro* and *in vivo*. *J. Cell Biol.* **90,** 507–514.

Papermaster, D. S., and Dreyer, W. J. (1974). Rhodopsin content in the outer segment membranes of bovine frog retinal rods. *Biochemistry* **13,** 2438–2444.

Papermaster, D. S., and Schneider, B. G. (1982). Biosynthesis and morphogensis of outer segment membranes in vertebrate photoreceptor cells. *In* "Cell Biology and the Eye" (D. S. McDevitt, ed.), pp. 475–531. Academic Press, New York.

Papermaster, D. S., Converse, C. A., and Siu, J. (1975). Membrane biosynthesis in the retina: Opsin transport in the photoreceptor cell. *Biochemistry* **14,** 1343–1352.

Papermaster, D. S., Converse, C. A., and Zorn, M. (1976). Biosynthetic and immunochemical characterization of a large protein in frog and cattle rod outer segment membranes. *Exp. Eye Res.* **23,** 105–115.

Papermaster, D. S., Schneider, B. G., Zorn, M. A., and Kraehenbuhl, J. P. (1978a). Immunocytochemical localization of opsin in outer segments of golgi zones of frog photoreceptor cells. *J. Cell Biol.* **77,** 196–210.

Papermaster, D. S., Schneider, B. G., Zorn, M. A., and Kraehenbuhl, J. P. (1978b). Immunocytochemical localization of a large intrinsic membrane protein to the incisures and margins of frog rod outer segment disks. *J. Cell Biol.* **78,** 415–425.

Papermaster, D. S., Schneider, B. G., and Besharse, J. C. (1979). Assembly of rod photoreceptor membranes: Immunocytochemical and autoradiographic localization of opsin in smooth vesicles of the inner segment. *J. Cell Biol.* **83,** 275a.

Papermaster, D. S., Burnstein, Y., and Schecter, I. (1980). Opsin mRNA isolation from bovine retina and partial sequence of the *in vitro* translation product. *Ann. N.Y. Acad. Sci.* **343,** 347–355.

Papermaster, D. S., Reilly, P., and Schneider, B. G. (1982). Cone lamellae and red and green rod outer segment disks contain a large intrinsic membrane protein on their margins: An ultrastructural immunocytochemical study of frog retinas. *Vision Res.* **22,** 1417–1428.

Papermaster, D. S., Schneider, B. S., and Besharse, J. C. (1985). Vesicular transport of newly synthesized opsin from the Golgi aparatus toward the rod outer segment: Immunocytochemical and autoradiographic evidence. *Invest. Ophthalmol. Visual Sci.* **26,** 1386–1404.

Peters, K. R., Palade, G. E., Schneider, B. S., and Papermaster, D. S. (1983). Fine structure of a periciliary ridge complex of frog retinal rod cells revealed by ultrahigh resolution scanning electron microscopy. *J. Cell Biol.* **96,** 265–276.

Peyman, G. A., and Bok, D. (1972). Peroxidase diffusion in the normal and the laser coagulated primate retina. *Invest. Ophthalmol.* **11,** 35–45.

Pfeffer, B. A., and Fisher, S. K. (1981). Development of retinal pigment epithelial surface structures ensheathing cone outer segments in the cat. *J. Ultrastruct. Res.* **76,** 158–172.

Philp, N. J., and Bernstein, M. H. (1981). Phagocytosis by retinal pigment epithelium explants in culture. *Exp. Eye Res.* **33,** 47–53.

Pierce, M. E., and Besharse, J. C. (1985). Circadian regulation of retinomotor movements: I. Interaction of melatonin and dopamine in the control of cone length. *J. Gen. Physiol.* **86,** 671–689.

Pierce, M. E., Iuvone, P. M., and Besharse, J. C. (1984). Melatonin and photoreceptor metabolism: Regulation of cone retinomotor movement by melatonin and dopamine. *Soc. Neurosci. Abstr.* **10,** 19.

Plantner, J. J., and Kean, E. L. (1976). Carbohydrate composition of bovine rhodopsin. *J. Biol. Chem.* **251,** 1548–1552.

Plantner, J. J., Poncz, L., and Kean, E. L. (1980). Effect of tunicamycin on the glycosylation of rhodopsin. *Arch. Biochem. Biophys.* **201,** 527–532.

Poo, M.-M., and Cone, R. R. (1974). Lateral diffuction of rhodopsin in the photoreceptor membrane. *Nature (London)* **247,** 438–441.

Pu, G. A., and Anderson, R. E. (1983). Alteration of retinal choline metabolism in an experimental model for photoreceptor cell degeneration. *Invest. Ophthalmol. Visual Sci.* **24,** 288–293.

Pu, G. A., and Masland, R. H. (1984). Biochemical interruption of membrane phospholipid renewal in retinal photoreceptor cells. *J. Neurosci.* **4,** 1559–1576.

Raviola, G. (1977). The structural basis of the blood–ocular barriers. *Exp. Eye Res. (Symp. Suppl.)* **25,** 27–63.

Reich-D'Almeida, F. B., and Hockley, D. J. (1975). *In situ* reactivity of retinal pigment epithelium. II. Phagocytosis in the dystrophic rat. *Exp. Eye Res.* **21,** 347–357.

Richardson, T. M. (1969). Cytoplasmic and ciliary connections between the inner and outer segments of mammalian visual receptors. *Vision Res.* **9,** 727–731.

Robinson, W. E., Gordon-Walker, A., and Bownds, D. (1972). Molecular weight of frog rhodopsin. *Nature (London) New Biol.* **235,** 112–114.

Röhlich, P. (1975). The sensory cilium of retinal rods is analogous to the transitional zone of motile cilia. *Cell Tissue Res.* **161,** 421–430.

Roof, D. J., and Heuser, J. E. (1982). Surfaces of rod photoreceptor disk membranes: Integral components. *J. Cell Biol.* **95,** 487–500.

Rosenstock, T., Basu, R., Basu, P. K., and Ranadive, N. S. (1980). Quantitative assay of phagocytosis by retinal pigment epithelium: An organ culture model. *Exp. Eye Res.* **30,** 719–729.

Saari, J. C., and Bunt, A. H. (1980). Fucosylation of rabbit photoreceptor outer segments: Properties of the labeled components. *Exp. Eye Res.* **30,** 231–244.

Sandoz, D., Gounon, P., Karsenti, E., and Sauron, M.-E. (1982). Immunocytochemical localization of tubulin, actin, and myosin, in axonemes of ciliated cells from quail oviduct. *Proc. Natl. Acad. Sci. U.S.A.* **79,** 3198–3202.

Schecter, I., Burstein, Y., Zemell, R., Ziv, E., Kantor, F., and Papermaster, D. S. (1979). Messenger RNA of opsin from bovine retina: Isolation and partial sequence of the *in vitro* translation product. *Proc. Natl. Acad. Sci. U.S.A.* **76,** 2654–2658.

Schmidt, S. Y. (1983a). Light enhances the turnover of phosphatidylinositol in rat retinas. *J. Neurochem.* **40,** 1630–1638.

Schmidt, S. Y. (1983b). Phosphatidylinositol synthesis and phosphorylation are enhanced by light in rat retinas. *J. Biol. Chem.* **258,** 6863–6868.

Schmidt, S. Y. (1983c). Cytidine metabolism in photoreceptor cells of the rat. *J. Cell Biol.* **97,** 824–831.

Schmidt, S. Y. (1983d). Light and cytidine-dependent phosphatidylinositol synthesis in photoreceptor cells of the rat. *J. Cell Biol.* **97,** 832–837.

Shimazaki, H., Karwoski, C. J., and Proenza, L. M. (1984). Aspartate-induced dissociation of proximal from distal retinal activity in the mudpuppy. *Vision Res.* **24,** 587–595.

Silverstein, S. C., and Loike, J. D. (1980). Phagocytosis. *In* "Mononuclear Phagocytes Functional Aspects Part II" (R. Van Furth, ed.), pp. 895–917. Nijhoff, The Hague.

Slaughter, M. M., and Miller, R. F. (1983). The role of excitatory amino acid transmitters in the mudpuppy retina: An analysis with kainic acid and *N*-methyl aspartate. *J. Neurosci.* **3,** 1701–1711.

Spitznas, M., and Hogan, M. (1970). Outer segments of photoreceptors and the retinal pigment epithelium. *Arch. Opthalmol.* **84,** 810–819.

Steinberg, R. H., and Miller, S. S. (1979). Transport and membrane properties of the pigment epithelium. *In* "The Retinal Pigment Epithelium" (M. F. Marmor and K. M. Zinn, eds.), pp. 205–225. Harvard Univ. Press, Cambridge, Massachusetts.

Steinberg, R. H., Wood, I., and Hogan, M. J. (1977). Pigment epithelial ensheathment and phagocytosis of extrafoveal cones in human retina. *Phil. Trans. R. Soc. Ser. B* **277,** 459–474.

Steinberg, R. H., Fisher, S. K., and Anderson, D. H. (1980). Disc morphogenesis in vertebrate photoreceptors. *J. Comp. Neurol.* **190,** 501–518.

Steinemann, A., and Stryer, L. (1973). Accessibility of the carbohydrate moiety of rhodopsin. *Biochemistry* **12,** 1499–1502.

Stirling, C. E., and Lee, A. (1980). (^{3}H) ouabain autoradiography of frog retina. *J. Cell Biol.* **85,** 313–324.

Stossel, T. P. (1974). Phagocytosis (first of three parts). *N. Engl. J. Med.* **290,** 717–723.

Streb, H., Irvine, R. F., Berridge, M. J., and Schulz, I. (1983). Release of Ca^{++} from nonmitochondrial intracellular store in pancreatic acinar cells by inositol 1,4,5-trisphosphate. *Nature (London)* **306,** 67–69.

Struck, D. K., and Lennarz, W. J. (1980). The function of saccharide-lipids in synthesis of glycoproteins. *In* "The Biochemistry of Glycoproteins and Proteoglycans" (W. J. Lennarz, ed.), pp. 35–83. Plenum, New York.

Tabas, I., Schlesinger, S., and Kornfeld, S. (1978). Processing of high mannose oligosaccharides to form complex type oligosaccharides on the newly synthesized polypeptides of vesicular stomatitis virus G protein and the IgG heavy chain. *J. Biol. Chem.* **253,** 716–722.

Tabor, G. A., Anderson, D. H., Fisher, S. K., and Hollyfield, J. G. (1982). Circadian rod and cone disc shedding in mammalian retina. *In* "The Structure of the Eye" (J. G. Hollyfield, ed.), pp. 67–73. Elsevier, Amsterdam.

Takahashi, J. S., DeCoursey, P. J., Bauman, L., and Menaker, M. (1984). Spectral sensitivity of a novel photoreceptor system mediating entrainment of mammalian circadian rhythms. *Nature (London)* **308,** 186–188.

Tamai, M., and O'Brien, P. J. (1979). Retinal dystrophy in the RCS rat: *In vivo* and *in vitro* studies of phagocytic action of the pigment epithelium on shed outer segments. *Exp. Eye Res.* **28,** 399–411.

Tamai, M., Tierstein, P., Goldman, A., O'Brien, P., and Chader, G. (1978). The pineal gland does not control rod outer segment shedding and phagocytosis in the rat retina and pigment epithelium. *Invest. Opthalmol. Visual Sci.* **17,** 558–562.

Tannenbaum, J., Tannenbaum, S. W., and Godman, G. C. (1977). The binding sites of cytochalasin D. II. Their relationship to hexose transport and to cytochalasin B. *J. Cell. Physiol.* **91,** 239–248.

Tierstein, P. S., Goldman, A. I., and O'Brien, P. J. (1980). Evidence for both local and central regulation of rat rod outer segment disc shedding. *Invest. Ophthalmol. Visual Sci.* **19,** 1268–1273.

Tkacz, J. S., and Lampen, J. O. (1975). Tunicamycin inhibition of polyisoprenyl *N*-acetylglucosaminyl pyrophosphate formation in calf liver microsomes. *Biochem. Biophys. Res. Commun.* **65,** 248–257.

Tsukamoto, Y., and Yamada, Y. (1982). Light-related changes of outer segment membranes from lamellae to tubules and two kinds of wavy configurations in frog visual cells. *Exp. Eye Res.* **34,** 675–694.

Tsunasawa, S., Narita, K., and Shichi, H. (1980). The *N*-terminal residue of bovine rhodopsin is acetylmethionine. *Biochim. Biophys. Acta* **624,** 218–225.

Tilney, L. G., and Kallenbach, N. (1979). Polymerization of actin. VI. The polarity of the actin filaments in the acrosomal process and how it might be determined. *J. Cell Biol.* **81,** 608–623.

Tilney, L. G., Hatano, S., Ishikawa, H., and Mooseker, M. S. (1973). The polymerization of actin: Its role in the generation of the acrosomal process of certain echinoderm sperm. *J. Cell Biol.* **59,** 109–126.

Waechter, C. J., and Lennarz, W. J. (1976). The role of polyprenol-linked sugars in glycoprotein synthesis. *Annu. Rev. Biochem.* **45,** 95–112.

Wald, G. (1968). The molecular basis of visual excitation. *Science* **162,** 230–239.

Wald, G., Brown, P. K., and Smith, P. H. (1955). Iodopsin. *J. Gen. Physiol.* **30,** 623–679.

Watkins, J. C. (1978). Excitatory amino acids. *In* "Kainic Acid as a Tool in Neurobiology" (E. G. McGeer, J. W. Olney, and P. L. McGeer, eds.), pp. 37–69. Raven, New York.

Wey, C. L., Cone, R. C., and Edidin, M. A. (1981). Lateral diffusion of rhodopsin in photoreceptor cells measured by fluorescence photobleaching and recovery. *Biophys. J.* **33,** 225–232.

Wickner, W. (1979). The assembly of proteins into biological membranes: the membrane trigger hypothesis. *Annu. Rev. Biochem.* **48,** 23–45.

Wilden, U., and Kuhn, H. (1982). Light-dependent phosphorylation of rhodopsin: number of phosphorylation sites. *Biochemistry* **21,** 3014–3022.

Williams, D. S., Wilson, C., and Fisher, S. (1984). Effect of Na^+ substitution, ouabain and strophanthidin on shedding of rod outer segment discs. *J. Comp. Physiol. A.* **155,** 763–770.

Wood, J. G., and Napier-Marshall, L. (1985a). Cytochemical analysis of oligosaccharide processing in frog photoreceptors. *Histochem. J.* **17,** 585–594.

Wood, J. G., and Napier-Marshall, L. (1985b). Differential effects of protease digestion on photoreceptor lectin binding sites. *J. Histochem. Cytochem.* **33,** 642–646.

Wu, S. M., and Dowling, J. E. (1978). L-Aspartate: Evidence for a role in cone photoreceptor synaptic transmission in the carp retina. *Proc. Natl. Acad. Sci. U.S.A.* **75,** 5205–5209.

Young, R. W. (1967). The renewal of photoreceptor outer segments. *J. Cell Biol.* **33,** 61–72.

Young, R. W. (1969). Passage of newly formed protein through the connecting cilium of retinal rods in the frog. *J. Ultrastruct. Res.* **23,** 462–473.

Young, R. W. (1971a). The renewal of rod and cone outer segments in the rhesus monkey. *J. Cell Biol.* **49,** 303–318.

Young, R. W. (1971b). An hypothesis to account for a basic distinction between rods and cones. *Vision Res.* **11,** 1–5.

Young, R. W. (1976). Visual cells and the concept of renewal. *Invest. Ophthalmol. Visual Sci.* **15,** 700–725.

Young, R. W. (1977). The daily rhythm of shedding and degradation of cone outer segment membranes in the lizard retina. *J. Ultrastruct. Res.* **61,** 172–185.

Young, R. W. (1978). The daily rhythm of shedding and degradation of rod and cone outer segment membranes in the chick retina. *Invest. Ophthalmol. Visual Sci.* **17,** 105–116.

Young, R. W., and Bok, D. (1969). Participation of the retinal pigment epithelium in the rod outer segment renewal process. *J. Cell Biol.* **42,** 392–403.

Young, R. W., and Droz, B. (1968). The renewal of protein in retinal rods and cones. *J. Cell Biol.* **39,** 169–184.

Zimmerman, W. F. (1984). Enzymes of phospholipid metabolism in bovine rod outer segments. *Invest. Opthalmol. Visual Sci. Suppl.* **25,** 113 (Abstr.).

Zisapel, N., Egozi, Y., and Laudon, M. (1982). Inhibition of dopamine release by melatonin: Regional distribution in the rat brain. *Brain Res.* **246,** 161–163.

INDEX

D

S